HOT TOPICS IN
INFECTION AND IMMUNITY
IN CHILDREN III

ADVANCES IN EXPERIMENTAL MEDICINE AND BIOLOGY

HOT TOPICS IN INFECTION AND IMMUNITY IN CHILDREN III

Edited by

Andrew J. Pollard

University of Oxford
Oxford, United Kingdom

Adam Finn

University of Bristol
Bristol, United Kingdom

 Springer

Andrew J. Pollard
University of Oxford
Level 4, John Radclif Hospital
Oxford, UK OX3 9DU

Adam Finn
University of Bristol
Tyndall Avenue
Bristol, UK B58 1TH

Library of Congress Control Number: 2005939184

ISBN-10: 0-387-31783-X
ISBN-13: 978-0387-31783-0

Printed on acid-free paper.

Printed in the United States of America.   (BS/MVY)

9 8 7 6 5 4 3 2 1

springer.com

Preface

This book is based on the course "Infection and Immunity in Children 2005" which was held at St Catherine's College Oxford, UK in June 2005. This is the third book in this series covering topics in infection and immunity during childhood and based on the Oxford courses. These courses, and their companion books, are aimed at encouraging excellence in clinical practice and raising the profile of paediatric infectious disease with a particular eye on the needs of trainees in the specialty. At the time of writing a fourth course is already at an advanced stage of planning for June 2006 with a completely new programme once again.

You will find in this book a wealth of state of the art information about various aspects of paediatric infectious diseases written by leading authorities in the field. We hope this volume will bring new insights into the management of children with infectious diseases and improve the health of children.

Andrew J Pollard and Adam Finn
January 2006

Acknowledgments

We are indebted to all of the contributors to this text who have generously provided their carefully written manuscripts in good time for editing and publication. We are also grateful to the staff of St Catherine's College, Oxford, UK who hosted *Infection and Immunity in Children 2005* on which the text is based.

We are particularly grateful to Sue Sheaf for the administration of the Course and who assisted with the preparation of the edited manuscripts and to our colleagues and families who supported our own activities in this venture. Our gratitude also for the support provided by the European Society for Paediatric Infectious Disease (ESPID), the British Paediatric Allergy Infection and Immunity Group (BPAIIG), and the Royal College of Paediatrics and Child Health (RCPCH). Lastly, we are pleased to acknowledge the generous and unrestricted financial support of Sanofi Pasteur MSD, Chiron Vaccines, GlaxoSmithKline Vaccines, and Wyeth Vaccines who made the meeting possible.

Contents

8. *Chlamydia trachomatis* Genital Infection in Adolescents and Young Adults

Toni Darville

12. Diagnosis and Prevention of Pneumococcal Disease
Hanna Nohynek

13. New Antibiotics for Gram-Positive Infections
John S. Bradley

17. Prevention of Transmission of HIV-1 from Mothers to Infants in Africa

Hoosen Coovadia and Derseree Archary

18. Practical Aspects of Antiretroviral Treatment in Children

Sam Walters

19. Antibiotic Treatment for Acute Otitis Media in Children
Matthew J. Thompson and Paul Glasziou

20. Antibiotic Prophylaxis for the Prevention of Recurrent Urinary Tract Infections in Children
Elliot Long, Samantha Colquhoun, and Jonathan R. Carapetis

21. Human Metapneumovirus: An Important Cause of Acute Respiratory Illness
Adilia Warris and Ronald de Groot

Contributing Authors

Derseree Archary
Department of Paediatrics and Child
  Health
University of KwaZulu-Natal
Nelson R Mandela School of Medicine
2nd Floor Doris Duke Medical Research
  Institute
Private Bag 7
Congella
Durban 4013, South Africa

John S. Bradley
Division of Infectious Diseases
Children's Hospital
San Diego
CA 92123
USA

Joseph Bresee
Viral Gastroenteritis Section
CDC
1600 Clifton Road
Atlanta, GA 30333
USA

Jonathan R. Carapetis
Director, Centre for International Child
  Health
Consultant in Paediatric Infectious
  Diseases
University of Melbourne Department of
  Paediatrics
Royal Children's Hospital
Parkville VIC 3052
Australia

Julia Clark
Paediatric Infectious Diseases Unit
Newcastle General Hospital
Newcastle upon Tyne NE4 6BE
UK

Samantha Colquhoun
Research Coordinator
Centre International Child Health
University of Melbourne, Department of
  Paediatrics, and
Murdoch Childrens Research Institute
Royal Children's Hospital
Parkville. Vic
Australia

Hoosen Coovadia
Centre for HIV and AIDS Networking
  (HIVAN)
University of KwaZulu-Natal
Nelson R Mandela School of Medicine
2nd Floor Doris Duke Medical Research
  Institute
Private Bag 7
Congella
Durban 4013, South Africa

Nigel Curtis
Department of Paediatrics
University of Melbourne and
Paediatric Infectious Diseases Unit
Department of General Medicine and
Murdoch Children's Research Institute
Royal Children's Hospital
Parkville VIC 3052
Australia

Toni Darville
Division of Pediatric Infectious Diseases
Department of Pediatrics and
Microbiology/Immunology
University of Arkansas for Medical
  Sciences
Little Rock
Arkansas
USA

Philip L. Davies
Department of Child Health
Cardiff University
Cardiff
CF14 4XN
UK

B. Keith English
The University of Tennessee Health
    Science Center
Memphis
Tennessee
USA

Alan Fenwick
Director of Schistosomiasis Control
    Initiative
Department of Infectious Disease
    Epidemiology
Imperial College
London, UK

Andrew R. Gennery
Paediatric Immunology Department
Newcastle General Hospital
Newcastle upon Tyne
NE4 6BE
UK

Jon Gentsch
Viral Gastroenteritis Section
CDC
1600 Clifton Road
Atlanta, GA 30333
USA

Roger Glass
Viral Gastroenteritis Section
CDC
1600 Clifton Road
Atlanta, GA 30333
USA

Paul Glasziou
Department of Primary Health Care
University of Oxford
Old Road Campus
Oxford OX3 7LF UK

Ronald de Groot
Pediatric Infectious Diseases Specialist
Professor and Head of the Pediatric
    Department
Radboud University Medical Center
    Nijmegen
P.O. Box 9101
6500 HB Nijmegen, the Netherlands

Robert S. Heyderman
Department of Cellular & Molecular
    Medicine
School of Medical Sciences
University of Bristol
University Walk
Bristol BS8 1TD, UK

Nico G. Hartwig
Paediatric Infectious Diseases Specialist
Department of Paediatrics
Erasmus MC-Sophia
Rotterdam
The Netherlands

Peter Hotez
Department of Microbiology, Immunology
    and Tropical Medicine
The George Washington University and
    Sabin Vaccine Institute
Washington DC 20037
USA

Baoming Jiang
Viral Gastroenteritis Section
CDC
1600 Clifton Road
Atlanta, GA 30333
USA

Eileen Lau
Viral Gastroenteritis Section
CDC
1600 Clifton Road
Atlanta, GA 30333
USA

David B. Lewis
Department of Paediatrics and the Program
 in Immunology
Stanford University
CCSR Building
Room 2115
Stanford University School of
 Medicine
269 Campus Drive
Stanford, CA 94305-5164
USA

Elliot Long
Paediatric Registrar
Royal Children's Hospital
Parkville Victoria 3052 Australia

Sailesh Kotecha
Department of Child Health
Cardiff University
Cardiff CF14 4XN

Kathryn Maitland
The Centre for Geographic Medicine
 Research
Coast, Kemri, Kenya
PO Box 230
Kilifi, Kenya
and
Department of Academic Paediatrics
Imperial College, London, UK

Nicola C. Maxwell
Department of Child Health
Cardiff University
Cardiff
CF14 4XN, UK

Tim J. Mitchell
Division of Infection and Immunity
Glasgow Biomedical Research Centre
University of Glasgow
120 University Place
Glasgow G12 8TA
UK

David Molyneux
Director of the Lymphatic Filariasis
 Support Centre
Liverpool School of Tropical
 Medicine
Liverpool, UK

Samantha J. Moss
Neonatal Department
Royal Victoria Infirmary
Newcastle upon Tyne
NE1 4LP
UK

Hanna Nohynek
Department of Vaccines
Clinical Unit
National Public Health Institute
Mannerheimintie 166
FIN-00300
Helsinki, Finland

Eric Ottesen
Task Force for Child Survival
Emory University
Atlanta, GA,
USA

Umesh Parashar
Viral Gastroenteritis Section
CDC
1600 Clifton Road
Atlanta, GA 30333
USA

David A. Randolph
Department of Paediatrics and Division of
 Immunology
Stanford University
CCSR Building
Room 2115
Stanford University School of Medicine
269 Campus Drive
Stanford, CA 94305-5164
USA

Frank Shann
Intensive Care Unit
Royal Children's Hospital
Melbourne Australia
and
University of Melbourne Australia

Matthew J. Thompson
Department of Primary Health Care
University of Oxford
Old Road Campus
Oxford OX3 7LF UK

Sam Walters
Senior Lecturer Pediatric Infectious
    Diseases
Imperial College of Science and Medicine
London
and
Honorary Consultant Pediatrician
Department of Paediatrics
St Mary's Hospital
Paddington
London W2 1NY, UK

Adilia Warris
Pediatric Infectious Diseases Specialist
Radboud University Medical Center
    Nijmegen
P.O. Box 9101
6500 HB Nijmegen, the Netherlands

Pablo Yagupsky
Clinical Microbiology Laboratories
Soroka University Medical Center
Ben-Gurion
University of the Negev
Beer-Sheva 84101
Israel

Frank Shann

Kathryn Maitland

Peter J. Hotez

Nigel Curtis

Roger Glass

David B. Lewis

Andrew Gennery

Toni Darville

Sailesh Kotecha

Tim J. Mitchell

B. Keith English

Hanna Nohynek

John S. Bradley

Robert S. Heyderman

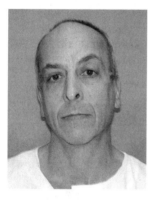

Pablo Yagupsky

Nico Hartwig

Jonathan R. Carapetis

Hoosen Coovadia

Sam Walters   Paul Glasziou

Adam Finn   Andrew J. Pollard

Warfare and the State of the World's Children

Frank Shann

1. Introduction

In the year 2000, 132 million children were born in the world, and 10.9 million children died before they reached 5 years of age (UNICEF, 2001). 99.4% of the 10.9 million deaths occurred in developing countries – a very high proportion indeed. If the whole world had had the same child mortality rate as the industrialized countries, there would have been only 0.8 million deaths before 5 years of age. Thus, there were 10.9 minus 0.8 million, or 10.1 million unnecessary deaths before 5 years of age in the world in the year 2000. This is 27,670 unnecessary deaths every day, or 1153 every hour.

Communities affected by war have very substantial increases in mortality. In a review of 37 studies, Guha-Sapir (2004) found that armed conflict increased mortality 1.8 fold in children less than 5 years of age, from 188 deaths per 1000 per year before a conflict, to 338 deaths per 1000 per year during the conflict. In people over 5 years of age, mortality increased 7.5 fold from 17 deaths per 1000 per year to 128 deaths per 1000 per year. The relative increase was greater over 5 years of age, but absolute mortality rates were much higher in children less than 5 years of age.

The nature of war has changed greatly in the last hundred years. In the first world war, the opposing armies fought mainly on battlefields, and only 19% of the casualties were civilians (Goldson, 1996). In the second world war, 48% of casualties were civilians (Goldson, 1996). Since 1980, many wars have been between opposing ethnic or religious groups, and wars have been fought in villages and communities with deliberate destruction of houses, schools, crops and wells – and 85% of war deaths have been civilians (Goldson, 1996).

Warfare causes immense physical and psychological morbidity, as well as killing people directly (through the effects of guns, bombs, mines and chemicals; the use of child soldiers; and systematic genocide) and indirectly (through infections

Hot Topics in Infection and Immunity in Children, edited by Andrew J. Pollard and Adam Finn.
Springer, New York, 2006

Table 1.1.   Direct and indirect effects of war

Direct effects of war

Guns, bombs, mines, chemicals:
   0.2 million children die per year
   0.5 million children disabled per year
Land mines kill 1000 children per year
300,000 child soldiers <18 years old
Rape, torture, slavery
Psychological trauma
Disruption of families and societies

Indirect effects of war

Diarrhoea, ARI, malaria, measles, HIV
Malnutrition contributes to mortality
2 million per year homeless or orphaned
Disruption of health and education services
Family and community breakdown

Money wasted (could be used for development)

World military expenditure US$1035 bn in 2004

---

that occur because of disruption to social structures and community services)
(Table 1.1). However, even greater harm is caused by the economic effects of the
huge amount of money spent on military activities – resources that should be used
to improve the state of the world's people, rather than causing suffering.

2. Direct Effects of War

It is estimated that, on average, guns, bombs, mines and chemicals kill 0.2
million children and disable 0.5 million children every year (Machel, 2000). In
addition, at any given time, 300,000 children less than 18 years of age are fighting
as soldiers (Machel, 2000) – many are forced to fight against their will, and many
are brutalized and tortured.

Land mines are still found in 105 countries. They are estimated to kill 1000
children every year, and cause serious injury to 20,000 children (Machel, 2000).
Although 20 million land mines have been destroyed, 250 million remain, and they
will pose a threat for many years to come.

Rape, brutality and slavery have become increasingly common in modern
warfare. 20,000 girls and women were raped in Yugoslavia in 1992, and 250,000
were raped in Rwanda (Machel, 2000).

3. Indirect Effects of War

As a rule, infections and malnutrition kill many more children than direct injuries during a war (Connolly, 2004). Recent exceptions to this have been the conflicts in Yugoslavia, Chechnya, Georgia and Iraq. The commonest infectious causes of death are diarrhoea, acute respiratory infections, measles, malaria, HIV and tuberculosis. Contributory factors are the destruction of houses, crowding, poverty, unsafe water supplies, poor sanitation, malnutrition, and a lack of health services (Connolly, 2004). Occasionally, many deaths are caused by the outbreak of an epidemic, for example, enteric disease in Bosnia in 1993, diphtheria in Tajikistan in 1993, cholera in Zaire in 1994, typhus in Burundi in 1996, and poliomyelitis in Angola in 1999 (Smallman-Raynor, 2004).

A study in the Democratic Republic of Congo in 2002 found that the under 5 mortality rate was 108 per 1000 per year in the war zone in the east, and 53 per 1000 per year in the west (where there was no war) (CDC, 2003). Only seven of the 443 deaths in the war zone were from war injuries – most of the excess deaths were caused by diarrhoea, measles, malnutrition, febrile illness (of unknown cause), and acute respiratory infection.

The Complex Emergencies Programme of the World Health Organization has suggested that the most important ways to reduce the incidence of infections during emergencies are the provision of carefully planned centres and camps, clean water and sanitation, food and other treatments for malnutrition, immunization, vector control, bed nets and health education (Connolly, 2004).

4. Sustainable Development

The direct effects of war attract a disproportionate amount of attention. Although it is a tragedy that an average of 0.2 million children die every year as a result of war (Machel, 2000), this is only a small proportion of the total of 10.1 million unnecessary deaths amongst children aged less than 5 years. Most of the 10.1 million deaths are caused by common childhood infections, such as pneumonia and diarrhoea, in communities that are not involved in war.

It is important to provide urgent help to people affected by wars and other disasters, but there is a danger that this will distract from development work – which is more mundane, but far more productive in the long-term.

The most harm from war comes from the enormous amount of money spent on military activity (Table 1.2). This waste of resources does far more damage than

Table 1.2. Annual expenditure on military and other activity

World expenditure on military activity	US$1035 billion
US and EU agricultural subsidies	US$325 billion
Total OECD development aid	US$60 billion
Developing countries health expenditure	US$382 billion

the direct effects of war. World poverty could be dramatically reduced if the money spent on destructive military activity were instead spent constructively on sustainable development.

5. Military Expenditure

The marked inequality between rich and poor countries is very dangerous. Widespread, severe poverty contributes to overpopulation, and threatens the environment. It breeds anger, promotion of violence, and greatly increases the risk of world pandemics of infectious diseases. Both humanitarian considerations and the self-interest of developed countries demand that poverty be reduced.

World military expenditure totalled US$1035 billion in 2004 (SIPRI, 2005). The annual budget for the World Health Organization is equivalent to about 3 hours of military expenditure (Goldson, 1996). The total expenditure on development aid by OECD countries was $68 billion in 2003 (OECD, 2004) – only one fifteenth of military expenditure. Developing countries spent a total of $382 billion on health in 2002 (World Bank, 2005) – only 37% of the world expenditure on military activity.

The United States of America is responsible for half the world's military expenditure of $1035 billion a year. The permanent members of the United National Security Council account for 90% of weapons sales to developing countries – the United States 41%, United Kingdom 20%, Russia 17%, France 8%, and China 5% (Chanaa, 2004). And yet these are the countries that have accepted responsibility for world security.

However, developing countries also carry a heavy responsibility for the suffering caused by war (Table 1.3). Ninety percent of the 150 wars from 1945 to 1995 were in developing countries (Chanaa, 2004). Thirty-three developing countries spend more on military activity than on either education or health, and seven spend more on military activity than on education and health combined (Chanaa, 2004). Loans incurred to finance arms purchases are responsible for 20% of developing country debt (Chanaa, 2004). Seven of the ten developing countries with the highest child mortality rates are at war, or have recently been at war. The developing countries of Africa, Asia, the Middle East and Latin America spend $22 billion a year on weapons (Chanaa, 2004) – which could be used to pay for universal primary

Table 1.3. Developing countries and war

90% of 150 wars were in developing countries in 1945–1995.
Developing countries spend $22 billion a year on weapons.
33 developing countries spend more on military activity than on either education or health.
7 developing countries spend more on military activity than on education and health combined.
20% of developing country debt is for arms purchases.
Of the 10 countries with highest child mortality, 7 are at war or have recently been at war.

school education ($10 billion a year) and the Millenium Development Goals Infant and Maternal Health Programme ($12 billion a year).

There are some bizarre examples of military spending in developing countries. In 2004, India spent $1.5 billion on a reconditioned Russian aircraft carrier – this amount of money would have paid for HIV drugs for 10 million people for a year. At the same time, India asked donor countries to contribute $50 million to help eradicate polio (Chanaa, 2004). In 1999, South Africa commenced a $6 billion programme to purchase weapons, but spent only $54 million on HIV.

6. What can be Done?

It will be very difficult to achieve real reductions in military activity and eradicate poverty (Table 1.4). Arms expenditure has increased in the last 20 years, development aid has decreased, and there has been little improvement in trade opportunities for developing countries.

To reduce military activity, the United Nations Security Council urgently needs radical reform, the UN Convention on the Rights of the Child must be enforced, and we need an effective arms trade treaty – such as the one suggested by Amnesty and Oxfam (Southall, 2002; Chanaa, 2004).

Arms dealing should be allowed only through the United Nations, solely for defence purposes (Southall, 2002). All other brokering in arms should be illegal (as with the drug trade). The numbers and kinds of weapons traded should be published – only three countries do this at present. Some progress has been made in enforcing ethical trading in arms by European Union countries, but not the United States (Southall, 2002).

Hundreds of illegal arms dealers, companies, banks and smugglers are already known, but there have been no convictions for violations of United Nations arms embargoes (Southall, 2002). The United Nations needs to develop an effective force to aggressively police improved controls in illegal arms trading and associated activities such as money laundering and smuggling.

Unfortunately, many developed countries make a lot of money from selling weapons, and some factions in developing countries seek to maintain or gain power by force whatever the cost to their societies. The five permanent members of the United Nations Security Council account for 90% of weapons sales to developing

Table 1.4.   Changes needed to reduce warfare

Urgent reform of UN Security Council
Permanent Members reduce military activity
Enforce UN Convention on the Rights of the Child
Effective arms trade treaty (eg Amnesty model)
Arms sales only for defence, and only via UN
Publication of number and type of weapons sold
Effective UN force to police illegal arms trading

countries (Chanaa, 2002), yet they are the very countries who have accepted the responsibility for ensuring world security. The United States of America made a very substantial contribution to Europe's recovery after the Second World War, and has the ability to substantially reduce military activity in the world. Unfortunately, as Laurence Korb, an Assistant Secretary of Defence in the Reagan administration, has pointed out, the United States has not tried to curb the arms trade because of greed and a reluctance to accept limitations on its sovereignty (Korb, 2003). The United States exports more military hardware than the rest of the world combined, is the source of 41% of weapons exports to developing countries, gave the second lowest percentage of its GDP as overseas aid in 2004, rejects the International Criminal Court, rejects the United Nations Convention on the Rights of the Child (Somalia is the only other non-signatory), refuses to sign the protocol to enforce the treaty banning biological weapons, and refuses to support plans to curtail the illegal trade in small arms (as this might impair its own citizens' "right to bear arms") (Southall, 2002).

7. Some Good News – Costa Rica

Costa Rica provides an extraordinary example of the effects of disarmament. In 1948, the presidential election was won by Otilio Ulate, but the incumbent refused to recognise the result. The resulting civil war was won by forces led by Jose Figueres Ferrer, who disbanded the army and then handed over power to Ulate in 1949. Infant mortality was almost 100 per 1000 live births. Since that time, Costa Rica has spent less than 0.05% of its GNI on military expenditure, and has concentrated on public health and education programs. In 1983, Costa Rica adapted a policy of active and permanent neutrality (Munos, 1995). Despite considerable long-term economic problems and a GNI per capita of only US$4,280 in 2003, Costa Rica had an infant mortality rate of only 9.95 per 1000 live births (by comparison, the United States had a GNI of $37,610 per capita, and an infant mortality rate of 6.50 per 1000 live births) (http://devdata.worldbank.org).

8. Conclusion

War is immensely harmful, but not primarily because of the horrible injuries, nor the large number of deaths it causes indirectly from infection, malnutrition and social and political disruption. By far the greatest harm comes from the diversion of huge amounts of money that could be used for beneficial development into harmful and destructive military activity.

The main problem is that many governments in both developed countries and developing countries do not want to limit military activity – the vested interests that support the military are too strong. We must do all we can to draw attention to the huge economic and human cost of war in the hope that this will force governments to take effective action to limit military activity.

References

Centres for Disease Control. (2003). Elevated mortality associated with armed conflict – Democratic Republic of Congo, 2002. *MMWR* 52:469–471.

Chanaa, J., Hillier, D., Powell, K., Epps, K., Hughes, H. (2004). *Guns or growth? Assessing the impact of arms sales in sustainable development.* Amnesty International, the International Action Network on Small Arms, and Oxfam International. www.controlarms.org.

Connolly, M.A., Gayer, M., Ryan, M.J., Salama, P., Spiegel, P., Heymann, D.L. (2004). Communicable diseases in complex emergencies: Impact and challenges. *Lancet* 364:1974–1983.

Goldson, E. (1996). The effect of war on children. *Child Abuse & Neglect* 20:809–819.

Guha-Sapir, D., Gijsbert, W. (2004). Conflict-related mortality: An analysis of datasets. *Disasters* 28:418–428.

Korb, L. (2003). Arms sales, health and security. A call for US leadership. *BMJ* 326:459–460.

Machel, G. (2000). *The Machel review, 1996–2000: A critical analysis of progress made and obstacles encountered in increasing protection for war-affected children.* Government of Canada: Canadian Centre of Foreign Policy Development, Ottawa. Summary available online at http://www.waraffectedchildren.gc.ca/machel_review-en.asp.

Munos, C., Scrimshaw, N.S. (1995). *The nutrition and health transition of democratic Costa Rica.* International Foundation for Developing Countries (INFDC), Boston.

Organization for Economic Cooperation and Development. (2004). Accessed 29th September 2005. http://www.oecd.org/dataoecd/42/61/31504039.pdf.

SIPRI Yearbook. (2005). *Armaments, disarmament and international security.* Stockholm International Peace Research Institute (SIPRI), Stockholm.

Smallman-Raynor, M.R., Cliff, A.D. (2004). Impact of infectious diseases on war. *Infect Dis Clin N Am* 18:341–368.

Southall, D.P., O'Hare, B.A.M. (2002). Empty arms: The effect of the arms trade on mothers and children. *BMJ* 325:1457–1461.

UNICEF. (2001). *The State of the World's Children 2001.* Oxford University Press, Oxford.

World Bank, Topic Indicators. (2002). Accessed 29th September 2005. http://www.devdata.worldbank.org/external/dgcomp.asp?rmdk+110&smdk=473886&w=0.

World Health Organization. (2002). Patterns of global health expenditures: Results for 191 countries. EIP/HFS/FAR Discussion paper No 51. WHO, Geneva.

2

How Do We Treat Children with Severe Malaria?

Kathryn Maitland

1. Introduction

Malaria imposes a profound burden on global public health; over one third of the world's population (~2 billion people) live in malaria-endemic areas, with ~1 billion people estimated to carry parasites at any one time (Guerin et al., 2002). The greatest burden of malaria falls on sub-Saharan Africa (SSA), where it causes between 200–450 million disease episodes each year and over 1 million deaths. Most of these deaths are in children <5 years. Throughout the world the burden of malaria is increasing, especially in sub-Saharan Africa, mainly attributed to drug and insecticide resistance and social and environmental changes. Couple this to the worldwide increase in international travel, increases in the numbers of legal and illegal economic migrants, as well as in refugees seeking asylum from wars or political unrest, it is not surprising that there is a parallel rise in imported malaria. Nevertheless, in the UK and other European countries malaria is still a rare cause of hospital admission, few paediatricians have ever seen a case of severe malaria. Management of severe malaria should not only focus upon the choice of antimalarial agents, but should include comprehensive management of the child (Maitland et al., 2005a).

2. Epidemiological Considerations

The resurgence of malaria in most tropical countries equates with an increased risk to travellers. Throughout the world, many countries are reporting an increasing number of cases of imported malaria. There are four species of malaria parasite known to naturally infect man. In the UK, other European countries and America there has been a steady rise in the number and proportion of the potentially lethal falciparum malaria. In the UK over two thousand cases of malaria are imported

Hot Topics in Infection and Immunity in Children, edited by Andrew J. Pollard and Adam Finn.
Springer, New York, 2006.

every year, however over two decades the predominant type has changed from vivax to falciparum malaria. Half of all cases occur in people visiting friends or relatives overseas, and increased migration from Africa (rather than from India) means that more of these visits are now to areas where falciparum malaria is common. Two thirds of UK falciparum cases occur in London, the vast majority in the African population. A large proportion of total imported cases are in those who have taken no prophylaxis in highly malarious areas. Approximately 13% of these cases were children <15 years (Travel, 2003).

3. Pre-hospital Management

Once infected by *P. falciparum*, without prompt diagnosis and appropriate treatment, progression to severe malaria is likely in a non-immune child. This should always be assumed even for children raised in malaria-endemic areas but now residing in non-malarious areas. Immunity requires years of repeated infection and rapidly wanes when the person is no longer exposed to infective mosquito bites. When a child presents with a possible travel-associated or tropical illness it is important to take a detailed travel and exposure history. Important points to focus upon include a full vaccination history; whether malaria chemoprophylaxis was taken correctly, if indicated; whether specific measures were employed to prevent insect bites. Nevertheless, as malaria prophylaxis and preventative measure are never 100% effective malaria should be considered in any patient presenting with a fever who has travelled to or come from an endemic area (for details visit: http://www.cdc.gov/travel/diseases.htm#malaria or http://www.who.int/topics/malaria/en/). The usual presentation is with fever within two weeks of exposure but in children the presentation can sometimes be with non-specific symptoms including cough, headache, malaise, vomiting and diarrhoea. Common supportive findings in malaria include splenomegaly, mild thrombocytopaenia, anaemia and mild jaundice. Although most patients present within a few weeks or months of their return, presentation may be delayed particularly in the semi-immune, those who have taken prophylaxis and in *P. vivax*, *P. ovale* and *P. malariae* infections. The assessment of the child should include a full clinical examination to exclude any of the severity features (see Box 2.1). Although most cases of *P. falciparum* malaria presenting to health services in the UK are uncomplicated, up to 10% become severe and life-threatening, principally due to delays in diagnosis and inadequate treatment (Bradley, 2003; Ladhani et al., 2003). In uncomplicated disease the clinical features of malaria are similar in both children and adults, however in severe disease the clinical spectrum, complications and management differ. While the clinical symptoms and signs do not help distinguish the infecting Plasmodium species, the travel history is extremely helpful (especially if travelled to Africa where *P. falciparum* is the most common species) in imported malaria cases and in guiding drug selection. Non-infectious diseases specialists are more likely to make errors in therapy than are infectious diseases specialists so prompt referral or consultation is recommended. Specialist advice is recommended for children infected with *P. falciparum* returning from South East Asia, where multi-drug resistant malaria is endemic.

Box 2.1.　Clinical features and priorities for management

Recognition of Severe Malaria
High priority: emergency management

- Depressed conscious state (any degree)
- Intercurrent seizures
- Hypoglycaemia <3 mmols
- Evidence of shock
- Hypoxia (oxygen saturations <95%)
- Metabolic acidosis (base deficit >8)
- Severe hyperkalaemia (potassium >5.5 mmols/L)

Intermediate risk: need for high dependency care
Haemoglobin <10 g/dl
History of convulsions during this illness
Hyperparasitaemia >5%
Visible jaundice
P. falciparum in a child with sickle cell disease

- **Low risk: need admission for parenteral medication**

Vomiting Unable to take or comply with oral medication

- **Low Risk: need for observation**

None of the above
Admit and observe on oral treatment

Adapted from BMJ paper

4. Parasitological Diagnosis

In general any fever starting within 8 days of entry into a malaria endemic area is probably not malaria. However, the presumptive diagnosis of malaria should prompt urgent referral for *immediate* diagnosis and management. Failure to expedite appropriate referral may lead to the development of life-threatening disease. Thick and thin blood films, processed from an EDTA sample by the local haematology laboratory are the mainstay of diagnosis. Three thick films taken 12 hours apart excludes most malaria infections in any patient exposed to malaria. If clinical suspicion is high, further films are warranted. Thick films are necessary to diagnose malaria and thin film is required to define species and stages. Failure to prepare and rigorously examine a *thick film* may lead to a falsely negative malaria film especially in non-immune patients who often present with scant parasitaemia. Despite this, some scanty infections may escape detection. There are a number of other tests that are used to detect parasitaemia. The tests detect parasite derived histidine-rich protein 2 (HRP-2) e.g. *Para*Sight F®, ICT malaria P.f.®; parasite lactate dehydrogenase (OptiMal®) or parasite nuclear material (quantitative buffy coat (QBC)). In general, these are quick and simple to use, distinguish between the major forms of human malaria, and may have some advantages over microscopy, particularly in children with low-density parasitaemia, a characteristic that often applies to those who have been on prophylaxis (Moody and Chiodini, 2002; Palmer et al., 2003).

5. Assessment and Triage of Children with Severe Malaria

As for any other sick child presenting to hospital the initial management of a child with suspected malaria should be guided by a rapid, structured, triage assessment, aimed at identifying emergency and priority signs (1997; Group, 2001) (Box 2.1). The sequence of clinical assessment should remain (1) the early recognition of impending respiratory failure (2) the detection of shock and (3) a neurological assessment. This approach, will guide early management towards the complications that are the most commonly life-threatening. Emergency management should not be delayed while the diagnosis of malaria is confirmed. Unless there is likely to be an undue delay, the administration of specific antimalarial drugs can usually be deferred until resuscitation treatments have been given and the diagnosis confirmed. Nevertheless, if the clinical suspicion of malaria is high, an intravenous infusion of quinine should be started empirically following initial resuscitation, even if the results are delayed. Experimental treatments, such as exchange transfusion, have no role to play in the initial management of children with suspected malaria, and may distract attention from providing urgent and simple life-saving interventions.

It is now clear that severe malaria encompasses a complex syndrome affecting many organs resulting in biochemical and haematological derangements which have many features in common with the pathophysiological derangement seen in children with the sepsis syndrome (Maitland and Newton, 2005). Among these, metabolic acidosis (manifesting as respiratory distress) has emerged as a central feature of severe malaria, and is widely recognized as the best independent predictor of a fatal outcome in both adults (Day et al., 2000) and children (Allen et al., 1996; Newton et al., 1998; Taylor et al., 1993; Waller et al., 1995), mortality being greatest in children in whom acidosis and impaired consciousness co-exist. In malaria-endemic areas, most deaths occur within hours of admission, principally from the failure of the clinician to recognize impending circulatory collapse or respiratory compromise. The latter is particularly true in children with prolonged seizures. Raised intracranial pressure (ICP) may complicate cases presenting in coma (Newton et al., 1998), prompting a cautious approach to volume resuscitation in such children.

5.1. Generic Approached to Management

The airway must be secured, oxygenation optimised, vascular access established, hypoglycaemia, if present should be corrected (5 ml/kg of 10% dextrose). Seizures are common however, in African children with severe malaria around 25% of seizures are subtle or sub-clinical (demonstrated by EEG) (Crawley et al., 1996) The presence of respiratory depression, irregular breathing, drooling, eye deviation or occasionally bizarre posturing should alert the clinician to either the presence of complex seizures. After ensuring adequate airway and respiratory support specific management should follow the evidence-based consensus guideline advocated by the Advance Paediatric Life support Group (Group, 2001). Seizure prophylaxis is not recommended (Crawley et al., 2000).

5.2. Identification and Management of Shock

Severe falciparum malaria is frequently complicated by features of shock. In a retrospective review of cases presenting to Kilifi District hospital factors associated with a fatal outcome included deep breathing or acidosis (base deficit below −8), hypotension (systolic blood pressure <80 mmHg), raised plasma creatinine (>80 μmols/l), low oxygen saturation (<90%), dehydration and hypoglycaemia (<2.5 mmols/L). Shock was present in 212/372 (57%) children, of whom 37 (17.5%) died, and was absent in 160, of who only 7 (4.4%) died (χ^2 = 14.9; P = 0.001 (Table 2.1) (Maitland et al., 2003a). Delayed capillary refill time (≥2 s) is a reasonable prognostic indicator, especially in children with a decreased conscious level (Pamba and Maitland, 2004). These data suggest that impaired tissue perfusion may play a role in the mortality of severe malaria. Moreover, these results suggest that volume resuscitation, an important life saving intervention in children with hypovolaemia, should be considered in severe malaria with evidence of impaired tissue perfusion.

Table 2.1. Triage parameters and electrolyte derangement suggesting urgent need for intervention in the critically ill child

Parameter	n (%)	Case Fatality (%)
Airway and respiration:		
Hypoxia (O$_2$ sats <90%)[a]	86/501 (17.1)	30
Respiratory rate >60 br.pm	83/501 (16.6)	20
Deep "acidotic" breathing	104/515 (20.2)	31
Cardiovascular and hydration:		
Severe tachycardia (>160 bpm)	81/503 (16.1)	17
Hypotension (<80 mmHg)	66/507 (13)	26
Capillary refill time >2 secs	165/496 (33.3)	15
Neurological:		
Seizures	303/515 (58.8)	8.3
Prostration[b]	353/514 (68.7)	13
Blantyre Coma Score ≤2	266/509 (52.3)	16
Laboratory:		
Severe acidosis (pH < 7.2)	96/436 (22)	36
Raised creatinine (>80 μmols/L)	96/469 (20.4)	26
Hypoglycaemia (<2.5 mmols/L)	58/478 (12.1)	28
Hyperkalaemia (>5.5 mmols/L)	61/493 (12.3)	28
Hyponatraemia (<125 mmols/L)	21/493 (4.3)	14
Hypokalaemia (<3 mmols/L)	16/493 (3.2)	6

[a]Measured by pulse oximeter;
[b]inability to sit unsupported or breast feed.

We have demonstrated that volume depletion (measured by central venous pressure) was present on admission in the majority of children with severe malaria complicated by acidosis. We also demonstrated that volume expansion safely corrects the hemodynamic abnormalities and is associated with improved organ function and reduction in acidosis (Maitland et al., 2003b).

5.2.1. Volume Resuscitation

Several resuscitation fluids are available for the treatment of severe dehydration or shock in children. Simple electrolyte solutions are of proven benefit in most situations where excess water and electrolyte depletion has resulted from severe diarrhoea or vomiting. In conditions such as septic shock or severe malaria there is still debate over the optimum solution and, in the latter condition, over the safety of volume resuscitation. As in sepsis, some advocate the use of colloidal solutions, which will restore both water and electrolytes and plasma oncotic pressure, thus reducing the risk of potentiating, raised intracranial pressure. There is undoubtedly some risk that aggressive volume expansion would accentuate ICP. Non-colloidal containing solutions such as isotonic saline will move rapidly from the intravascular compartment into the tissues with the potential risk of accentuating raised ICP and pulmonary oedema. This situation is reminiscent of the similar dilemma in the treatment of meningitis, in which it was customary in the past to volume restrict (Herson and Todd, 1977; Powell et al., 1990). Recent studies in meningitis have shown that when compared to volume expansion modest fluid restriction is detrimental to outcome. We have shown in a Phase II randomised controlled trial (RCT), that volume resuscitation with between 20 and 40 mls/kg of either 0.9% saline or 4.5% human albumin solution (HAS), safely corrects the haemodynamic features of shock and improves renal function in Kenyan children with severe malaria (Maitland et al., 2005b; Maitland et al., 2005c). Pulmonary oedema was a rare complication of volume expansion (<0.5%) (Maitland et al., 2005c).

We advise the cautious use of fluid resuscitation when treating children presenting in coma. In the same trial children presenting in coma with the features of shock 11 out of 24 (46%) receiving saline died compared to only 1 out of 21 (5%) receiving HAS (relative risk 9.6; 95% CI 1.4 to 68; P = 0.002) (Maitland et al., 2005c). Until further data become available from larger trials we recommend that HAS should be considered the resuscitation fluid of choice in the subgroup of children who present with coma and features of shock. Volume resuscitation should proceed cautiously and be terminated once there is a satisfactory cardiovascular response to the volume challenge. A urine output of <1 ml/kg/hr, in the absence of urinary retention, indicates impaired renal perfusion secondary to hypovolaemia, and is a good non-invasive guide to fluid management. If the patient is still shocked, or if the shocked state returns, then treatment of shock should take priority, since cerebral perfusion depends on an adequate blood pressure.

Re-evaluation of the respiratory and circulatory status after each intervention is important. For any child with persisting features of shock after having received 40 mls/kg of fluid, we recommend elective tracheal intubation, mechanical ventilation and placement of a central venous catheter to guide further fluid management (Group, 2001). Patients with severe acidosis may self ventilate $pCO2$ to very low levels as compensation for the metabolic acidosis. Great care should be taken when initiating ventilation to avoid a sudden rise of $pCO2$, even to normal levels, before acidosis has been partially corrected.

5.3. The Child with Impaired Consciousness (GCS ≤ 8)

Rapid assessment of neurological function should include an assessment of the conscious level (AVPU or Children's GCS scale are adequate), pupillary size and reaction to light, in addition to observation of the child's posture and convulsive movements, if present. Other CNS infections or intracranial haemorrhage should be considered as an alternative diagnosis in a child with neck stiffness or a full fontanelle. The presentation of an acute neurological syndrome characterised by impaired consciousness, convulsions, abnormal neurological signs, and opisthotonic posturing are cardinal features of cerebral malaria (2000). However, these features may also suggest raised ICP in a small proportion of children (Newton et al., 1991a; Walker et al., 1992; Waller et al., 1991b). Initial management should include maintenance of the airway, support of breathing and immediate correction of hypoglycaemia and volume deficits. These interventions will correct hypoxia, hypoglycaemia or shock, which maybe potential contributors to the depressed conscious level. Children who remain unconscious (GCS ≤ 8) or have features suggestive of raised ICP warrant elective intubation and ventilation. For those with seizures, the decision to ventilate may be delayed if they are in a post-ictal state, as long as the airway is patent. Repeated seizures and motor posturing movements are commonly seen in severe malaria. In sub-Saharan Africa abnormal motor posturing (AMP) manifesting as decorticate, decerebrate or opisthotonus is often observed since few hospitals have facilities for paralysis and ventilation. However, the pathogenesis of these signs and prognosis of each type of posturing is unclear (Idro et al., 2004; Molyneux et al., 1989; Newton et al., 1997; Newton and Krishna, 1998; Schmutzhard and Gerstenbrand, 1984). Raised ICP, cerebral ischaemia, hypoxia, hypoglycaemia, and hyponatraemia are suggested as causes of AMP. The relationship AMP and raised ICP has yet to be established, nevertheless, owing to the potential risk of raised ICP ventilation should aim to optimise the $pCO2$ in the normal range, as there is no evidence that hyperventilation is beneficial in raised ICP. To date, the only clinical sign that was associated with the development of intermediate or severe intracranial hypertension, was a sluggish or absent pupillary response (Newton et al., 1997). Other signs (such as absent or extensor motor response, pupillary dilatation, decerebrate posturing, or absent oculocephalic reflexes) were not (Newton et al., 1997). Recent studies in children in Malawi have demonstrated a retinopathy that is peculiar to severe malaria, consisting of patchy whitening of the retina both in the macular and extra-macular areas, pale opacification of retinal vessels, and white-centred haemorrhages (Lewallen et al., 1999) (Figure 3). In children who died histopathological examinations of retina, parietal and cerebellar sections of the brains showed a correlation between the density of haemorrhages in the retina with their density in the brain (Lewallen et al., 2000; Lewallen, 2000; White et al., 2001). The presence of the retinopathy may help guide management (Taylor et al., 2004).

6. Monitoring

6.1. Blood Tests

Once the diagnosis is made, repeat blood films for parasite counts may be useful in following the progress of the disease. However, the paediatrician should be aware that, as quinine acts upon the later stages (schizonts), which are the generally sequestered in the microcirculation and thus not generally visible to the microscopist examining the blood film (that generally examines the circulating younger ring stages), peripheral parasitemia might continue to increase over the first 24 hours. In cases presenting from Africa (where quinine resistance is rarely encountered), this rarely indicates quinine resistance, and should not cause undue alarm. The addition or development of features indicating clinical severity should be used to guide initial and subsequent management rather than the parasitaemia alone.

6.2. Other Laboratory Investigations

To determine the range of biochemical and haematological derangements the following basic blood tests should be performed: full blood count, electrolytes (including potassium magnesium, phosphate and calcium) (Maitland et al., 2005d), urea (BUN) or creatinine, blood glucose, lactate and a blood gas. These should be measured serially, at least 12 hourly, if the child remains critically ill. Additional blood should be obtained for blood cultures and for blood grouping, in case a transfusion is required subsequently.

7. General Management

7.1. Antimalarial Medication

Parenteral quinine remains the first line anti-malarial drug for patients with severe falciparum malaria. Parenteral quinine, prescribed as quinine dihydrochloride (122 mg salt contains 100 mg quinine base) remains the first line treatment of severe falciparum malaria given as an initial loading dose of 20 mg salt/kg IV over 4 hours then 10 mg/kg IV tds. Intravenous quinine should be prescribed for any child with *P. falciparum* malaria who is unable to take oral medication or is vomiting. Mefloquine is a synthetic analogue of quinine and the *quinine loading dose should be omitted if the patient has taken mefloquine prophylaxis* in the previous 24 hours or received a treatment dose within the previous 3 days. Quinine should be prescribed for 7 days however once the child can take oral medication. Further treatment with another antimalarial is recommended (since its bitter taste precludes

Table 2.2. Antimalarial treatments

Drug	Uncomplicated Malaria (oral medication)	Complicated Malaria (parenteral medication)
Proguanil with Atovaquone (Malarone®)	Daily dose for 3 days: **Paediatric tablets** 5–8 kg: 2 paediatric 25 mg tablets, 9–10 kg: 3 paediatric 25 mg tablets given once daily	**Quinine hydrochloride:** Loading dose 20 mg salt/kg IV (over 4 hours) then 10 mg/kg IV (over 4 hours) 8 hourly for 7 days.
	Adult tablets 21–30 kg: 2 adult 100 mg tabs, 31–40 kg: 3 adult (100 mg) tabs, >40 kg: 4 adult (100 mg) tabs given once daily 15 mg base/kg followed by a second dose of 10 mg/kg 8–24 hours later[2]	Or if oral medications tolerated before 7 days then switch to *complete course* of oral medication (Malarone®, mefloquine or artemether with lumefantrine) as for uncomplicated malaria[1]
Mefloquine (Lariam®) **Artemether with lumefantrin (Riamet® or Coartem®)**	Given at 0, 8, 24, 36, 48 and 60 hours 5–<15 kg: 1 tab; 15–<25 kg 2 tabs; 25–<35 kg 3 tabs; ≥35 kg: 4 tabs (adult dose) Give with milk or similar as bioavailability significantly enhanced with fat.	

[1]Avoid prescribing oral quinine in children as the bitter taste may affect compliance.
[2]Repeat dose if vomited within 30 minutes.

its oral prescription to children). Mefloquine (Lariam®), Atovaquone-proguanil (Malarone®) or Lumefantrine-artemether (Coartem®, Novartis, Basel, Switzerland is currently the only fixed ratio artemether combination therapy that has been developed up to international standards) are potential follow on therapies. See Table 2.2 for doses. For those returning from Southeast Asia, where quinine resistance is problematic, advice regarding the use of intravenous artesunate should be sought from one of the regional centres. Other anticipated complications and management are covered in Box 2.2.

7.2. Role of Exchange Transfusion

Exchange transfusion has been advocated for hyper-parasitaemia (>10%) in adult ICU settings, despite no evidence to suggest that it confers a better outcome (Riddle et al., 2002). Even when parasitaemia exceeds 25%, the vast majority of children respond rapidly to the treatments outlined above. In those with persistent acidosis, and multi-organ impairment not responsive to the resuscitation treatments outlined above, exchange transfusion may be considered as a means of rapidly reducing the level of abnormal red cells, or parasite toxins. This treatment however remains experimental.

Box 2.2.   Anticipated complications

Very common

- **Hypoglycaemia (blood glucose <3 mmols):** correlates with disease severity.
- **Hyperpyrexia:** increases the risk of convulsions in children and should be treated with antipyretics/tepid sponging. Ibuprofen superior to paracetamol fever reduction, dose should be reduced in cases complicated by impaired renal function.
- **Seizures and Posturing** – see under emergency management

Common

- **Electrolytes derangement:**
  - hyperkalaemia complicates cases with severe metabolic acidosis at admission.
  - hypokalaemia, hypophosphataemia and hypomagnesaemia only apparent postadmission. Serial monitoring of plasma electrolytes is suggested.
  Treatment should follow standard APLS guidelines
- **Metabolic Acidosis:**
  - Resolves with the correction of hypovolaemia and treatment of anaemia by blood transfusion
  - No evidence to support the use of sodium bicarbonate.
- **Severe malaria anaemia:**
  - Most cases will experience some reduction of haemoglobin, and do not require transfusion. Decision to transfuse influenced by the parasitaemia level and clinical condition of the child.
  - Transfuse if the Hb falls below an absolute value of 10 g/dl.

Uncommon

- **Secondary bacterial infection** may occur and empiric broad-spectrum antibiotics are warranted
- **Coagulation activation:** Bleeding is rare despite the customary thrombocytopenia of severe malaria (platelet counts often $<50 \times 10^9$/L)

---

8. Outcome

Once recognised and adequately treated the outcome from severe malaria is generally favourable. Experience from the non-immune or semi-immune adults with severe malaria suggest that the complication of jaundice, renal failure, pulmonary oedema and adult respiratory distress syndrome are more frequent. Whilst transient renal impairment, secondary to hypovolaemic shock, is common in children with severe malaria from hyperendemic areas. Renal function generally corrects with the provision of volume expansion. There have been case reports from India demonstrating that children of an older age are more likely to have the adult phenotype of severe malaria including renal failure and ARDS suggesting that more data are needed to determine whether the management prescribed above is appropriate for children with severe malaria who are not ordinarily resident in malaria endemic areas.

References

(1997). "Pediatric Advanced Life Support 1997–1999." American Heart Association, Dallas, Texas.

(2000). Severe falciparum malaria. World Health Organization, Communicable Diseases Cluster. *Trans R Soc Trop Med Hyg* **94 Suppl 1**, S1–90.

Allen, S.J., O'Donnell, A., Alexander, N.D., and Clegg, J.B. (1996). Severe malaria in children in Papua New Guinea. *Q J Med* **89**, 779–88.

Bradley, D. (2003). Illness in England, Wales, and Northern Ireland associated with foreign travel: A baseline report to 2002. pp. 48–55. Travel Health Surveillance Section, London.

Crawley, J., Smith, S., Kirkham, F., Muthinji, P., Waruiru, C., and Marsh, K. (1996). Seizures and status epilepticus in childhood cerebral malaria. *Q J Med* **89**, 591–7.

Crawley, J., Waruiru, C., Mithwani, S., Mwangi, I., Watkins, W., Ouma, D., Winstanley, P., Peto, T., and Marsh, K. (2000). Effect of phenobarbital on seizure frequency and mortality in childhood cerebral malaria: a randomised, controlled intervention study. *Lancet* **355**, 701–6.

Day, N.P., Phu, N.H., Mai, N.T., Chau, T.T., Loc, P.P., Chuong, L.V., Sinh, D.X., Holloway, P., Hien, T.T., and White, N.J. (2000). The pathophysiologic and prognostic significance of acidosis in severe adult malaria. *Crit Care Med* **28**, 1833–40.

Group, A.L.S. (2001). "Advanced Paediatric Life support: The practical approach." BMJ Publishing Group, London.

Guerin, P.J., Olliaro, P., Nosten, F., Druilhe, P., Laxminarayan, R., Binka, F., Kilama, W.L., Ford, N., and White, N.J. (2002). Malaria: current status of control, diagnosis, treatment, and a proposed agenda for research and development. *Lancet Infect Dis* **2**, 564–73.

Herson, V.C., and Todd, J.K. (1977). Prediction of morbidity in *Hemophilus influenzae* meningitis. *Pediatrics* **59**, 35–9.

Idro, R., Karamagi, C., and Tumwine, J. (2004). Immediate outcome and prognostic factors for cerebral malaria among children admitted to Mulago Hospital, Uganda. *Ann Trop Paediatr* **24**, 17–24.

Ladhani, S., El Bashir, H., Patel, V.S., and Shingadia, D. (2003). Childhood malaria in East London. *Pediatr Infect Dis J* **22**, 814–19.

Lewallen, S., Harding, S.P., Ajewole, J., Schulenburg, W.E., Molyneux, M.E., Marsh, K., Usen, S., White, N.J., and Taylor, T.E. (1999). A review of the spectrum of clinical ocular fundus findings in P. falciparum malaria in African children with a proposed classification and grading system. *Trans R Soc Trop Med Hyg* **93**, 619–22.

Lewallen, S., White, V.A., Whitten, R.O., Gardiner, J., Hoar, B., Lindley, J., Lochhead, J., McCormick, A., Wade, K., Tembo, M., Mwenechanyana, J., Molyneux, M.E., and Taylor, T.E. (2000). Clinical-histopathological correlation of the abnormal retinal vessels in cerebral malaria. *Arch Ophthalmol* **118**, 924–8.

Lewallen, S.W., Whitten, V.A., Gardiner, R.O., Hoar, J., Lindley, B., Lochhead, J., McCormick, J., Wade, A., Tembo, K., Mwenechanyana, M., Molyneux, J., Taylor, ME., TE. (2000). Clinical-histopathological correlation of the abnormal retinal vessels in cerebral malaria. *Arch Ophthalmol* **118**, 924–8.

Maitland, K., Levin, M., English, M., Mithwani, S., Peshu, N., Marsh, K., and Newton, C.R. (2003a). Severe *P. falciparum* malaria in Kenyan children: evidence for hypovolaemia. *Q J med* **96**, 427–34.

Maitland, K., Nadel, S., Pollard, A.J., Williams, T.N., Newton, C.R., and Levin, M. (2005a). Management of severe malaria in children: proposed guidelines for the United Kingdom. *BMJ* **331**, 337–43.

Maitland, K., and Newton, C.R. (2005). Acidosis of severe falciparum malaria: heading for a shock? *Trends Parasitol* **21**, 11–16.

Maitland, K., Pamba, A., English, M., Peshu, N., Levin, M., Marsh, K., and Newton, C.R. (2005b). Pre-transfusion management of children with severe malarial anaemia: a randomised controlled trial of intravascular volume expansion. *Br J Haematol* **128**, 393–400.

Maitland, K., Pamba, A., English, M., Peshu, N., Marsh, K., Newton, C.R.J.C., and Levin, M. (2005c). Randomized trial of volume expansion with albumin or saline in children with severe malaria: preliminary evidence of albumin benefit. *Clin Infect Dis* **40**, 538–45.

Maitland, K., Pamba, A., Fegan, G., Njuguna, P., Nadel, S., Newton, C.R., and Lowe, B. (2005d). Perturbations in electrolyte levels in kenyan children with severe malaria complicated by acidosis. *Clin Infect Dis* **40**, 9–16.

Maitland, K., Pamba, A., Newton, C.R., and Levin, M. (2003b). Response to volume resuscitation in children with severe malaria. *Pediatr Crit Care Med* **4**, 426–31.

Molyneux, M.E., Taylor, T.E., Wirima, J.J., and Borgstein, A. (1989). Clinical features and prognostic indicators in paediatric cerebral malaria: a study of 131 comatose Malawian children. *Q J Med* **71**, 441–59.

Moody, A.H., and Chiodini, P.L. (2002). Non-microscopic method for malaria diagnosis using OptiMAL IT, a second-generation dipstick for malaria pLDH antigen detection. *Br J Biomed Sci* **59**, 228–31.

Newton, C.R., Crawley, J., Sowumni, A., Waruiru, C., Mwangi, I., English, M., Murphy, S., Winstanley, P.A., Marsh, K., and Kirkham, F.J. (1997). Intracranial hypertension in Africans with cerebral malaria. *Arch Dis Child* **76**, 219–26.

Newton, C.R., Kirkham, F.J., Winstanley, P.A., Pasvol, G., Peshu, N., Warrell, D.A., and Marsh, K. (1991a). Intracranial pressure in African children with cerebral malaria. *Lancet* **337**, 573–6.

Newton, C.R., Kirkham, F.J., Winstanley, P.A., Pasvol, G., Peshu, N., Warrell, D.A., and Marsh, K. (1991b). Intracranial pressure in African children with cerebral malaria. *Lancet* **337**, 573–6.

Newton, C.R., and Krishna, S. (1998). Severe falciparum malaria in children: current understanding of pathophysiology and supportive treatment. *Pharmacol Ther* **79**, 1–53.

Newton, C.R., Taylor, T.E., and Whitten, R.O. (1998). Pathophysiology of fatal falciparum malaria in African children. *Am J Trop Med Hyg* **58**, 673–83.

Palmer, C.J., Bonilla, J.A., Bruckner, D.A., Barnett, E.D., Miller, N.S., Haseeb, M.A., Masci, J.R., and Stauffer, W.M. (2003). Multicenter study to evaluate the OptiMAL test for rapid diagnosis of malaria in U.S. hospitals. *J Clin Microbiol* **41**, 5178–82.

Pamba, A., and Maitland, K. (2004). Capillary refill: prognostic value in Kenyan children. *Arch Dis Child* **89**, 950–5.

Powell, K.R., Sugarman, L.I., Eskenazi, A.E., Woodin, K.A., Kays, M.A., McCormick, K.L., Miller, M.E., and Sladek, C.D. (1990). Normalization of plasma arginine vasopressin concentrations when children with meningitis are given maintenance plus replacement fluid therapy. *J Pediatr* **117**, 515–22.

Riddle, M.S., Jackson, J.L., Sanders, J.W., and Blazes, D.L. (2002). Exchange transfusion as an adjunct therapy in severe *Plasmodium falciparum* malaria: a meta-analysis. *Clin Infect Dis* **34**, 1192–8.

Schmutzhard, E., and Gerstenbrand, F. (1984). Cerebral malaria in Tanzania. Its epidemiology, clinical symptoms and neurological long term sequelae in the light of 66 cases. *Trans R Soc Trop Med Hyg* **78**, 351–3.

Taylor, T.E., Borgstein, A., and Molyneux, M.E. (1993). Acid-base status in paediatric *Plasmodium falciparum* malaria. *Q J Med* **86**, 99–109.

Taylor, T.E., Fu, W.J., Carr, R.A., Whitten, R.O., Mueller, J.G., Fosiko, N.G., Lewallen, S., Liomba, N.G., and Molyneux, M.E. (2004). Differentiating the pathologies of cerebral malaria by postmortem parasite counts. *Nat Med* **10**, 143–5.

Travel, H.S.S. (2003). Illness in England, Wales, and Northern Ireland associated with foreign travel: a baseline report to 2002. Health Protection Agency, London.

Walker, O., Salako, L.A., Sowunmi, A., Thomas, J.O., Sodeine, O., and Bondi, F.S. (1992). Prognostic risk factors and post mortem findings in cerebral malaria in children. *Trans R Soc Trop Med Hyg* **86**, 491–3.

Waller, D., Crawley, J., Nosten, F., Chapman, D., Krishna, S., Craddock, C., Brewster, D., and White, N.J. (1991b). Intracranial pressure in childhood cerebral malaria. *Trans R Soc Trop Med Hyg* **85**, 362–4.

Waller, D., Krishna, S., Crawley, J., Miller, K., Nosten, F., Chapman, D., ter Kuile, F.O., Craddock, C., Berry, C., Holloway, P.A., et al. (1995). Clinical features and outcome of severe malaria in Gambian children. *Clin Infect Dis* **21**, 577–87.

White, V.A., Lewallen, S., Beare, N., Kayira, K., Carr, R.A., and Taylor, T.E. (2001). Correlation of retinal haemorrhages with brain haemorrhages in children dying of cerebral malaria in Malawi. *Trans R Soc Trop Med Hyg* **95**, 618–21.

3

The Neglected Tropical Diseases: The Ancient Afflictions of Stigma and Poverty and the Prospects for their Control and Elimination

Peter Hotez, Eric Ottesen, Alan Fenwick, and David Molyneux

1. Introduction

The World Health Organizations and other international health agencies identify a select group of 13 tropical infections as the neglected tropical diseases (NTDs). These diseases, which include leprosy, kala-azar, river blindness, guinea worm, schistosomiasis, hookworm and lymphatic filariasis, strike the world's poorest people living in remote and rural areas of low-income countries in Sub-Saharan Africa, Asia and the Americas. They inflict suffering by causing life-long disabilities, disfigurement, reduced economic productivity, and social stigma (WHO, 2003). Unlike better-known global health threats such as HIV-AIDS, malaria, and tuberculosis, the NTDs do not receive enough international attention. Instead, they are neglected diseases among forgotten people found only in the setting of geographic isolation and intense poverty (Molyneux, 2004).

Impoverished and marginalized populations with the NTDs represent the lowest priority markets for U.S. and European pharmaceutical manufacturers. The NTDs do not occur in the industrialized world or even among the substantial wealthy and middle-classes in developing countries. They are not a significant health risk for foreign travelers or the military. This is in contrast to the more substantial commercial markets for HIV-AIDS, malaria and tuberculosis ("the big three"). The recent creation of massive funding schemes for the big three, such as The Global Fund to Fight AIDS, Tuberculosis, and Malaria, and the U.S. President's Emergency Plan for AIDS Relief provides additional financial incentives, as well as a certain amount of panache and luster. In contrast, the commercial market for NTD drug

Hot Topics in Infection and Immunity in Children, edited by Andrew J. Pollard and Adam Finn. Springer, New York, 2006

and vaccine development is essentially zero, and there is as yet no Bono or equivalent celebrity to champion its cause.

Since the millennium, the hopeless outlook for those afflicted with the NTDs has undergone a surprising turnaround. Through unprecedented advocacy and the creation of new and innovative public-private-partnerships, often based on generous pharmaceutical company donations of effective products, several NTDs have been targeted for control or elimination. Although not well known by the lay public, a silent revolution is gathering, which could alleviate some of our planet's greatest health disparities.

2. The NTDs as the Ancient Afflictions of Stigma and Poverty

The NTDs are caused by parasitic worms, protozoa, and the bacterial agents of leprosy, Buruli ulcer, and trachoma (Table 3.1). Unlike the well-publicized newly emerging infections, such as Ebola, West Nile fever, or avian influenza, the NTDs have burdened humanity since the beginning of recorded history (WHO, 2003). Descriptions of leprosy, schistosomiasis, guinea worm, hookworm, trachoma and other NTDs are found in the Bible (Dirckx, 1985; Hulse, 1971; Ceccarelli, 1994), the Talmud (Ostrer, 2002), *Papyrus Ebers* (c.1550 B.C.), *Kahun papyrus*

Table 3.1. The Neglected Tropical Diseases

Parasitic Diseases (Protozoan)	Etiologic Agent
Kala-azar (Visceral leishmaniasis)	*Leishmania donovani*
African Sleeping Sickness (African trypanosomiasis)	*Trypanosoma gambiense* and *T. rhodesiense*
Chagas Disease (American trypanosomiasis)	*Trypanosoma cruzi*

Parasitic Diseases (Helminth)	
Schistosomiasis	*Schistosoma haematobium, S. mansoni, S. japonicum*
Lymphatic Filariasis (elephantiasis)	*Wuchereria bancrofti* and *Brugia malayi*
Onchocerciasis (river blindness)	*Onchocerca volvulus*
Dracunculiasis (guinea worm)	*Dracunculiasis medinensis*
Soil-transmitted Helminthiases	
Ascariasis (roundworm)	*Ascaris lumbricoides*
Trichuriasis (whipworm)	*Trichuris trichiura*
Hookworm	*Necator americanus* and *Ancylostoma duodenale*

Bacterial Diseases	
Leprosy	*Mycobacterium leprae*
Buruli Ulcer	*Mycobacterium ulcerans*
Trachoma	*Chlamydia trachomitis*

Table 3.2. Annual Deaths from the Neglected
Tropical Diseases[1]

Kala-azar	51,000
African Trypansomiasis	48,000
Schistosomiasis	15,000
Chagas disease	14,000
Soil-transmitted Helminthiases[2]	12,000
Leprosy	6,000
Lymphatic Filariaisis	0
Onchocerciasis	0
Guinea Worm	0
Total Neglected Diseases	146,000

[1]World Health Report 2004 (Annex Table 2).
[2]Ascariasis, Hookworm, and Trichuriasis.
[3]Some estimates indicate that African trypanosomiasis causes 100,000 deaths, leishmaniasis 100,000 deaths, hookworm 65,000 deaths, and schistosomiasis 150,000–280,000 deaths annually. Therefore more than 500,000 deaths annually may result from NTDs.

(c.1900 B.C.), the writings of Hippocrates and other ancient texts (Grove, 1990). Joshua's curse and the abandonment of Jerico's walls have been attributed to schistosomiasis (Hulse, 1971), while guinea worms (*Dracunculiasis medinensis*) are believed to be the "fiery serpents" that attacked the Israelites in the desert during their exodus from Egypt (Grove, 1990).

Every year, the NTDs annually kill at least as many people as were killed as a result of the 2004 Christmas tsunami (Table 3.2). Some estimates suggest that more than 500,000 people die annually from NTDs. However, their overall toll cannot be measured by mortality alone. Generally speaking, the NTDs cause much more chronic disability and morbidity rather than death. For instance, Chagas disease causes a chronic and disabling heart condition; hookworm, chronic intestinal blood loss and anemia; onchocerciasis, blindness and intense itching that results in chronic skin changes; guinea worm, localized pain that prevents temporary disuse of a lower limb.

Whereas it is relatively easy to understand the significance of death rates and mortality figures, it is more difficult to translate chronic disability and morbidity into a value that is readily understood by public health officials and health advocates. The human toll from the NTDs is better explained in terms of their disease burden using the disability-adjusted life year (DALY) as a metric. When measured in DALYs (Murray, 1996), the NTDs account for approximately one-quarter of the global disease burden from HIV-AIDS and almost the same burden as malaria (Table 3.3). Even these high DALY figures are probably underestimates based on new studies pointing out the hidden morbidity and mortality of chronic infections with schistosomiasis, onchocerciasis and other NTDs (King et al., 2005; Little et al., 2004; Hotez et al., 2004a).

The NTDs also have a huge social impact due to lost educational potential, reduced economic productivity, and stigma. Schistosomiasis and hookworm impair the ability of school-aged children to learn in school (King et al., 2005; Hotez

Table 3.3.   The Neglected Tropical Diseases Ranked by
Disease Burden (DALYs 000)[1]

Lymphatic Filariasis	5,654
Soil-transmitted Helminthiases[2]	4,706
Kala-azar	2,357
Trachoma	2,329
Schistosomiasis	1,760
African Sleeping Sickness	1,598
Onchocerciasis	987
Chagas Disease	649
Leprosy	177
Buruli Ulcer	<100
Guinea Worm	<100
Total Neglected Diseases	20,217
Total HIV-AIDS[3]	84,458

[1]WHO, World Health Report 2002.
[2]Ascariasis, Hookworm, Trichuriasis.
[3]WHO, World Health Report 2004.

et al., 2004a), while lymphatic filariasis, guinea worm and river blindness result in missed days of work for adults, especially the family breadwinner (Little et al., 2004). This includes the loss of US $1 billion annually from lymphatic filariasis in India alone (Ramaiah et al., 2000), and $5.3 billion from blinding trachoma (Frick et al., 2003), and substantial reductions in future wage-earning capacity as a result of chronic hookworm infection in childhood (Bleakley, 2003). Therefore, the NTDs not only occur in the context of poverty, but through their adverse social impact they also promote poverty (WHO, 2002). The stigmatizing nature of the NTDs often causes afflicted individuals to shun social contact or seek medical attention (WHO, 2002). There is also an intimate link between the incidence of NTDs and conflict. The interruption of public health services and forced human migrations resulting from decades of wars in Angola, Congo and Sudan, for example, has produced resurgence in African sleeping sickness, kala-azar, and guinea worm (Molyneux, 1997; Ekwanzala et al., 1996; Hotez, 2001).

3. The Role of the Pharmaceutical Industry

By the year 2015 each of the 191 United Nations member states have pledged to meet a set of Millennium Development Goals for global health and human rights (Table 3.4). Considering the NTDs probably impact on more of these goals than any other single entity it is astonishing that the NTDs still retain their neglected status. The global pharmaceutical industry failure to invest in new drugs for these diseases has produced particularly serious consequences. Of the 1,233 drugs commercialized worldwide between 1975 and 1997 only 4 were developed specifically for the NTDs (Pecoul et al., 1999; Kremer and Glennerster, 2004). The Pharmaceutical Manufacturers of America similarly reports that only a single drug for the NTDs was in the

Table 3.4. Reducing Neglected Tropical Disease Burden Impacts on 7 of the 8 UN Millennium Development Goals

---
1. *Eradicate extreme poverty and hunger*
2. *Achieve universal primary education*
3. *Promote gender equality and empower women*
4. *Reduce child mortality*
5. *Improve maternal health*
6. *Combat HIV/AIDS, malaria and other diseases*
7. Ensure environmental sustainability
8. *Develop a global partnership for development*

---

pipeline for 2000, compared with 8 for erectile dysfunction, 7 for obesity, and 4 for sleep disorders (Medicins Sans Frontieres, 2001).

The absence of drug research and development (R&D) means that we have only a handful of available drugs for the NTDs and no vaccines. Many of the existing drugs were developed more than 50 years ago, and are themselves highly toxic. For example, the drug of choice used to treat the late stages of African trypanosomiasis is an arsenic-containing compound known as melarsoprol developed in the 1940s. The treatment goal is essentially to use the arsenic to poison the trypanosomes before seriously affecting the patient. A toxic antimony-containing compound is still the treatment of choice for kala-azar (visceral leishmaniasis) in many parts of the world. The absence of R&D and interest in manufacturing new drugs for the NTDs reflects their market failure. The last 50 years have been a dark age for the R&D and manufacture of NTDs drugs and vaccines.

4. New Promise Through Public Private Partnerships

Are we emerging from the dark ages? In 1988, the pharmaceutical giant, Merck & Co., Inc., and later, GlaxoSmithKline and Pfizer, inaugurated a series of extraordinary NTDs drug donation programs. The Ivermectin Donation Program (Merck & Co., Inc.), a partnership with the WHO, the Task Force for Child Survival, and other non-governmental organizations (www.mectizan.org) began after the announcement by Roy Vagelos, Merck CEO, that his company would donate the drug free of charge for as long as needed to anyone with river blindness (Levine, 2004a). Over the last 17 years the Donation Program has provided more than 300 million treatments at roughly $1.50 per dose (Levine, 2004a).

Ivermectin was not in fact developed by Merck to treat river blindness or LF, but rather as part of its veterinary discovery efforts to identify promising new compounds for parasites of livestock. Unlike the NTDs of humans, the parasitic diseases of cattle and sheep represent a market that exceeds US$ 1 billion annually. In 1987, Ivermectin was the ranked as Merck's second best selling drug (Levine, 2004a). When Ivermectin was shown to be safe and effective for the treatment of human river blindness in Africa Merck & Co., Inc. appropriately realized that its small commercial human market was not worth efforts to sell the drug as a medical product. They chose instead to donate it in an extraordinary and at the time unique gesture of good will and corporate philanthropy. However, the fact that the drug required donation for widespread medical use illustrates the reluctance of the pharmaceutical industry to embark on R&D for the NTDs. Despite the clear success of the Mectizan® Donation Program in relieving human suffering, it did not overcome the hurdles that block R&D and manufacture of new products.

The barriers in NTD product R&D started to erode beginning in the late 1990s through the establishment of a new type of non-profit entity. Known as the Product Development Private-Public-Partnership (PD-PPPs), these non-profits embrace industry practices and collaboration, using private sector approaches to attack R&D challenges, and develop products for developing countries (Widdus and White, 2004). New private public partnerships include vaccine development initiatives for leishmaniasis, hookworm, and infectious diarrheas, and drug development initiatives for leishmaniasis, African trypanosomiasis, and Chagas disease (Table 3.5). This is resulting in an unprecedented involvement of the non-profit sector in the manufacture and testing of new products. A major stimulus to establish PPPs was a quantum leap in funding (referred to as "push mechanisms") that began in 1999 (Hotez, 2004a), including key contributions from the Bill and Melinda Gates Foundation. The U.S. Orphan Drug Act, which provides tax credits and grants for diseases of fewer than 200,000 in the U.S. (Kremer and Glennerster, 2004), and NIH small business innovation grants, which foster academic partnering to manufacture new products, also provided incentives.

Concurrent with the opportunities created by private push mechanisms, many middle-income countries have dramatically increased their contributions to government-sponsored R&D. The term *innovative developing country* (IDC) has been applied to nations such as Brazil, China, Cuba, Egypt, India, South Africa, and South Korea, which have modest economic strength but are relatively advanced with respect to their sophistication in health biotechnology and government-supported research and development (Morel et al., 2005). The IDCs have capacity for producing their own pharmaceuticals and vaccines; at the same time, all except South Korea suffer from endemic NTDs. This creates a unique mix that encourages these

Table 3.5. Major Public Private Partnerships Committed to the Neglected Tropical Diseases

---

Drugs for Neglected Diseases Initiative – DNDi (Geneva)
WHO Partnership for Parasite Control – PPC (Geneva)
Institute for One World Health (IOWH) (San Francisco CA)
Diseases of the Most Impoverished – DOMI (Seoul Korea)
Human Hookworm Vaccine Initiative – HHVI (Wash DC)
International Trachoma Initiative (New York NY)
Infectious Diseases Research Institute (Seattle WA)
Global Alliance to Eliminate LF (Liverpool UK)
Schistosomiasis Control Initiative (London UK and Boston MA)
African Programme for Onchocerciasis Control (APOC)
Onchocerciasis Elimination Programme in the Americas (OEPA)

---

nations to begin solving their own NTDs problems with only modest technical or financial assistance from more developed countries.

5. Vertical Control Efforts

The new PPPs and IDCs are providing a new generation of products (often referred to as "tools") to control the NTDs. However, in the meantime, another group of PPPs, which include the Global Alliance to Eliminate Lymphatic Filariasis (GAELF), the Schistosomiasis Control Initiative (SCI), the International Trachoma Initiative (ITI), the African Programme for Onchocerciasis Control (APOC), and several initiatives focused on leprosy, are directly working to apply existing and new tools to disease control. To date, these PPPs are having an enormous impact on reducing NTD disease prevalence and burden through vertical control programs. For example, through APOC, a 15-year partnership of the endemic countries World Bank, WHO, NGDO's and other international organizations, more than 67 million doses of Mectizan® have been distributed and administered through a system of community-directed treatment – ultimately the program aspires to scale up to about 90 million yearly treatments and to prevent 43,000 cases of blindness annually (Levine et al., 2004a). Similarly, through widespread use of two drug combinations comprised of either diethylcarbamazine and albendazole, or Mectizan® and albendazole the GAELF is making dramatic progress in reducing the global disease burden of LF (Molyneux and Zagaria, 2002). The ITI is reducing the global burden of blinding trachoma through use of the Pfizer-donated antibiotic Zithromax®, surgery for trichiasis, and complementary hygienic measures (environmental and sanitation and face washing SAFE strategy) (Mecaskey et al., 2003), and the number of new cases of leprosy has been dramatically reduced through combination antibiotic therapy (Lockwood and Suneetha, 2005). Through the use of a synthetic long-acting pyrethroid insecticide, an initiative known as the Southern Cone Initiative to Control/Eliminate Chagas Disease has dramatically reduced or halted the incidence of new cases of Chagas Disease in Argentina, Bolivia, Brazil, Chile, Paraguay, Uruguay and Peru since it began in 1991 (Levine et al., 2004b). Most dramatic of all, is the progress in guinea worm control with a 98 percent reduction in the number of cases worldwide since control efforts began in the early 1980s (Hopkins et al., 2002). Finally, LF in China and schistosomiasis in Egypt have been controlled beyond expectations by long-term investment in government sponsorship and World Bank investment in control activities. The reasons for these successes include donor commitment, regional approaches, specific technical needs, and the absence of emerging drug or insecticide resistance (WHO, 2003).

If these trends continue the WHO projects that some important NTDs, such as lymphatic filariasis, river blindness, Chagas disease, blinding trachoma, guinea worm, and leprosy, may be eliminated by the year 2020. The term *elimination* refers to the reduction of disease incidence to zero but requiring ongoing public health measures to prevent reemergence (Hotez et al., 2004b; Molyneux et al., 2004). Elimination of some of the NTDs is considered feasible because disease transmission could be interrupted with specific control tools, and does not require sanitation and clean water, which are far more costly (WHO, 2003).

6. Integrated Control

There is additional excitement in the NTD scientific and public health community about the possibility of bundling some of these vertical control efforts in a program of integrated control (Molyneux and Nantulya, 2004; Molyneux et al., 2005). With four drugs, albendazole, Mectizan®, Zithromax®, and praziquantel, it is possible to target seven of the NTDs, namely lymphatic filariasis, onchocerciasis, hookworm, trichuriasis, ascariasis, schistosomiasis, and trachoma in areas where they geographically overlap, such as in Sub-Saharan Africa and focal regions of the Americas (Fenwick et al., 2005; Molyneux et al., 2005). In poor and rural areas in Africa it is not uncommon to identify individuals who are polyparasitized with three or more NTD etiologic agents. The four drugs are mutually compatible in terms of delivery approaches. Through integrated control efforts the international community could potentially control or eliminate morbidity and blindness from these NTDs, and at a fraction of the cost needed to control other diseases (Fenwick et al., 2005).

The Commission for Africa, established by the Britain's Prime Minister Tony Blair recognized the importance of NTDs in their 2005 report (www.commissionforafrica.org). In response to the Commission's calls to implement pro-poor strategies, Fenwick et al. (2005) determined that approximately 500 million people at risk for NTDs in Africa could be treated with all four drugs at a cost of $200 million annually, or $0.40 per patient if these resources were allocated as a package. These figures take into consideration that the multinational corporations Merck, GSK, and Pfizer have committed to donating three of the four drugs, Mectizan®, albendazole, and Zithromax®, respectively.

7. New Generation Control Tools

Despite their promise, the NTD vertical and integrated control efforts are still fragile owing to their one- dimensional reliance on a single chemical agent or drug combination to achieve their intended goals (Hotez, 2004b). Therefore unanticipated drug resistance or other product failures could reverse their success. An example is the near eradication of malaria in India in the early 1960s, which was undermined by the rising costs of DDT to control mosquito populations, outright DDT resistance, and chloroquine anti-malarial drug resistance (Harrison, 1978). As a result, malaria returned to previous levels. Had ongoing R&D continued to generate new control tools, malaria could possibly have been eliminated in India. Similarly, vertical and integrated control efforts directed against hookworm are not likely to be effective in reducing transmission because of the variable efficacy of benzimidazole anthelminthic agents (e.g., albendazole or mebendazole), the high rates of post-treatment hookworm re-infection that occurs in areas of high transmission, and the diminishing efficacy of benzimidazoles with frequent and periodic use (Hotez et al., 2005). These lessons raise the stakes for the activities of the PPPs and IDCs. Unless new control tools are developed the promise for eliminating the NTDs may not be realized.

Successful elimination will therefore require further R&D investment. In 2003, the NIH Director established a Roadmap for the development of new therapies (www.nih.gov/roadmap). Although primarily intended for disease in the industrialized world, this also affords an opportunity to target the NTDs (Hotez, 2004b). In addition, the Bill and Melinda Gates Foundation has made significant investments in the development of new generation vaccines for NTDs including leishmaniasis and hookworm, and new drugs for African sleeping sickness, leishmaniasis, Chagas disease (www.gatesfoundation.org) (Hotez, 2001; Hotez et al., 2005). These new tools could become urgently needed in the face of emerging drug resistance. We might be facing once-in-a-lifetime opportunities to forever eliminate humankind's most ancient scourges. Never has the time been so propitious for ending centuries of neglect.

Acknowledgements

Peter Hotez is supported by the Human Hookworm Vaccine Initiative of the Sabin Institute and the Bill and Melinda Gates Foundation (BMGF). Alan Fenwick is supported by the Schistosomiasis Control Initiative of the BMGF. The authors thank Dr. Gavin Yamey of Public Library of Science for his thoughtful review of the manuscript.

References

Bleakley, H. (2003). Disease and development: evidence from the American South. *Journal of the European Economic Association.* **1**, 376–386.

Ceccarelli, G. (1994). Infettivologia e parasssitologia nella Bivvia. *Minerva Medica.* **85**, 417–422.

Dirckx, J.H. (1985). The biblical plague of "hemorrhoids". An outbreak of bilharziasis. *American Journal of Dermatopathology.* **7**, 31–36.

Ekwanzala, M., Pepin, J., Khonde, N., Molisho, S., Bruneel, H., and De Wals, P. (1996). In the heart of darkness: sleeping sickness in Zaire. *Lancet.* **348**, 1427–1430.

Fenwick, A., Molyneux, D., Nantulya, V. (2005). Achieving the millennium development goals. *Lancet.* **365**, 1029–1030.

Frick, K.D., Hanson, C.L., and Jacobson, G.A. (2003). Global burden of trachoma and economics of the disease. *American Journal of Tropical Medicine and Hygiene.* **69**, 1–10.

Grove, D.I. (1990). *A History of Human Helminthology.* CAB International, pp. 209, 469, 521, and 616.

Harrison, G. (1978). *Mosquitoes, Malaria & Man: A History of the Hostilities since 1880.* New York: E.P. Dutton, pp. 242–54.

Hopkins, D.R., Ruiz-Tiben, E., Diallo, N., Withers, P.C. Jr., and Maguire, J.H. (2002). Dracunculiasis eradication: and now, Sudan. *American Journal of Tropical Medicine and Hygiene.* **67**, 415–422.

Hotez, P.J. (2001). Vaccines as instruments of foreign policy. *EMBO reports.* **2**, 862–868.

Hotez, P.J. (2004a). Should we establish a North American school of global health sciences? *American Journal of Medical Sciences.* **328**, 71–77.

Hotez, P.J. (2004b). The National Institutes of Health roadmap and the developing world. *Journal of Investigative Medicine.* **52**, 246–247.

Hotez, P.J., Brooker, S., Bethony, J.M., Bottazzi, M.E., Loukas, A., and Xiao, S.H. (2004a). Hookworm infection. *New England Journal of Medicine.* **351**, 799–807.

Hotez, P.J., Remme, H., Buss, P., Alleyne, G., Morel, C., and Breman, J. (2004b). Combating tropical communicable diseases: workshop report of the disease control priorities project. *Clinical Infectious Diseases.* **38**, 871–878.

Hotez, P.J., Bethony, J., Brooker, S., and Albonico, M. (2005). Eliminating neglected diseases in Africa. *Lancet.* **365**, 2089.

Hulse, E.V. (1971). Joshua's curse and the abandonment of ancient Jericho: schistosomiasis as a possible medical explanation. *Med. Hist.* **15**, 376–386.

King, C.H., Dickman, K., Tisch, D.J. (2005). Reassessment of the cost of chronic helmintic infection: a meta-analysis of disability-related outcomes in endemic schistosomiasis. *Lancet.* **365**, 1561–1569.

Kremer, M. and Glennerster. (2004). *Strong Medicine, Creating Incentives for Pharmaceutical Research on Neglected Diseases.* Princeton University Press. pp. 25–64.

Levine, R. and the What Works Working Group. (2004a). *Millions Saved, Proven Successes in Global Health,* Case 6, Controlling Onchocerciasis in Sub-Saharan Africa, Center for Global Development, pp. 57–64.

Levine, R. et al. (2004b). *Millions Saved,* Case 11, Controlling Chagas Disease in the Southern Cone of South America, pp. 99–104.

Little, M.P., Breitling, L.P., Basanez, M.G., Alley, E.S., and Boatin, B.A. (2004). Association between microfilarial load and excess mortality in onchocerciasis: an epidemiological study. *Lancet.* **363**, 1514–1521.

Lockwood, D.N. and Suneetha, S. (2005). Leprosy: too complex a disease for a simple elimination paradigm. *Bulletin of the World Health Organization.* **83**, 230–235.

Mecaskey, J.W., Knirsch, C.A., Kumaresan, J.A., and Cook, J.A. (2003). The possibility of eliminating blinding trachoma. *Lancet.* **3**, 728–734.

Medecins Sans Frontieres. (2001). *Fatal Imbalance, The Crisis in Research and Development for Drugs for Neglected Diseases.* Medecins Sans Frontieres Access to Essential Medicines Campaign and the Drugs for Neglected Diseases Working Group, p. 12.

Molyneux, D.H. (1997). Patterns of change in vector-borne diseases. *Annals of Tropical Medicine and Parasitology.* **91**, 827–839.

Molyneux, D.H. (2004). "Neglected" diseases but unrecognized successes – challenges and opportunities for infectious disease control. *Lancet.* **364**, 380–383.

Molyneux, D.H. and Nantulya, V.M. (2004). Linking disease control programmes in rural Africa: a pro-poor strategy to reach Abuja targets and millennium development goals. *BMJ.* **328**, 1129–1132.

Molyneux, D.H. and Zagaria, N. (2002). Lymphatic filariasis elimination: progress in global programme development. *Annals of Tropical Medicine and Parasitology.* **96** Suppl 2, S15–40.

Molyneux, D.H., Hopkins, D.R., and Zagaria, N. (2004). Disease eradication, elimination and control: the need for accurate and consistent usage. *Trends in Parasitology.* **20**, 347–351.

Molyneux, D.H., Hotez, P.J., and Fenwick, A. (2005). "Rapid-impact Interventions": How a policy of integrated control for Africa's neglected tropical diseases could benefit the poor. *PLoS Medicine,* **2**: e336.

Morel, C.M., Acharya, T., Broun, D., Dangi, A., Elias, C., Ganguly, N.K., Gardner, C.A., Gupta, R.K., Haycock, J., Heher, A.D., Keusch, G.T., Hotez, P.J., Krattiger, A.F., Kreutz, F.T., Lall, S., Lee, K., Mahoney, R., Martinez-Palomo, A., Mashelkar, R.A., Matlin, S., Min, H.K., Mzimba, M., Oehler, J., Pick, W., Ridley, R.G., Senanayake, P., Singer, P., and Yun, M.Y. (2005). Health innovation: the neglected capacity of developing countries to address neglected diseases. *Science.* **309**, 401–404.

Murray, C.J.L. (1996). Rethinking DALYs. In: *The Global Burden of Disease* (Murray, C.J.L. and Lopez, A.D., eds.), Global Burden of Disease and Injury Series, Harvard School of Public Health on behalf of the WHO and World Bank, Harvard University Press.

Ostrer, B.S. (2002). Leprosy: medical views of Leviticus Rabba. *Early Science and Medicine.* **7**, 138–154.

Pecoul, B., Chirac, P., Trouiller, P., and Pond, J. (1999). Access to essential drugs in poor countries: a lost battle? *Journal of the American Medical Association.* **281**, 361–367.

Ramaiah, K.D., Das, P.K., Michael, E., and Guyatt, H. (2000). The economic burden of lymphatic filariasis in India. *Parasitol Today.* **16**, 251–253.

Widdus, R. and White, K. (2004). *Combating Diseases Associated with Poverty,* Financing Strategies for Product Development and the Potential Role of Public-Private-Partnerships, The Initiative on Public-Private-Partnerships for Health (IPPH), pp. 1–10.

World Health Organization. (2003). *Communicable diseases 2002: global defence against the infectious disease threat* (ed. Kindhauser, M.K.), Geneva, pp. 106–107.

4

Viral Haemorrhagic Fevers Caused by Lassa, Ebola and Marburg Viruses

Nigel Curtis

1. Introduction

The possibility of a viral haemorrhagic fever (VHF) in a patient presenting with fever and haemorrhagic symptoms, who has recently returned from an area endemic for VHF viruses, is likely to strike fear into many clinicians. An understanding of the epidemiology, pathogenesis and clinical features of VHFs, as well as potential treatment and preventive strategies, is important for the effective recognition and management of such patients.

VHFs are caused by a number of different encapsulated single-stranded RNA viruses. The survival of these viruses is dependent on their carriage in a natural reservoir (rodents, bats, mosquitos or, in the case of Ebola, an unknown host). Consequently the geographic distribution of VHFs is restricted by the distribution of the host species that harbours their causative virus. Humans are not the natural reservoir of VHF viruses but rather are incidentally infected from the natural reservoir hosts. Outbreaks of VHFs typically occur sporadically, irregularly and unpredictably. VHFs comprise a number of similar clinical syndromes although the specific diseases as well as their pathogenesis and treatment are different. Another shared and important feature of VHF viruses is the potential infectious hazard they pose to health workers and to laboratory workers handling samples.

There are four families of VHF viruses. The viruses within these families and the diseases they cause are detailed in Table 4.1. Only Lassa, Ebola and Marburg VHFs will be considered in this chapter. There are excellent reviews on these and other VHFs available elsewhere (Borio et al., 2002; Richmond and Baglole, 2003; Ndayimirije and Kindhauser, 2005; Feldmann et al., 2003; Geisbert and Hensley, 2004; Salvaggio and Baddley, 2004) in addition to comprehensive, up to date, internet-based resources from the UK Health Protection Agency (HPA) (HPA,

Hot Topics in Infection and Immunity in Children, edited by Andrew J. Pollard and Adam Finn.
Springer, New York, 2006

Table 4.1. Viruses causing haemorrhagic fever

Family	Genus	Virus	Disease	Natural Reservoir	Geographic Distribution
Filoviridae	Filovirus	Ebola subtype Zaire subtype Sudan subtype Ivory Coast subtype Reston	Ebola hemorrhagic fever	Unknown	Africa
		Marburg	Marburg hemorrhagic fever	Unknown	Africa
Arenaviridae	Arenavirus	Lassa	Lassa fever	Rodent	West Africa
		New World Arenaviridae Machupo Junin Guanarito Sabia	New world hemorrhagic fevers Bolivian Argentinian Venezuelan Brazilian	Rodent	Americas
Bunyaviridae	Nairovirus	Crimean-Congo hemorrhagic fever	Crimean-Congo hemorrhagic fever	Tick	Africa, central Asia, eastern Europe, Middle East
	Phlebovirus	Rift Valley fever	Rift Valley fever	Mosquito	Africa, Saudi Arabia, Yemen
	Hantavirus	Agents of hemorrhagic fever with renal syndrome	Hemorrhagic fever with renal syndrome	Rodent	Asia, Balkans, Europe, Americas
Flaviviridae	Flavivirus	Dengue	Dengue fever, Dengue hemorrhagic fever and Dengue shock syndrome	Mosquito	Asia, Africa, Pacific, Americas
		Yellow fever	Yellow fever	Mosquito	Africa, tropical Americas
		Omsk hemorrhagic fever	Omsk hemorrhagic fever	Tick	Central Asia
		Kyasanur Forest disease	Kyasanur Forest disease	Tick	India

2005b), EU Eurosurveillance website (Eurosurveillance, 2002), US Centers for Disease Control and Prevention (CDC) (CDC, 2005b) and World Health Organization (WHO) (WHO, 2005b). The potential deliberate release of VHF viruses as agents of biowarfare and bioterrorism has spawned additional interest and resources (CDC, 2005a; WHO, 2005a; HPA, 2005a; Eurosurveillance, 2004).

2. Epidemiology

2.1. Lassa Fever

The Arenavirus Lassa was first discovered in 1969 when two missionary nurses in Nigeria developed a VHF illness in the village after which the virus is now named. Lassa virus is endemic in parts of West Africa and outbreaks of the disease, typically occurring in the dry season between January and April and affecting all age groups and both sexes, have occurred in Guinea, Sierra Leone, Liberia, Nigeria and Central African Republic. Serological studies in Senegal, Mali and Congo suggest the disease also occurs in these countries. Lassa virus is unique amongst Old World Arenaviruses in causing secondary human to human transmission with the potential for large nosocomial outbreaks with a high case fatality rate. Lassa fever is said to be responsible for between 100,000 and 500,000 infections per year and around 5,000 deaths.

Lassa virus is zoonotic in rodents, in particular the "multimammate rat" (*Mastomys* species complex). These rodents are widely distributed over large parts of sub-Saharan Africa. The virus is transmitted and amplified in rodents and subsequently shed in urine, faeces and other excreta. The multimammate rat is characterised by frequent breeding with large numbers of offspring, and is a frequent coloniser of homes.

2.2. Ebola and Marburg Fever

Ebola virus is said to be the most virulent pathogen known to infect humans. It is named after the Ebola river in the Democratic Republic of Congo (DRC) (formerly Zaire), where the first known outbreak of Ebola VHF occurred in 1976. Since then, nearly 2,000 cases with over 1,200 deaths have been documented from this disease.

Marburg virus was first described in 1967 when an outbreak of VHF occurred in laboratory workers simultaneously in Marburg, Frankfurt and Belgrade. These workers were subsequently shown to have been infected from African green monkeys imported from Uganda for polio vaccine manufacture. Of the 31 infected lab workers, 7 died (all of whom belonged to the sub group with primary infections).

Subsequent to these original descriptions of Ebola and Marburg VHF there have been a number of well documented outbreaks of both these VHFs. Four subtypes of Ebola virus (Zaire, Sudan, Ivory Coast, and Reston) have been described and these have been responsible for outbreaks in the DRC, Gabon, Republic of Congo, Sudan, Uganda and Ivory Coast. Most recently, an outbreak in the Republic of Congo from April to June 2005 was responsible for 12 cases including 9 deaths. The Reston subtype of Ebola virus is unique in causing disease only in non-human primates. It was first described in Reston, Virginia in the USA in 1989 in Cynomolgus monkeys (*Macacca fascicularis*) imported from the Philippines. Although serological evidence suggested four scientists had become infected, none become sick. Outbreaks of Marburg VHF have been described in South Africa, Kenya and the DRC. Most recently, an outbreak in Uige province in Angola, that started in October 2004 continuing through to July 2005, was responsible for 374 cases with 329 deaths. The high case fatality ratio (88%) in this outbreak included at least 16 doctors and nurses. A characteristic of this outbreak was the high proportion of children under five affected, although this may reflect the fact that the outbreak seemed to have initiated amongst young children and mothers in a paediatric ward in a provincial hospital.

The reservoir for Ebola and Marburg viruses is unknown despite intensive investigation and searches following a number of outbreaks. Both viruses are believed to be zoonotic but the animal carrier host remains unknown. Although primates are infected by these viruses, monkeys are not believed to be the reservoir host because they die too rapidly from the disease to become carriers. Rather, it is thought that the carrier is a rare species or a species that only rarely comes into contact with humans and monkeys, or alternatively that the virus is not easily transmitted from the carrier host. It has been speculated that bats are the natural reservoir of these viruses. This is an attractive proposition because a number of outbreaks have been associated with caves and mines, and bats experimentally infected with the virus do not die. However, the virus has not been isolated from bats in the wild. The natural reservoir host therefore remains unknown.

3. Transmission

3.1. Lassa Fever

Lassa virus is transmitted from rodents to humans by the inhalation of aerosolised virus, the ingestion of food contaminated by infected rodent excreta or when the *Mastomys* is eaten (the rodent is a delicacy in many parts of West Africa). Human to human transmission occurs by direct contact with blood, tissue, secretions or excretions from infected patients, and also by needlestick injury.

3.2. Ebola and Marburg Fever

As with Lassa virus, Ebola and Marburg viruses are transmitted by direct contact with blood or secretions (saliva, respiratory, urine, faeces, vomitus), organs or semen from infected patients. The virus can also be transmitted by handling or eating ill or dead infected animals. Transmission through contaminated syringes and needles has also been shown to be particularly efficient. Finally, transmission has also been strongly associated with practices associated with burial ceremonies in Africa. In contrast, transmission through casual contact is rare. Further, airborne (aerosol) spread is believed *not* to occur naturally although has been demonstrated in research conditions. Most importantly, Ebola and Marburg (in common with Lassa) viruses are not transmitted during the incubation period.

4. Virology

4.1. Lassa Virus

Lassa virus is an enveloped, round or pleomorphic 50 to 300 nanometre diameter, single stranded RNA virus which possesses two genes with an ambisense orientation on each of two segments encoding 5 proteins. The virus is inactivated by heating to 56°C, pH <5.5 or >8.5, ultraviolet or gamma irradiation, or detergents.

4.2. Ebola and Marburg Viruses

Ebola and Marburg viruses are filamentous or bacillus form with a uniform diameter of 80 nanometres but a variable length between 860 and 1086 nanometres that can help distinguish Marburg and the different subtypes of Ebola viruses. These viruses contain non-segmented negative-strand RNA with a 19 kb genome that encodes 7 structural (and in the case of Ebola also one non-structural) proteins. Ebola and Marburg viruses have 55% nucleotide homology. The viruses are coated with an envelope that contains glycoprotein peplomers on the surface, which is believed to be an important virulence factor.

5. Clinical Features

5.1. Lassa Fever

Most infections with Lassa virus are mild and/or subclinical. Zoological studies in Guinea, Nigeria and Sierra Leone suggest that up to half the population have previously been exposed to this virus. Hospitalised cases therefore represent only the tip of the iceberg and severe symptomatic disease is said to occur in approximately 20% of infected individuals. In Sierra Leone infection rates of up to 20% of the population each year have been documented and it has been shown that up to 20% of febrile patients have illnesses that are attributable to this virus.

After an incubation period of between 5 and 21 days there is the gradual onset of an illness that lasts between one and four weeks. The clinical manifestations of this and other VHFs are protean but typically cases start with non-specific symptoms of fever, headache, arthralgia and myalgia. Patients may also suffer from a sore throat with exudative pharyngitis and cough. This is followed by abdominal symptoms (vomiting, abdominal pain and tenderness, diarrhorea). Severe illness with haemorrhagic symptoms and signs, confusion, neck and facial swelling, and progression to shock with multisystem organ failure occurs in only a minority (fewer than 20%) of patients. Other features may include encephalopathy, encephalitis, meningitis, cerebellar syndrome, pericarditis, uveitis and orchitis. A characteristic feature of Lassa fever, unrelated to the severity of the acute illness, is a unilateral or bilateral sensorineural deafness that occurs in one third of patients, typically during early convalescence, and which is permanent in up to a third of those affected by this complication.

The case fatality in hospitalised patients is between 15 and 25% but has been documented up to 50% in outbreaks. However, as a result of to the larger number of mild or asymptomatic infections, the overall case fatality in Lassa fever is only about 1%. Poor prognosis has been shown to be associated with high viraemia, AST greater than 150 IU/L, bleeding, encephalitis, oedema, and infection during the third trimester of pregnancy. The presence of a high viraemia and a raised AST alone were associated with a mortality of 78% (compared with 17% without these features) in one study.

5.2. Lassa Fever in Pregnancy and Children

Lassa VHF is associated with particularly severe disease in pregnant women and their infants and is associated with high levels of viraemia. Infection during the third trimester of pregnancy is associated with maternal mortality greater than 30% and increased foetal and neonatal mortality (greater than 85%). Delivery improves the mother's chance of survival.

Lassa virus infection is a significant cause of paediatric admissions in some areas of West Africa and the clinical features can be more variable and clinically confusing in this age group. The "swollen baby syndrome" comprising oedema, jaundice, abdominal distension and bleeding is associated with a particularly poor prognosis.

5.3. Ebola and Marburg Fever

The incubation period for Ebola virus is between 3 and 21 days and for Marburg virus between 3 and 9 days. In contrast to Lassa fever, the onset of VHF caused by these viruses is abrupt, with onset of a prodrome lasting less than 7 days. This comprises non-specific symptoms including fever, chills, severe headache, malaise, myalgia and a maculopapular rash (onset day 2 to 7; typically on the face, neck, trunk and arms). This is followed by a rapid and progressive deterioration with severe, watery diarrhoea, abdominal pain and cramps, nausea and vomiting. Patients typically have ghost-like, drawn features with deep-set eyes and an expressionless facies. This phase is characterised by extreme lethargy. Other clinical features include chest pain, sore throat, hiccups, conjunctivitis, haematemesis, cough, photophobia and back pain.

Haemorrhagic manifestations occur in about 70% of patients with Ebola and 45% of patients with Marburg virus, usually around day 5 to 7 of the illness. Widespread bleeding occurs internally and from body orifices, and is associated with disseminated intravascular coagulation. Patients may have petechiae, epistaxis, haematemesis, melaena, bleeding gums, and bleeding from puncture sites. At this time there is a sustained high fever and there may be associated jaundice, pancreatitis, severe weight loss and delirium. Central nervous system involvement may include confusion and irritability. Death from Ebola and Marburg VHF occurs from massive blood loss and shock with associated organ systemic failure. This occurs between day 7 and 14 in Ebola VHF and between day 8 and 9 in Marburg VHF (although earlier haemorrhagic manifestations with death between day 3 and 7 was a feature of the Angola outbreak in 2005). Occasionally death is as early as day 2 after onset of symptoms. Case fatalities between 50 and 90% are typical.

6. Pathogenesis

The pathogenesis of VHFs is complex but involves widespread cytolytic infection with associated inhibition of both innate and adaptive immune responses. Endothelial cell infection and cytokine-mediated immune mechanisms are responsible for the vascular leak, coagulopathy and apoptosis that underlie the clinical features. Detailed reviews of the pathogenesis of VHFs have recently been published (Mahanty and Bray, 2004; Geisbert and Jahrling, 2004).

7. Differential Diagnosis

As a result of the early non-specific features of VHFs, and the multiple and variable subsequent clinical features, the differential diagnosis for VHFs is large. Specific clinical diagnosis is often difficult (Lowenstein, 2004). However, the presence of the combination of fever, pharyngitis, retrosternal chest pain and proteinuria predicted 70% of subsequently laboratory-confirmed Lassa VHF cases in one series (McCormick et al., 1987).

Laboratory testing of samples is potentially dangerous for scientists and requires biosafety laboratory category 4 testing facilities (Klietmann and Ruoff, 2001). In suspected cases, laboratory tests should be minimised and possibly restricted to the exclusion of malaria as a differential diagnosis only. When tested, patients with VHF typically have leucopenia (but leucoytosis in some Lassa VHF patients) followed by neutrophilia, thrombocytopenia, anaemia or haemoconcentration, coagulopathy with disseminated intravascular coagulation, and increased hepatic transaminases. Specific tests for VHF include ELISA for both viral antigen and antibody, RT-PCR, viral culture and immunohistocytochemistry.

8. Treatment

Supportive measures are the mainstay of treatment of VHFs. There is no effective treatment for Ebola and Marburg VHF. In contrast, ribavirin is effective in Lassa VHF when started within 7 days. In one study in Sierra Leone it was associated with a reduction in mortality from 55% to 5% in patients with severe disease associated with a high AST (McCormick et al., 1986). Other drugs, such as cysteine protease inhibitors, are in developmental phase only. The use of convalescent plasma has shown mixed results. Although highly effective in New World Arenavirus infection (e.g., Argentine VHF), convalescent plasma has not been effective in Lassa VHF. This is believed to be due to the fact that neutralising antibodies take weeks to develop and also show low titres and low avidity in animal experiments. Of more concern is the fact that convalescent plasma has been associated with enhanced viral replication in experimentally-infected animal models. This treatment is also limited by practical considerations.

9. Prevention and Control Measures

9.1. Vaccine

A number of strategies have been used for the development of both Lassa and Ebola virus vaccines (Baize et al., 2001; Fisher-Hoch and McCormick, 2004; Sullivan et al., 2000). One novel approach involving immunisation with a DNA priming vaccine followed by a booster immunisation containing recombinant adenovirus expressing glycoprotein was associated with the development of both humoral and cellular immunity against Ebola virus in Cynomolgus monkeys (Sullivan et al., 2000). Recently, the threat of bioterrorism using VHF agents has underpinned renewed impetus and enthusiasm for the development of effective vaccines (Baize, 2005; Jones et al., 2005; Peters, 2005).

9.2. Community Awareness, Education and other Control Measures

Education, particularly in concert with the support of local traditional healers and religious leaders, has proven a valuable component of outbreak control in the African setting as some cultural and traditional practices can be counterproductive, particularly those associated with certain burial practices. After death, cadavers should be buried or burned immediately and sheets, gloves and clothing that have been in contact with patients also need to be burned (CDC, 2005c). Finally, an integral part of the control of Lassa fever is effective control of the rodents that harbour the virus.

10. Recognition and Management of Supected Cases of Viral Haemorrhagic Fever

Comprehensive and detailed guidelines for the recognition and management of patients with suspected VHF are readily available for use in the UK (1996; HPA, 2005b; HPA, 2005a), Europe (Eurosurveillance, 2002; Eurosurveillance, 2004) and North America (CDC, 2005b; CDC, 2005a). They share a similar approach that emphasises the importance of strict infection control measures including appropriate isolation facilities (see below). They also highlight the importance of minimising 'routine' blood tests.

Typical criteria for the identification of suspected index cases of VHF in an individual from an endemic area include the presence of a fever greater than 38.3°C of less than 3 weeks duration together with a severe illness that includes haemorrhagic features (haemorrhagic or purpuric rash, epistaxis, haematemesis, haemoptysis, blood in stools) with no predisposing factors for haemorrhagic manifestions and no alternate diagnosis.

Reporting of cases to the relevant health department as well as to local infection control and laboratory personnel is clearly a priority. Contact tracing and surveillance is another critical aspect of controlling the spread of VHFs. This generally involves twice daily monitoring of temperature for 21 days with, in the case of Lassa fever, the instigation of ribavirin prophylaxis if fever develops.

11. Infection Control

Isolation and strict barrier nursing of patients with suspected VHF are critical in all settings. Strict disinfection procedures for all items that come in contact with patients are mandatory. Additional measures should ideally include the use of gloves, gowns, masks, shoe covers and eye/face protection. Where available resources permit they may also involve the use of airborne precautions such as HEPA-filtered masks or respirators, negative pressure isolation rooms and positive pressure protective respirators. Dedicated facilities for the care of patients with suspected VHF are available in many developed countries, such as the High Security Infectious Diseases Units (HSIDU) in Coppett's Wood (London) and Newcastle General Hospitals in the UK where patients are managed in High Security Isolator Beds ('Trexler units').

The effectiveness of routine infection control measures has been confirmed in many documented cases of VHF that have been imported into Europe and North America from endemic areas overseas. In practically all cases there has been no secondary nosocomial spread from these cases, despite the fact that in many instances the patient was managed using only standard infection control precautions by large numbers (often hundreds) of health care workers before the diagnosis was made, or in many cases even considered.

Acknowledgements

I am grateful to Dr Penelope Bryant for helpful comments on the manuscript.

References

Advisory Committee on Dangerous Pathogens. (1996). Management and control of viral haemorrhagic fevers. http://www.hpa.org.uk/infections/topics_az/VHF/ACDP_VHF_guidance.pdf.
Baize, S. (2005). A single shot against Ebola and Marburg virus. *Nat Med* 11:720–1.

Baize, S., Marianneau, P., Georges-Courbot, M.C. and Deubel, V. (2001). Recent advances in vaccines against viral haemorrhagic fevers. *Curr Opin Infect Dis* 14:513–8.

Borio, L., Inglesby, T., Peters, C.J., Schmaljohn, A.L., Hughes, J.M., Jahrling, P.B., Ksiazek, T., Johnson, K.M., Meyerhoff, A., O'toole, T., Ascher, M.S., Bartlett, J., Breman, J.G., Eitzen, E.M., Jr., Hamburg, M., Hauer, J., Henderson, D.A., Johnson, R.T., Kwik, G., Layton, M., Lillibridge, S., Nabel, G.J., Osterholm, M.T., Perl, T.M., Russell, P. and Tonat, K. (2002). Hemorrhagic fever viruses as biological weapons: medical and public health management. *Jama* 287:2391–405.

Centers for Disease Control and Prevention (2005a). http://www.bt.cdc.gov/agent/vhf/index.asp.

Centers for Disease Control and Prevention (2005b). http://www.cdc.gov/ncidod/dvrd/spb/mnpages/dispages/vhf.htm.

Centers for Disease Control and Prevention (2005c). http://www.cdc.gov/ncidod/dvrd/spb/mnpages/vhfmanual.htm.

Eurosurveillance (2002). http://www.eurosurveillance.org/em/v07n03/v07n03.pdf.

Eurosurveillance (2004). http://www.eurosurveillance.org/em/v09n12/0912-235.pdf.

Feldmann, H., Jones, S., Klenk, H.D. and Schnittler, H.J. (2003). Ebola virus: from discovery to vaccine. *Nat Rev Immunol* 3:677–85.

Fisher-Hoch, S.P. and Mccormick, J.B. (2004). Lassa fever vaccine. *Expert Rev Vaccines* 3:189–97.

Geisbert, T.W. and Hensley, L.E. (2004). Ebola virus: new insights into disease aetiopathology and possible therapeutic interventions. *Expert Rev Mol Med* 6:1–24.

Geisbert, T.W. and Jahrling, P.B. (2004). Exotic emerging viral diseases: progress and challenges. *Nat Med* 10: S110–21.

Health Protection Agency (2005a). http://www.hpa.org.uk/infections/topics_az/deliberate_release/VHF/homepage.asp.

Health Protection Agency (2005b). http://www.hpa.org.uk/infections/topics_az/VHF/menu.htm.

Jones, S.M., Feldmann, H., Stroher, U., Geisbert, J.B., Fernando, L., Grolla, A., Klenk, H.D., Sullivan, N.J., Volchkov, V.E., Fritz, E.A., Daddario, K.M., Hensley, L.E., Jahrling, P.B. and Geisbert, T.W. (2005). Live attenuated recombinant vaccine protects nonhuman primates against Ebola and Marburg viruses. *Nat Med* 11:786–90.

Klietmann, W.F. and Ruoff, K.L. (2001). Bioterrorism: implications for the clinical microbiologist. *Clin Microbiol Rev* 14:364–81.

Lowenstein, R. (2004). Deadly viral syndrome mimics. *Emerg Med Clin North Am* 22:1051–65, ix–x.

Mahanty, S. and Bray, M. (2004). Pathogenesis of filoviral haemorrhagic fevers. *Lancet Infect Dis* 4:487–98.

Mccormick, J.B., King, I.J., Webb, P.A., Johnson, K.M., O'sullivan, R., Smith, E.S., Trippel, S. and Tong, T.C. (1987). A case-control study of the clinical diagnosis and course of Lassa fever. *J Infect Dis* 155:445–55.

Mccormick, J.B., King, I.J., Webb, P.A., Scribner, C.L., Craven, R.B., Johnson, K.M., Elliott, L.H. and Belmont-Williams, R. (1986). Lassa fever. Effective therapy with ribavirin. *N Engl J Med* 314:20–6.

Ndayimirije, N. and Kindhauser, M.K. (2005). Marburg hemorrhagic fever in Angola–fighting fear and a lethal pathogen. *N Engl J Med* 352:2155–7.

Peters, C.J. (2005). Marburg and Ebola–arming ourselves against the deadly filoviruses. *N Engl J Med* 352:2571–3.

Richmond, J.K. and Baglole, D.J. (2003). Lassa fever: epidemiology, clinical features, and social consequences. *Bmj* 327:1271–5.

Salvaggio, M.R. and Baddley, J.W. (2004). Other viral bioweapons: Ebola and Marburg hemorrhagic fever. *Dermatol Clin* 22:291–302, vi.

Sullivan, N.J., Sanchez, A., Rollin, P.E., Yang, Z.Y. and Nabel, G.J. (2000). Development of a preventive vaccine for Ebola virus infection in primates. *Nature* 408:605–9.

World Health Organization (2005a). http://www.who.int/csr/disease/en/.

World Health Organization (2005b). http://www.who.int/topics/haemorrhagic_fevers_viral/en/.

5

Rotavirus and Rotavirus Vaccines

**Roger I. Glass, Joseph Bresee, Baoming Jiang,
Umesh Parashar, Eileen Yee, and Jon Gentsch**

1. Introduction to Diarrhea in Children

We expect diarrhea to be a problem for children in the developing world where poor hygiene and sanitation, lack of clean water and proper food storage all favor the transmission of a host of different enteric pathogens. In these settings, diarrhea remains the first or second most common cause of hospitalizations and death among children <5 years. Nonetheless, in the developed world, despite all the improvements made in sanitation and the provision of clean food and water to our population, childhood diarrhea still remains one of the most common illnesses leading to doctor visits and hospitalizations. Why then does diarrhea continue to be a problem in the industrialized world and what can we do to prevent it?

The answer to this seeming paradox has only been elucidated in the past 30 years with improvements in our understanding of the microbiology of gastrointestinal infections. In 1970, only a few organisms such as Salmonella, Shigella, Vibrio cholerae O1, Entamoeba histolytica, and Giardia lamblia, were recognized to cause diarrhea and with the diagnostics available, fewer than 10% of all diarrheal diseases could be linked to a known infectious agent. The remainder were attributed to conditions such as weaning, malnutrition, or food allergy and the largest single diagnostic group was called "idiopathic", a term indicating that no cause could be identified. Since 1970, a scientific explosion has occurred with the discovery of more than 25 different pathogens that can cause diarrhea in children – novel bacteria, viruses and parasites. The field has become so rich with new agents and so diagnostically complicated that outside of a research setting, few laboratories or groups are capable of adequately establishing an etiologic diagnosis. Consequently, physicians and even research laboratories remain with an incomplete ability to detect the cause of diarrhea in their patients.

CDC – required disclaimer: The findings and conclusions in this paper are those of the authors and do not necessarily represent the views of the funding agency.

Hot Topics in Infection and Immunity in Children, edited by Andrew J. Pollard and Adam Finn.
Springer, New York, 2006

Despite these diagnostic shortcomings, some major advances have been made in our understanding of the causes of childhood diarrheas, their treatment and prevention. First, with economic development, improvement in water, sanitation and hygiene, the incidence of diarrheal illness has declined so that while children in the developing world may have 3–7 episodes of diarrhea per year for the first few years of life, children in the industrialized countries count on 1–2 episodes per year. Furthermore, the agents responsible for their diarrhea have changed. In the developing world, children succumb to infections with a wide range of bacteria, parasites, and viruses but with economic development, the bacteria and parasites have become rare except in small outbreaks settings such as day care centers. This evolution in the profile of enteric pathogens reflects the observation that the bacteria and parasites are infections spread by sewage, contaminated food or water, and their transmission is aided by lapses in hygiene – practices such as the lack of refrigeration, close contact with animals in the home, the absence of toilet paper and good hygienic practices, and the limited access to clean running water in the home.

Unlike bacteria and parasites, most of the viruses seem to have a mode of transmission that is independent of these lapses of sanitation and hygiene. The viral agents of gastroenteritis have been termed "democratic" because all children are infected in their first few years of life and are therefore not likely to be spread by poor hygiene and sanitation. An American or British child today will have 5–8 episodes of diarrhea in its first few years of life and for the most part, these will be caused by viruses. The "rogues gallery" of viral pathogens includes rotaviruses, enteric adenoviruses, astroviruses, and several caliciviruses of both the norovirus and sapovirus genogroups. Yet even today with all of our diagnostic advances, it would be difficult for even a research lab to adequately detect each of these agents. Rotavirus has become the flagship virus for this group because it is the easiest to detect by simple diagnostic assays, is the agent most likely to bring a child to the hospital for severe disease, and is now closest to control through the use of vaccines. This review will focus on the epidemiology of rotavirus and the prospects of its control though the use of vaccines.

2. The Role of Rotavirus

2.1. The Epidemiology of Rotavirus and the Natural History of the Disease

The epidemiology and natural history of rotavirus seems quite straightforward. All children are infected in their first few years of life, these first infections are usually symptomatic, and subsequent infections are silent, rarely causing symptoms and associated with very little if any viral shedding. Humans appear to be the main reservoir of infection but rare or novel strains can creep into humans, often representing strains from animals. In temperate climates, rotavirus has a winter peak whereas in the tropics and the developing world, disease can occur year round and at a slightly earlier age. While outbreaks do occur in day care settings and in hospitals, these outbreaks probably reflect the density of immunologically naïve groups of children who are highly susceptible, the short incubation period of infection – 18–36 hours, and the low inoculum required to cause infection making spread rapid and efficient. In fact, children not exposed to these setting will be infected as well although perhaps, not as early or as part of an identifiable outbreak. We still have no idea where rotavirus resides in the long summer periods and how or why the annual season begins each fall or winter.

2.2. The Pathogenesis of Rotavirus Diarrhea

Following the ingestion of rotavirus, the virus attaches to the epithelial surface of the small intestine, enters the cell and begins it replication cycle. Within 18–36 hours, a few viruses have multiplied so many times that as many as 10^{12} particles are shed per gram of stool. Illness often begins with vomiting, perhaps reflecting the early release of cytokines acting centrally. The virus first elaborates a toxin called NSP4 (Non structural protein 4) which by itself can cause diarrhea in mice. As the disease progresses, epithelial cells of the small intestine begin to slough and the child will suffer profuse watery diarrhea. This diarrhea can be enhanced by increased activity of the enteric nervous system. After several days, diarrhea stops, the intestine recovers, and shedding rapidly falls off to levels not detectable by immunoassays but detectable by PCR for 2 weeks or more. While the disease is primarily localized in the intestine, recent studies suggest that many children may experience a mild viremia. Extra-intestinal manifestations of this viremia are rare and rotavirus RNA has been identified in the CSF of children with seizures and in the liver of children with severe immunodeficiency syndromes. The importance of infections in these extra-intestinal sites is not fully understood.

The clinical presentation of rotavirus diarrhea is indistinguishable from most acute watery diarrheas of children. Rotavirus diarrhea is usually mild but about 20% of children will experience more severe disease with moderate dehydration requiring a clinic visit. About 1.5% to 3% of children will have such severe diarrhea and dehydration that they require hospitalization and more aggressive rehydration. In the developing world, children with severe disease accompanied by malnutrition, limited access to care, and other intercurrent infections become so dehydrated that they can go into shock and die from their infection.

2.3. Diagnostics and Virology

The diagnosis of rotavirus should be suspected in a child under the age of 3–5 years who has diarrhea preceded by vomiting at the time of the usual winter seasonal epidemic. At this time, the child will be shedding so much virus (i.e., $>10^{12}$) that diagnosis is easy. The most commonly available test is an immunoassay for either an enzyme immunoassay (e.g., ELISA) or a latex assay which can be quicker but is slightly less sensitive. The assays will normally remain positive for the first few days of the disease. RT-PCR is useful for confirmation and characterization of strains but, this is only used in research laboratories. The virus is easily visible by electron microscopy with dark stain and by polyacrilamide gel electrophoresis which displays the 11 segments of RNA.

Rotavirus belongs to the family Reoviridae and is characterized by the presence of 11 segments of double stranded RNA that are encased by 3 shells, an inner core, an internal capsid and an outer coat. The outer coat is composed of 2 proteins that are seen by the child's immune system. These outer capsid proteins, called Viral Proteins 4 and 7 (VP4 and VP7) and also labeled the Protease-cleaved protein and the Glycoprotein respectively (P and G proteins) are the basis to characterize the diversity of this virus group and the targets to mount an immune response. Four or five strains of rotavirus ([P8], G1; [P8], G3; [P8], G4; [P8], G9, and [P4], G2) predominate worldwide and these have become the prime targets for the vaccines in development. At the same time, in developing countries, many new strains arise although they rarely become the predominant strains over time.

2.4. The Mechanism of Immunity

Studies of the natural history of rotavirus in Australia, Mexico and India all confirm that first infections with rotavirus can lead to good immunity against subsequent disease. While first infections are often symptomatic and associated with much viral shedding and a robust immune response, subsequent infections are usually asymptomatic or lead to very mild disease with little shedding but a boost in immunity. Mexican children are frequently infected and 20% of children followed for the first 2 years of life have 5 documented infections evidenced by four fold rises in their antibody titers. The fact that a child gets only one symptomatic infection implies that this infection can protect the child against symptoms from infections with the other common serotypes. The strategy to develop vaccines has therefore been to work with live, attenuated strains of rotavirus that can mimic the immune response of natural infection without causing disease.

The mechanism of immunity to rotavirus has never been fully determined. Since rotavirus infections are localized in the gut, local immunity in the small intestine mediated by IgA is believed to be critical. However, since infants become infected but rarely develop symptoms in the first few months of life, humoral antibody IgG, passed from the mother may be protective as well. Recent studies in primates indicate that infection with rotavirus can be prevented with the passive transfer of circulating IgG antibodies but the action of these antibodies might well be through transudation into the gut lumen. For vaccine development, IgA seroresponses have been the best indicator of vaccine "take". However, to date, we have no reliable serologic proxy for vaccine efficacy – the presence of a good antibody response to a vaccine does not correlate fully with protection and the lack of an antibody response to vaccine does not rule out the efficacy of the vaccine. Until such a marker is identified, field trials of these vaccines will be the only acceptable way to demonstrate the proper efficacy of a new candidate.

2.5. The Burden of Rotavirus Disease

The rationale for global programs to develop rotavirus vaccines has rested upon the enormous burden of rotavirus diarrhea in children. In the developing world, an estimated 610,000 children die each year from rotavirus representing about 5% of all deaths among children <5 years. Every child is infected in the first few years of life, about 1 in 5 to 1 in 8 will require a medical visit for the treatment of moderate dehydration, 1 in 30 to 1 in 80 will be hospitalized for severe disease, and 1 in 250 will die from their infection. In the developed world, mortality is low but similar rates of doctor visits and hospitalizations apply. Furthermore, the push for rotavirus vaccines in the developed world has been based upon arguments of cost-effectiveness of vaccines compared to the direct and indirect costs of the disease. Consequently, a vaccine for rotavirus would have global usefulness, preventing death in the developing world, and severe disease and providing economic benefit in the industrialized world.

3. Vaccines as an Approach to Control

3.1. The Earliest Vaccine

To date, all of the vaccines against rotavirus that have entered clinical trials have been live attenuated rotavirus strains that have been administered orally with buffers. Attenuation has been achieved either through the repeated passaging of human rotavirus strains or the natural attenuation of animal strains administered to humans. The first candidate rotavirus vaccine was a bovine strain of rotavirus, RIT 4237 prepared by SmithKline Rixensart, that was administered to infants just before the winter rotavirus season. The unexpected high efficacy of this animal strain to protect infants in Finland against rotavirus diarrhea, often in the absence of a measurable immune response, provided the motivation to pursue this line of research. However, the variable efficacy of this vaccine in subsequent trials in other settings led to the demise of this program.

3.2. Rhesus Tetravalent Vaccine-Rotashield

The rhesus tetravalent rotavirus vaccine became the next major candidate for development. The monovalent rhesus vaccine suffered the same variable efficacy in field trials, as the original RIT borne vaccine, so a tetravalent vaccine was developed. This vaccine was based upon the assumption that animal strains that shared no neutralization epitopes with human strains would need to carry genes encoding these key neutralization targets – principally the VP7 genes of the most common rotavirus strains. Reassortant strains were developed that were composed of 10 genes with the attenuation properties from the parent rhesus vaccine strain and a single gene encoding the VP7 capsid protein from each of the 4 serotypes of rotavirus in common circulation. In clinical trials, this tetravalent vaccine was significantly more effective than it monovalent predecessor. In 1998, this vaccine, developed at the National Institutes of Health by Dr. A. Kapikian and prepared by Wyeth Lederle as RotaShield, became the first rotavirus vaccine to be licensed.

In the United States, the vaccine was immediately introduced into the routine program of childhood immunization. In the first 9 months after licensure, more than 600,000 children received one to three doses of the vaccine. Trials of the vaccine were just beginning in the developing world when in July, 1999, 15 children were

identified who developed intussusception in the 2 week period immediately following vaccine administration. When investigations linked this severe adverse event to
the vaccine, the recommendation for use of the vaccine was withdrawn, and the
company removed their product from the market. Studies to identify the mechanism
for this statistical association were never definitive and the actual risk of intussusception following vaccination ranged from early estimates of 1 event per 2500 vaccines to less than 1 event in 30,000 to 50,000. However, the risk was particularly
low in those infants who had received their first dose of vaccine at the proper time,
ie. less than 90 days of age. Therefore, developers of the next generation of vaccines
were all advised to limit administration of their vaccines to children in this restricted
age window and to conduct very large safety trials to ensure that the next generation
of vaccine was safer than the first.

3.3. The Current Generation of Vaccines

The problem of intussusception from the rhesus vaccine changed the way that
the next generation of rotavirus vaccines would have to be tested. The rhesus vaccine
had demonstrated that live oral vaccines could be highly effective and protect children against severe rotavirus diarrhea. However, new vaccines also had to jump the
hurdle of safety and demonstrate greater safety than Rotashield. While trials of
several thousand children were adequate to demonstrate vaccine efficacy, trials
exceeding 60,000 children would be required to demonstrate safety if the risk of
intussusception were about 1 event per 5000 vaccines.

Both GlaxoSmithKline and Merck proceeded to develop alternative live oral
vaccines. The GSK vaccine, called Rotarix^{TM}, was based on a single strain of rotavirus derived from a patient and attenuated by multiple passaging in cell culture.
The vaccine comes as a lyophilized powder that is reconstituted with a buffer and
administered orally in a 2 dose schedule. The Merck vaccine, called Rotateq^{TM}, was
based on a bovine strain of rotavirus that, like Rotashield, was reassorted to yield
5 bovine-human reassortant strains that carried the single VP7 or VP4 genes of the
most common humans serotypes, G1, G2, G3, G4 and P1A. These strains were
combined into a single vaccine that is administered orally in 3 doses in a liquid
premixed with buffer.

In large scale studies, both vaccines have demonstrated a high level of efficacy
(i.e., >85%) against severe rotavirus diarrhea. In depth analysis of the trial results
indicate that each vaccine protects against a range of serotypes although full results
are not available for all serotypes. Furthermore, in safety trials each involving more
than 60,000 infants, no evidence of intussuception could be linked to either vaccine,
an observation suggesting that the problem of intussusception was associated primarily with the rhesus strain.

The GSK vaccine was first licensed in Mexico and has been licensed in more
than 25 developing countries and the European Union (EMEA). The Merck vaccine
was licenced by the Food and Drug Administration in the United States in February
2006 and has been licenced in several countries of Latin America. While these vaccines may soon enter the vaccination programs of many industrialized and middle
income countries, their efficacy in preventing severe rotavirus diarrhea among chil-

dren in the poorest developing countries in Asia and Africa has never been tested and remains to be established. Live oral vaccinees against polio, cholera, and typhoid have all had to be modified when tested and used in the developing world. Unlike parenteral vaccinees, these live oral vaccines have to replicate and be processed in the gut of the vaccines, and differences in the bacterial and viral flora of these children, nutritional constitution and levels of maternal and breast-milk antibodies could all effect the way that a live oral rotavirus vaccine would behave. Past dosing studies with rotavirus vaccines have indicated that the immune response of children is clearly dose-related and if bacterial flora or breast milk in the guts of infants in the developing world were to reduce the titer of the vaccine substantially, we might expect to see an efficacy that was substantially reduced. Consequently, there is a clear need to test these vaccines in those populations where they would be life-savers.

3.4. Future Vaccines in the Pipeline

It is unlikely that vaccines produced by these two multinationals will be able to satisfy the global need for rotavirus vaccines at an affordable cost. A number of other candidate vaccines are in the pipeline and all are likely to be produced by emerging manufacturers in the developing world.

In China, a rotavirus vaccine based on a single strain of lamb rotavirus and produced by the Lanzhou Institute of Biological Products has been licensed and is currently being sold. The efficacy of this vaccine has never been clearly established and a second generation of vaccine based upon reassortant strains of this vaccine is in development. The vaccine is unlikely to be sold outside of China.

Neonatal strains of rotavirus are being developed by groups in Australia and India. Newborns have been infected with rotavirus strains in neonatal units in many countries. These neonatal strains are genetically unusual because they can replicate in infants despite the presence of high titers of maternal antibody, cause no symptoms, but can induce an immune response. Infants infected with these strains are naturally immunized and are protected from severe disease upon reinfection. An Australian strain, RV3, developed by Ruth Bishop and Graeme Barnes has been immunogenic in low dose in about 50% of infants who are then protected against subsequent disease. More work is needed to raise the titer of this candidate vaccine and determine if the level of immunogenicity and efficacy can be improved. In India, two candidate neonatal strains are being developed in Hyderabad and no results from clinical studies have been reported to date.

The NIH has developed a bovine-based reassortant vaccine (UK strain) that has demonstrated its efficacy in a study in Finland. This vaccine is similar to the reassortant vaccine of Merck but has been licensed to manufacturers in Brazil, China and India. It will be several years before the immunogenicity and efficacy of this vaccine can be assessed in the target populations of infants in these countries.

Other approaches to rotavirus vaccines are being pursued although enthusiasm for alternative vaccines will wane until the success of the live oral vaccines can be further assessed. These alternative approaches include parenteral vaccines derived from killed virus or virus-like particles (VLP) constructed in Baculovirus vectors, DNA vaccines, single expressed antigens or subunit vaccines, and VLP vaccines administered orally. None of these has approached clinical trial but all have demonstrated some proof of concept in animal studies.

4. Looking into the Future

4.1. The Goal for Rotavirus Vaccines

The goal for rotavirus vaccines remains the prevention of rotavirus diarrhea in children worldwide. In each country, this goal should be measurable within several years of the vaccine's introduction and will differ in the developed and developing world. In industrialized countries, a rotavirus vaccine program should decrease hospitalizations and clinic visits for diarrhea in children by 30–50% within 2–3 years of the vaccine's introduction. This would lead to substantial economic savings in hospital and clinic care as well as social costs – such as savings from parents' loss of work to care for sick children. In developing countries, the economic benefits will be dwarfed by improvements in child survival, especially from diarrheal deaths.

Monitoring the impact of a rotavirus vaccine should be straightforward since severe disease can be monitored in hospitalized patients, the diagnosis from fecal specimens is cheap and easy, and infection presents uniquely as diarrhea, and the impact of the vaccine should prevent diarrhea hospitalizations within 1–2 years of the introduction of the vaccine.

4.2. Challenges for the Future

The next major hurdle for rotavirus vaccines will be establishing their efficacy for children in poor developing countries. Unlike parenteral vaccines which induce a good immune responses in most children, live oral vaccines against cholera, polio, and typhoid were not fully immunogenic or efficacious when first tested in populations in the developing world. Similarly, early candidate oral rotavirus vaccines failed in several trials in Africa. Because of this experience, a WHO consensus group determined that the efficacy of live oral rotavirus vaccines would need to be established in children in poor settings in Africa and Asia before WHO would offer a global recommendation. To date, neither Rotarix™ nor Rotateq™ has yet been tested in such settings although these trials are beginning or in planning.

The first rotavirus vaccine left a difficult legacy of intussusception that future immunization programs will have to address. While both vaccines to be introduced have demonstrated their safety against intussusception in trials of more than 60,000 children each, the age group of infants <3 months in whom the vaccines were tested would have had very low rates of intussusception given the past experience with Rotashield. Consequently, post-licensure surveillance will need to be put in place to monitor the safety of these vaccines in developed countries against a risk that could be as low as 1 in 30,000 from Rotashield, a level that led the vaccine to be withdrawn in the United States. This data could take years to develop and advisory groups have not made a determination of the level of risk that might be acceptable,

Table 5.1. The Next Generation of Live Oral Rotavirus Vaccines

Vaccine	Rotateq™ (Merck)	Rotarix™ (GSK)
Principle	Pentavalent (bovine-human reassortants)	Monovalent human G1 attenuated strain
Schedule	3 oral doses	2 oral doses
Efficacy	>90% severe disease	>90% severe disease
Safety	No IS risk in 70 K trial	No IS risk in 60 K trial
Licensure	Licenced with FDA (US), EMEP	Licensed by the EMEA (European Union) and in >20 countrie
Inventors	H Fred Clark, Stanley Plotkin, and Paul Offit, Park Offit Children's Hospital of Pennsylvania	Richard Ward, David Bernstein Gamble Institute, Cincinnati

a level that would vary greatly by country and setting and perception of the severity of rotavirus disease itself.

A third challenge will be arriving at a price for vaccines that will make them accessible to all. This is a particular problem in developing countries where vaccines need to be inexpensive if programs are to be sustainable. Efforts by international donors including the Global Alliance for Vaccines and the International Finance Facility may aid with vaccine introduction in the short run. Later on, experience with past vaccines has demonstrated that multiple manufacturers, including some in the developing world, may need to make vaccines to increase competition and drive down the price. To date, a number of emerging manufacturers in Brazil, China, India and Indonesia are considering rotavirus vaccine programs but it will be several years before this generation of vaccines comes on line.

Finally, few people around the world recognize rotavirus by name so pediatricians and policy makers have little idea of the importance of the disease or the potential for prevention with vaccines. A global network of surveillance activities has been set up to introduce diagnostics into many countries so that the burden of disease can be assessed and the cost-effectiveness of vaccines can be established on a country by country basis. This effort should provide the data that decision makers will need when they consider introduction of rotavirus vaccines in the future.

5. Summary

Two new rotavirus vaccines have recently been licensed that will provide the intervention needed to diminish the huge burden of rotavirus disease among all children. In many upper and middle income countries, these vaccines will soon be available for the routine immunization of children. The impact should be a rapid and measurable reduction in hospitalizations and doctor visits for acute diarrhea in children, especially in the winter rotavirus season. While few deaths occur in these settings, the illness has consequences in terms of both medical costs and indirect costs, including parents work time lost. In the developing world, clinical trials are still needed to ensure that the vaccines being licensed will work as expected in children living in poor settings. In these settings, other enteric flora, micronutrient malnutrition, higher titers of maternal antibody and other factors still poorly defined have compromised other live oral vaccines and have required the developers to alter vaccine formulation, dose, or schedule. Until these trials are completed, we can only hope that the efficacy will be comparable and that the vaccine will prove to be life-savers. Once the efficacy and safety have been established, rotavirus vaccines could provide a major boost to programs for child survival.

References

Bresee, J.S., Parashar, U.D., Widdowson, M.A., Gentsch, J.R., Steele, A.D., and Glass, R.I. (2005). *Ped. Infect. Dis.* **24**, 11.

Clark, H.F., Offit, P., Glass, R.I., and Ward, R.L. (2004). Rotavirus vaccines. In: Plotkin S.A., Orenstein W.A., eds. Vaccines. 4th ed. Philadelphia: Saunders-Elsevier. 1327–1345.

Glass, R.I., Bresee, J.S., Parashar, U.D., Jiang, B., and Gentsch, J.R. (2003). *Lancet* **363**, 1547–1550.

Glass, R.I., Bresee, J.S., Turcios, R., Fischer, T.K., Parashar, U.D., and Steele, A.D. (2005). *J. Infect. Dis.* **192**, 160–166.

Murphy, B.R., Morens, D.M., Simonsen, L., Chanock, R.M., La Montagne, J.R., and Kapikian, A.Z. (2003). *J. Infect. Dis.* **187**, 1301–1308.

Murphy, T.V., Gargiullo, P.M., Massoudi, M.S., et al., (2001). *New Engl. J. Med.* **344**, 564–572.

Parashar, U.D., Hummelman, E.G., Bresee, J.S., Miller, M.A., and Glass, R.I. (2003). *Emerg. Infect. Dis.* **9**, 565–572.

Ruiz-Palacios, G.M., Perez-Schael, I., Velazquez, F.R., et al. (2006). New Engl. J. Med. **354**, 11–22.

Vesikari T., Matson, D.O., Dennehy, P. (2006). New Engl. J. Med. **354**, 23–33.

6

Transient Deficiencies of T-Cell-Mediated Immunity in the Neonate

David A. Randolph and David B. Lewis

1. Introduction

Cell-mediated immunity driven by $\alpha\beta$-T cells is essential for protection against intracellular pathogens. Compared to adults, neonates demonstrate increased susceptibility to a number of intracellular infections, including those by certain bacteria, viruses, fungi, and protozoa. In part, the relatively immuno-compromised state of the neonate is due to the lack of memory T-cell responses, which tend to be stronger and more rapid than initial responses. Except in the case of congenital infection, all pathogen encounters in the newborn period are first-time encounters that are characterized by primary adaptive T-cell and B-cell immune responses that are less robust and of slower onset than memory immune responses. However, even when compared to first-time responses in adults, neonates appear to have developmental limitations in their T-cell responses. While this has been long recognized clinically, the cellular and molecular basis for this phenomenon has only recently begun to be understood. This review will summarize the cellular and molecular requirements for effective T-cell-mediated immunity and discuss what is known about the relative deficiencies of these components in infants.

2. Overview of T-Cell-Mediated Immunity

$\alpha\beta$-T cells represent more than 90% of all T cells and are defined by their surface expression of antigen-specific, heterodimeric T-cell receptors (TCRs) composed of a TCR-α and TCR-β chain (Chaplin, 2003). $\alpha\beta$-T cells are further divided into subsets based on their surface expression of different proteins, which have been assigned cluster of differentiation (CD) numbers. T cells expressing CD4 are called

Hot Topics in Infection and Immunity in Children, edited by Andrew J. Pollard and Adam Finn.
Springer, New York, 2006

helper T cells and are particularly important in orchestrating the overall immune response by providing stimulatory signals to other cells of the immune system in the form of secreted cytokines and certain cell surface molecules. CD8$^+$ T cells are also called cytolytic T lymphocytes (CTLs) or killer T cells and are important for killing virally infected cells.

2.1. CD4$^+$ T Cells

The TCRs of CD4$^+$ T cells recognize peptides bound in a molecular complex to major histocompatibility complex (MHC) class II proteins on the surface of specialized antigen presenting cells (APCs), such as myeloid dendritic cells (Figure 6.1A) (Watts, 2004). MHC class II peptides are mainly derived from extracellular proteins or pathogens that have entered into lipid bilayer bound compartments in the APC, such as endosomes, by phagocytosis, pinocytosis, or internalization of the cell membrane. After antigen-specific naïve CD4$^+$ T cells are activated via their TCR, they undergo clonal expansion and differentiation into various types of effector cells, depending on cues in their local environment.

CD4$^+$ T cells that are activated in the presence of cytokines, such as interleukin (IL)-12 (Hunter, 2005), differentiate into T helper 1 (Th1) cells that can secrete large amounts of interferon (IFN)-γ, IL-2, and tumor necrosis factor (TNF)-α when appropriately triggered. Th1 immunity is particularly important for the control of intracellular infections that occur in APCs, such as mononuclear phagocytes. CD4$^+$ T cells that are activated in the presence of IL-4 differentiate into T helper 2 (Th2) cells which secrete large amounts of IL-4 and IL-5 but not IFN-γ. They are particularly important for humoral immunity, especially IgE, and cellular immunity by eosinophils, basophils, and mast cells, to extracellular parasites, such as helminths. Th2 responses are also important in classic allergic disease.

2.2. T Helper 1 Cells

Th1 immunity is essential to control infection by certain intracellular bacteria (e.g., Mycobacteria, Salmonella, Listeria) [reviewed in Picard and Casanova (2004)], viruses (e.g., herpesviruses), fungi (e.g., Candida, Pneumocystis), and protozoa (e.g., Toxoplasma, Plasmodium). A typical primary Th1 immune response to an intracellular pathogen, such as *Mycobacterium tuberculosis*, begins when a myeloid dendritic cell takes up cellular material that includes pathogen proteins and loads peptides from these proteins onto MHC class II molecules. For example, the pathogenic proteins may be contained in the debris of infected cells. Entry of intact organism into the dendritic cell, e.g., by phagocytic uptake or infection, is not necessary for such loading. The dendritic cell surface presents pathogen-derived peptides/ MHC complexes to the TCRs of a naïve antigen-specific CD4$^+$ T cell (Figure 6.1A). Dendritic cells constitutively express low amounts of B7 molecules and the CD40 protein. The B7 proteins deliver a co-stimulatory signal through the CD28 receptor on the T cell that is required for effective T-cell activation. The activated CD4$^+$ T cell rapidly produces de novo cell surface CD154 (CD40-ligand). T-cell CD154 binds to CD40 on the dendritic cell and induces increased dendritic cell expression

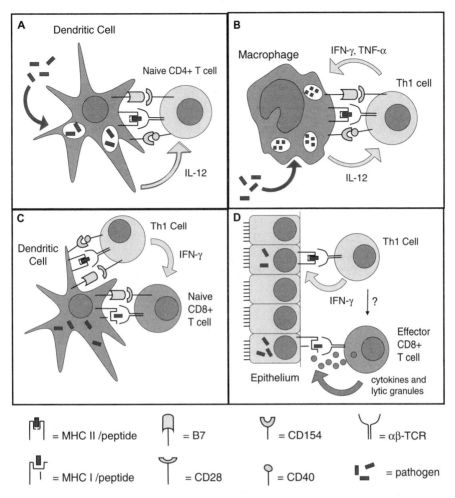

A Dendritic Cell

Naive CD4+ T cell

IL-12

B Macrophage

IFN-γ, TNF-α

Th1 cell

IL-12

C Th1 Cell

Dendritic Cell

IFN-γ

Naive CD8+ T cell

D

Th1 Cell

IFN-γ ?

Effector CD8+ T cell

Epithelium

cytokines and lytic granules

= MHC II /peptide = B7 = CD154 = αβ-TCR

= MHC I /peptide = CD28 = CD40 = pathogen

Figure 6.1. Activation of naïve T cells into effectors that control intracellular infection. For the generation of Th1 effector cells (A) myeloid dendritic cells take up antigens and present peptide fragments bound to MHC class II to naïve CD4[+] T cells. Full activation of naïve CD4[+] cells requires co-stimulatory signals delivered through B7-CD28 interactions in addition to antigen-specific TCR signals. This results in the rapid appearance on the CD4[+] T-cell surface of CD154, which is not found on the resting CD4[+] T cell. The CD154-CD40 interaction enhances dendritic cell cytokine production, e.g., IL-12 and IL-27, and B7 surface expression, which re-enforces T-cell activation. Together, these events drive differentiation of the naïve CD4[+] T cell into Th1 effector cells. (B) Differentiated Th1 cells promote killing of intracellular pathogens, such as M. tuberculosis, that attempt to elude the immune system by inhabiting intracellular microvesicular compartments of mononuclear phagocytes. Pathogen-derived peptides are bound by MHC class II and are displayed on the infected cell surface where they engage the Th1 TCRs. The Th1 cytokines bind to specific surface receptors on the infected cell surface and increase its microbicidal activity to help eliminate the pathogen. The generation of CD8[+] T-cell effectors involves (C) activation of the naïve CD8[+] T cell by engagement of its TCR by cognate peptide/MHC class I complexes on the myeloid DC. This signal in conjunction with B7/CD28 co-stimulation and, possibly, cytokines from neighboring Th1 cells, results in clonal expansion and differentiation into CD8[+] T-cell effectors. (D) These CD8[+] effectors have pre-formed granules of perforin and granzymes and a capacity to express death receptors ligands and kill target cells infected with virus and that display viral peptide/MHC complexes on their cell surface. They can also produce cytokines with direct anti-viral activity and that enhance antigen presentation to T cells. Th1 cells may also assist in anti-viral immunity by producing cytokines and in helping maintain functional memory/effector CD8[+] T-cell populations in cases of persistent viral infection.

of B7 molecules and the secretion of cytokines, such as IL-12, that promote further Th1 differentiation (Quezada et al., 2004).

Once generated, effector Th1 cells are especially important in the control of intracellular infection of MHC class II bearing APCs, such as macrophages. Many organisms, such as Mycobacteria, are able to enter into mononuclear phagocytes or other cell types and persist intracellularly in microvesicular compartments away from the potentially lethal extracellular environment. However, pathogen-derived peptides from these microvesicular compartments enter into the MHC class II pathway, allowing their detection on the infected cell surface by the TCR of the Th1 cell, resulting in Th1 cytokine secretion (Figure 6.1B). Th1 cytokines, particularly IFN-γ, increase the microbicidal activity and antigen presentation capacity of the infected cell helping it to eliminate these intracellular pathogens (Figure 6.1B). Activated Th1 cells also express CD154, and the CD154/CD40 interaction between Th1 cells and mononuclear phagocytes appears to be particularly important in the control of infection by *Pneumocystis jiroveci* (Quezada et al., 2004).

2.3. CD8⁺ T Cells

CD8⁺ T cells recognize peptides bound to MHC class I, which is ubiquitously expressed. In most cell types, the peptides bound to MHC class I originate from the cytosol and are derived from either endogenous proteins or viral proteins if the cell is infected. As for naïve CD4⁺ T cells, myeloid dendritic cells are an important APC for the activation of naïve CD8⁺ T cells. In contrast to most cell types, myeloid dendritic cells have the ability to load MHC class I with peptides that are derived from the extracellular space, a function known as cross-presentation (Cresswell et al., 2005). Thus, activation of naïve CD8⁺ T cells by myeloid dendritic cells typically involves the same pathways of antigenic uptake from the extracellular space as is the case for naïve CD4⁺ T cells. The activation events are similar to those for CD4⁺ T cells and include co-stimulation via the CD28 molecule (Figure 6.1C). In contrast to CD4⁺ T cells, CD8⁺ T cells express little CD154 after activation, and, therefore, the CD154/CD40 interaction between CD8⁺ T cells and myeloid dendritic cells has at most only a minor role in CD8⁺ T-cell activation. In contrast, the CD154/CD40 interactions between Th1 cells and myeloid dendritic cells may indirectly enhance CD8⁺ T-cell activation by increasing dendritic cell surface expression of B7 molecules and the production of cytokines, such as IL-12, that promote CD8 T-cell proliferation and the acquisition of cytotoxic effector mechanisms.

Once antigen-specific effector CD8⁺ T cells are generated, they survey the tissues for cells bearing antigenic peptides bound to surface MHC class I molecules (Cresswell et al., 2005). Antigen-specific effector CD8⁺ T cells recognize their cognate ligands, and kill infected target cells (Figure 6.1D). Killing is usually achieved by secretion of cytolytic granules containing perforin and granzyme, which activate apoptotic pathways in the infected target cell. Alternatively, killing may be achieved using Fas/Fas-ligand interactions in which Fas-ligand expressed on the activated effector CD8⁺ T cell engages Fas on the target cell to induce apoptosis. The effector CD8⁺ T cell also secretes cytokines, particularly IFN-γ and

TNF-α, which have direct antiviral effects on infected cells, and also can enhance antigen presentation, e.g., by overcoming the inhibition of MHC class I antigen presentation by the immunoevasive proteins of herpesviruses. Th1 cells may also contribute to anti-viral immunity in non-lymphoid tissues, such as the mucosa infected with cytomegalovirus (Lucin et al., 1992; Polic et al., 1998), although the cellular sources of viral peptide/MHC class II molecules for this response are unclear. The apparent inability of neonates to mount strong T cell-mediated immune responses conceivably could arise from defects at any of the steps shown in Figure 6.1. As discussed below, both APC functions and CD4[+] T-cell helper functions differ from those of adults, whereas CD8[+] T-cell function appears to be relatively intact in neonates.

3. Clinical Evidence for Deficiencies of T-Cell-Mediated Immunity in the Neonate

Cytomegalovirus (CMV) is a ubiquitous herpes virus that ultimately infects 50–90% of the population. For the vast majority of children and adults, infection, which usually occurs after mucosal contact with bodily secretions, is either asymptomatic or results in a self-limited non-specific viral syndrome characterized by fever, hepatosplenomegaly, leukopenia, and myalgias (Gandhi and Khanna, 2004). During active infection, virus is shed from mucous membranes and is detectable in both urine and saliva. Cell-mediated immunity is essential for control of the disease, and onset of T-cell immunity in results in resolution of viremia, although latent virus can be detected in tissues for life (Harari, 2004). In adults, severe systemic disease is seen only in settings of substantial immunodeficiency, such as concurrent HIV infection or following hematopoietic stem cell transplantation, where infection can result in pneumonitis, hepatitis, retinitis, and other organ dysfunction (Gandhi and Khanna, 2004). Reconstitution of virus-specific CD8[+] T cells is sufficient for temporary control of the virus, but CD4[+] T-cell immunity appears necessary for long-term immunity (Peggs et al., 2003; Walter et al., 1995) CD4[+] T-cell immunity acts, at least in part, by helping maintain functional antigen-specific memory CD8[+] T cells, but the nature of this help remains controversial (Bevan, 2004).

In contrast, infection *in utero* can have dramatic, damaging effects on an otherwise healthy fetus (Brown and Abernathy, 1998; Gandhi and Khanna, 2004). Although the majority of infected infants are asymptomatic, 5–10% will suffer severe neurologic damage including microcephaly, seizures, deafness, and retardation. Additional infants will appear asymptomatic at birth but will progress to have significant hearing loss. Infection acquired after birth is usually asymptomatic, but interestingly both congenital infection and post-natal infection through the pre-school years result in prolonged shedding of the virus, while in adults such continuous shedding after primary CMV infection is limited to approximately 6 months after acquisition. This indicates an inability of the neonatal and infant immune system to control the virus compared to the immunocompetent adult (Stagno, 1983).

Recently we investigated the T-cell responses of infants and children infected with CMV (Tu, 2004). Compared to adults, children demonstrated impaired virus-specific Th1 responses but had relatively normal CD8$^+$ T-cell responses [Figure 6.2 and see (Chen et al., 2004)]. This was true even when primary infections of similar duration in both adults and children were compared. The delay in CD4$^+$ T-cell immunity correlated with prolonged viral shedding in the urine (Tu et al., 2004). Interestingly, CD8$^+$ T-cell immunity seems to be relatively intact even in the setting of congenital CMV infection. Marchant et al. (2003) studied 8 newborns with congenital CMV infection compared to 15 healthy controls. Using tetramer staining

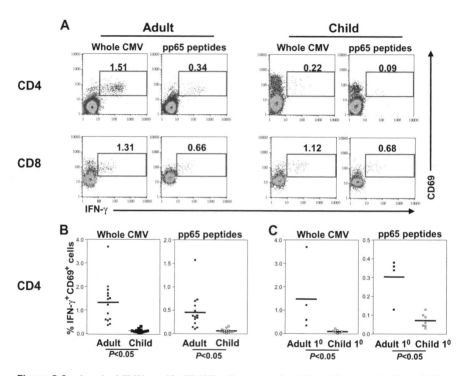

Figure 6.2. Impaired CMV-specific CD4$^+$ T-cell responses in children. Blood samples from children and adults with documented CMV infection were stimulated with lysates from CMV infected fibroblasts (whole CMV) or with a set of 138 overlapping peptides derived from the CMV protein pp65 protein (pp65 peptides), and T-cell responses were studied by flow cytometry. (A) Representative results of CMV-specific IFN-γ and CD69 expression by CD4$^+$ and CD8$^+$ T cells of a CMV Ab-seropositive adult with presumed chronic infection and a young child with infection of less than 2 years duration. The numbers represent the percentage of CMV-specific or pp65-specific CD4$^+$ or CD8$^+$ T cells as assessed by cells positive for both IFN-γ and CD69. (B) Frequency of antigen-specific CD4$^+$ T cells determined flow cytometrically after CMV or pp65 stimulation for a cohort of children with 8–29 months duration of primary infection (child) and a group of adults with presumed chronic CMV infection (adult). (C) CMV-specific CD4$^+$ T cells for children and adults with infections of similar duration (child 1° and adult 1°). Horizontal bars in (B) and (C) indicate group means. Statistical significance was evaluated using the two-tailed, unpaired Student's t-test (from Tu W. et al. Persistent deficiency of CD4 T cell immunity to cytomegalovirus in immunocompetent young children, *J. Immunol.*, 172:3260–67, 2004).

and flow cytometry, they showed that CMV-specific CD8$^+$ T cells could be detected pre-natally and early after birth, and these cells expressed IFN-γ, perforin, and granzyme A and could lyse target cells loaded with CMV peptides (Marchant et al., 2003). Although this detailed study does not exclude a potential lag in the onset of CD8$^+$ T-cell immunity in response to congenital CMV infection compared to post-natal infection, the robust nature of the immune response is very striking.

A robust CD8$^+$ but not CD4$^+$ T-cell response to congenital infection is not unique to CMV infection, as a similar pattern of immunity has been observed at birth in cases of congenital infection with *Trypanosoma cruzi* (Hermann et al., 2002). This congenital infection results in a marked expansion of CD8$^+$ T cells rather than CD4$^+$ T cells, with evidence of oligoclonality of the TCR repertoire indicating that this is the result of antigen-driven expansion. These CD8$^+$ T cells are enriched in markers for activation (HLA-DR^{high}), memory (CD45R0^{high}), and end-stage effector cells (CD28$^{-/low}$), and for cytotoxicity (perforin$^+$). They also have a markedly greater capacity to produce IFN-γ and TNF-α than CD8 T$^+$ cells from uninfected newborns. In comparison, the CD4$^+$ T cells in these congenitally infected newborns have undergone much less clonal expansion and acquisition of effector function (Hermann et al., 2002).

Newborns also are highly vulnerable to severe infection with herpes simplex virus (HSV)-1 and -2. Neonatal infection frequently results in death or severe neurological damage, even with the administration of high doses of anti-viral agents, such as acyclovir (Kimberlin, 2004). In contrast, death from disseminated primary HSV infection is distinctly unusual outside the newborn period, except in cases of T-cell immunodeficiency or in recipients of T-cell ablative chemotherapy or immunosuppression (Herget et al., 2005). The increased disease severity in infants correlates with delayed and diminished appearance of HSV-specific Th1 responses, including CD4$^+$ T-cell proliferation, IFN-γ and TNF-α secretion, and production of HSV-specific T-cell dependent antibody, compared to adults with primary infection (Burchett et al., 1992; Sullender et al., 1987). It is unknown whether CD8$^+$ T-cell immunity is similarly diminished and delayed. It is also unclear by what age HSV-specific CD4$^+$ T-cell immunity achieves a level equivalent to that of adults.

Developmental limitations in immunity are not restricted to anti-viral immunity. Infections with *Toxoplasma gondii*, an obligate intracellular protozoan, are common with an overall seroprevalence of 22.5% in the USA (Montoya and Liesenfeld, 2004). In adults, infections most often occur after ingestion of under-cooked meat or food or water contaminated with cat feces, and are usually asymptomatic or result in mild, non-specific symptoms including non-tender lymphadenopathy. In immunocompetent adults, primary toxoplasmosis rarely disseminates to cause other sites of disease, such as chorioretinitis. Cell-mediated immunity driven by Th1 cells is required for containment of the infection, and deficiency in CD4$^+$ T cells, IL-12, CD154, or IFN-γ results in increased severity of disease in animal models (Subauste et al., 1999). In contrast to infection in adulthood, congenital infection, which occurs when the organism is transmitted transplacentally to the fetus during an active maternal infection, can have dire consequences: While the majority of infants are asymptomatic at birth, a large number will progress to develop chorioretinitis and other neurologic complications

(Wilson et al., 1980). In one study, chorioretinitis was already present in 19% of otherwise well appearing infants in whom toxoplasmosis was diagnosed by neonatal screening, and even when treatment was initiated promptly after diagnosis, additional children subsequently developed retinal disease (Guerina et al., 1994).

Similarly, the neonatal immune system has difficulty containing both fungal infections, such as mucocutaneous candidiasis which is common in the first year of life, and intracellular bacterial infections, such as *M. tuberculosis*. Here the tendency of neonates to develop miliary disease and tuberculous meningitis is paralleled by decreased cell-mediated immunity as assessed by delayed-type hypersensitivity skin tests compared to older children and young adults (Smith et al., 1997). This susceptibility to disseminated tuberculosis appears to persist for at least one year after birth.

Thus, neonates and young infants clinically exhibit increased susceptibility to a number of pathogens that are normally contained by T cell-mediated immunity.

4. Cellular and Molecular Basis for Impaired Cell-Mediated Immunity

4.1. Impaired Dendritic Cell Function

Myeloid dendritic cells are particularly effective at presenting antigen to naïve T cells and therefore play a critical role in initiating both $CD4^+$ and $CD8^+$ responses [(Liu, 2001; Mellman and Steinman, 2001) and Figure 6.1]. In a typical immune response, myeloid dendritic cells take up antigen in infected or inflamed tissues and then migrate to draining lymph nodes via afferent lymphatics or to the spleen via the circulation where they take up residence near naïve T cells. Concurrent with this migration, myeloid dendritic cells increase expression of molecules involved in antigen presentation, including MHC class II and B7 molecules, and expression of the CCR7 chemokine receptor. CCR7 helps guide the dendritic cell to the T-cell zones of the peripheral lymphoid organs where the CCR7 ligands, CCL19 and CCL21, are expressed. This stereotyped alteration in phenotype and migration to lymphoid tissues can be initiated by ligation of a number of receptors on the dendritic cell, particularly receptors in the family of Toll-like receptors (TLRs). TLRs recognize molecules that are unique to pathogens, e.g., TLR4 specifically recognizes the lipopolysaccharide component of gram-negative bacteria, and promote immune responses by delivering intracellular "danger" signals (Takeda and Akira, 2005). These signals result in multiple pro-inflammatory alterations in cell function, including the secretion of cytokines.

A number of studies have found decreased function of APCs from infants and children compared to those from adults (Adkins et al., 2004; Clerici et al., 1993; Goriely et al., 2001; Langrish et al., 2002; Trivedi et al., 1997; Upham et al., 2002). Most of these have focused the ability of neonatal APCs to secrete IL-12 as a correlate of Th1 immunity. While there is some variability in the results, the overall trend is that APCs from infants secrete less IL-12 (Goriely et al., 2001; Langrish et al., 2002; Upham et al., 2002). For example, Langrish et al. (2002) derived dendritic cells from adult and cord blood by culturing PBMCs with GM-CSF and IL-4. They then matured the cells via a brief incubation with LPS and analyzed expression of cytokines and surface markers. Neonatal dendritic cells secreted significantly less IL-12 and had significantly decreased surface expression of MHC class II and B7 proteins.

These studies provide insight into the capabilities of circulating antigen presenting cells, but it is technically more difficult, particularly in humans, to assess the capabilities of dendritic cells that are resident in the peripheral tissues. To address this issue, we recently examined the impact of TLR4 signaling on myeloid dendritic function in young mice (Dabbagh et al., 2002). Myeloid dendritic cells from the spleens of 6–12 week-old TLR4-deficient (C3He/J) mice showed reduced capacity to increase expression of B7 proteins either in response to GM-CSF alone or together with CD40 engagement. Moreover, myeloid dendritic cells from TLR4-deficient mice also had significantly reduced capacity to produce IL-12 in response to CD40 engagement compared to those from wild-type mice, and were substantially less effective at priming naïve CD4$^+$ T cells into either Th1 or Th2 effector cells. It is interesting to speculate that dendritic cells from neonates, born from a sterile uterine environment, are functionally immature until they have had exposures to bacterial products in the extrauterine environment.

4.2. Impaired T Helper 1 Function

Numerous studies have demonstrated decreased Th1 function in neonatal CD4$^+$ T cells (Adkins, 1999; Ehlers and Smith, 1991; Gasparoni et al., 2003; Pirenne-Ansart et al., 1995; Trivedi et al., 1997; Wu et al., 1993) In part, these findings reflect the increased numbers of differentiated memory T cells of the Th1 subset in adult blood compared to neonatal blood. Naïve (CD45RA^{high}CD45R0^{low}) cells comprise approximately 60% of the CD4$^+$ T cells in most adults while they represent nearly 90% of cells in infants (Gasparoni et al., 2003). Thus any direct comparison of cord blood T cells to unfractionated adult cells will be comparing a relatively pure population of naïve cells to a mixed population containing both naïve and memory cells. However, even when purified naïve adult CD4$^+$ T cell are used for comparison, there appear to be differences. We recently investigated the capacity of neonatal T cells to mount Th1 responses (Chen et al., 2006). To avoid questions of inadequate antigen presentation, a pool of allogeneic adult dendritic cells was used as stimulators. Compared to purified adult naïve CD4$^+$ T cells, neonatal naïve CD4$^+$ T cells purified from cord blood secreted much less IL-2 and IFN-γ (Figure 6.3) and expressed less CD154 on their cell surface (Chen et al., 2006). There was also a decrease in the IL-12 detected in the culture supernatants, indicating decreased induction of IL-12 in the dendritic cells by the neonatal CD4$^+$ T cells, and this is likely due, in part, to decreased CD154/CD40 interactions. In addition, the neonatal cells seemed to be blocked from differentiating into Th1 cells since they expressed less STAT4 and had lower levels of STAT4 phosphorylation, which is required for IL-12 signaling (Chen et al., 2006).

There was no evidence of increased skewing towards Th2 cells, based on the low level of IL-4 produced in both neonatal and adult CD4$^+$ T-cell cultures with dendritic cells. There was also no evidence of an increased presence of regulatory CD4⁺ T cells, as the levels of the immunoregulatory cytokine, IL-10, were not increased in the cultures of neonatal CD4⁺ T cells and dendritic cells (Figure 6.3). Recently, we have also determined directly that neonatal CD4⁺ T cells have a relatively reduced number of regulatory CD4⁺ T cells, based on their intracellular expression of the Foxp3 transcription factor (Figure 6.4). Thus, the reduced ability of neonatal naïve CD4⁺ T cells to differentiate into Th1 cells in response to a potent allogeneic stimulus is not accounted for by an increased number regulatory T cells, a cell population that is able to inhibit CD4⁺ T-cell effector function (Schwartz, 2005).

Epigenetic mechanisms also regulate of IFN-γ in neonatal T cells. Early studies using methylation-sensitive restriction mapping demonstrated a hypermethylated CpG site in the IFN-γ promoter of neonatal and adult naïve (CD45RA^high) CD4 T cells compared to memory/effector (CD45RO^high) CD4⁺ T cells (Melvin, 1995). This correlated with decreased IFN-γ expression in the cells with the hypermethylated IFN-γ promoter. More recently White et al. (2002) used a more sensitive

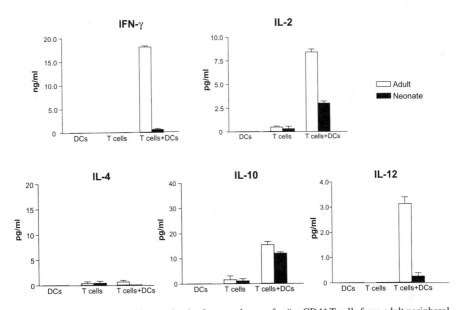

Figure 6.3. Reduced cytokine production by co-cultures of naïve CD4⁺ T cells from adult peripheral blood (ABP CD4, open bars) or cord blood (CB CD4, filled bars). Highly purified naïve CD4⁺ T cells were isolated from adult peripheral blood or cord blood. Monocyte-derived dendritic cells were prepared from three adult donors who were unrelated to the donors used for T-cell isolation by incubating blood monocytes with IL-4 and GM-CSF followed by maturation with TNF-α treatment. The dendritic cells were pooled together at an equal concentration to maximize the allogeneic stimulus. IFN-γ, IL-2, IL-4, IL-10, and IL-12 content in cell culture supernatants was determined by ELISA of cell supernatants collected at 48 hrs from cultures of dendritic cells alone (DCs), naïve CD4⁺ T cells alone (T cells), or both cell populations (T cells + DCs). *P < 0.05 compared to adult CD4⁺ T-cell/dendritic cell co-cultures (from Chen, L. et al. Impaired allogeneic activation and T helper 1 differentiation of human cord of blood naïve CD4 T cells, *Biol. Blood Marrow Transplant.*, 12: 160–171, 2006).

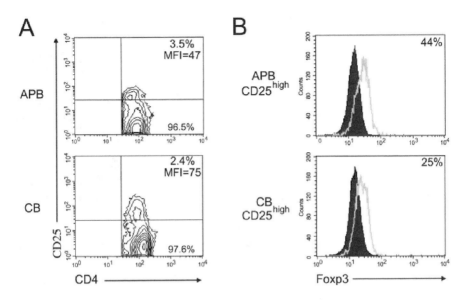

Figure 6.4. Naïve circulating CD4$^+$ T cells of cord blood (CB) are not enriched in regulatory T cells expressing Foxp3 compared to the naïve CD4$^+$ T-cells of the adult peripheral blood (APB). (A) Flow cytometric analysis comparing CD4 versus CD25 in APB naïve CD4$^+$ T cells (upper panel) versus naïve CD4 T cells of CB (lower panel) for CD25 expression. The percentage of positive cells and mean fluorescent index (MFI) are indicated. (B) Flow cytometric analysis of intracellular Foxp3 expression in freshly isolated naïve CD4$^+$ T cells that are CD25^{high} from APB (upper panel) or from CB (lower panel). Filled histogram (isotype control) is compared to solid gray line (Foxp3). The percentage of positive cells is indicated in the upper right hand panel (from Chen, L. et al. Impaired allogeneic activation and T helper 1 differentiation of human cord blood naïve CD4 T cells, *Biol. Blood Marrow Transplant.*, 12: 160–171, 2006).

bisulfite sequencing technique to show that the IFN-γ promoter is hypermethylated at a number of sites in neonatal CD4$^+$ T cells compared to adult CD4$^+$ CD45RA$^+$ T cells. Interestingly, the IFN-γ promoter in neonatal CD8$^+$ T cells did not show the same degree of hypermethylation, and indeed, stimulated neonatal CD8$^+$ T cells were capable of making significant amounts of IFN-γ, albeit not as much as adult CD8$^+$ T cells.

4.3. Reduced CD154 Expression

CD154, also known as CD40-ligand, is a member of the tumor necrosis factor superfamily and is an important co-stimulatory molecule. CD154-CD40 interactions between activated T cells and B cells are critical for inducing immunoglobulin class switching in B cells. Deficiency in CD154 results in the hyper-IgM syndrome in which patients are unable to efficiently produce IgG isotypes and are severely immuno-compromised (Etzioni and Ochs, 2004). CD154-CD40 interactions between T cells and dendritic cells are also important. CD40 ligation on dendritic cells induces maturation of the DC and IL-12 expression, both of which are necessary for full activation of cell-mediated immunity (Quezada, 2004). Hyper-IgM syndrome patients also lack vaccine-specific T cells that proliferate and produce IFN-γ, indicating the importance of the CD154/CD40 interaction in the accumulation of functional memory T cells, including those of the Th1 subset (Ameratunga et al., 1997; Jain et al., 1999).

Part of the defect in Th1 differentiation likely lies in the inability of neonatal T cells to upregulate CD154 which in turn results in decreased IL-12 production from dendritic cells. We and others have shown that stimulated neonatal T cells fail to upregulate CD154 in spite of upregulation of other activation markers such as CD69 [(Jullien et al., 2003; Nonoyama et al., 1995) and see Figure 6.5]. The decreased expression of CD154 was due to decreased transcription. This was linked to decreased calcium flux after TCR engagement, but even when the TCR was pharmacologically bypassed using ionomycin, transcription of CD154 remained low. As mentioned above, decreased CD154 expression was also observed when neonatal naïve CD4⁺ T cells were stimulated with fully mature allogeneic dendritic cells

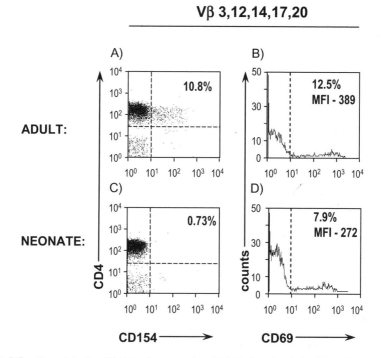

Figure 6.5.   Neonatal naïve CD4⁺ T cells express less CD154 after stimulation than adult naïve CD4⁺ T cells. Naïve CD4⁺ T cells were isolated from adult peripheral blood (A and B) or cord blood (C and D) and activated with a pool of anti-Vβ monoclonal antibodies (mAbs) and anti-CD28 mAb coated onto the wells of plastic tissue culture dishes. After 24 hours the cells were then stained for CD4 and either CD154 (A and C) or CD69 (B and D) and analyzed by flow cytometry. The percentage of positive cells for CD154 and CD69 and the mean fluorescent index of CD69⁺ cells are shown (from Jullien P., et al. Mechanisms for decreased CD40-ligand (CD154) expression by human neonatal CD4 T cells, *Int. Immunol.*, 15:1461–72, 2003).

(Chen, 2006). This indicates that the defect in CD154 expression is intrinsic to the neonatal T cell and is likely to apply to activation in vivo that occurs in response to foreign antigens.

5. Conclusion

In summary, neonates demonstrate striking deficiencies in cell-mediated immunity to a number of pathogens. A number of carefully orchestrated cellular and molecular events must occur before a strong cell-mediated immune response can be mounted, and newborns are transiently deficient at a number of points along the way. Neonatal antigen presenting function is less efficient than that of adults, and, in particular neonatal dendritic cells secrete less IL-12. Neonatal CD4 cells are prevented from differentiating into Th1 cells due to decreased levels of STAT4 and epigenetic regulation of the IFN-γ promoter. They also express less CD154. Consequently, the normal positive feedback loops for driving cell-mediated immunity, in which IFN-γ and CD154 from T cells induce dendritic cells to produce more IL-12, are interrupted. A better understanding of these mechanisms will be important in the care of newborns as well as in vaccine development.

References

Adkins, B. (1999). T-cell function in newborn mice and humans. *Immunol. Today* **20**, 330–335.

Adkins, B., Leclerc, C., and Marshall-Clarke, S. (2004). Neonatal adaptive immunity comes of age. *Nat. Rev. Immunol.* **4**, 553–564.

Ameratunga, R., Lederman, H.M., Sullivan, K.E., Ochs, H.D., Seyama, K., French, J.K. et al. (1997). Defective antigen-induced lymphocyte proliferation in the X-linked hyper-IgM syndrome. *J. Pediatr.* **131**, 147–150.

Bevan, M. (2004). Helping the CD8(+) T-cell response. *Nat. Rev. Immunol.* **4**, 595–602.

Brown, H.L., and Abernathy, M.P. (1998). Cytomegalovirus infection. *Semin. Perinatol.* **22**, 260–266.

Burchett, S.K., Corey, L., Mohan, K.M., Westall, J., Ashley, R., and Wilson, C.B. (1992). Diminished interferon-gamma and lymphocyte proliferation in neonatal and postpartum primary herpes simplex virus infection. *J. Infect. Dis.* **165**, 813–818.

Chaplin, D.D. (2003). 1. Overview of the immune response. *J. Allergy Clin. Immunol.* **111**, S442–459.

Chen, L., Cohen, A.C., and Lewis, D.B. (2006). Impaired allogeneic activation and T helper 1 differentiation of human cord blood naive CD4 T cells. *Biol. Blood Marrow Transplant.* **12**: 160–171.

Chen, S.F., Tu, W.W., Sharp, M.A., Tongson, E.C., He, X.S., and Greenberg, H.B., et al. (2004). Antiviral CD8 T cells in the control of primary human cytomegalovirus infection in early childhood. *J. Infect. Dis.* **189**, 1619–1627.

Clerici, M., DePalma, L., Roilides, E., Baker, R., and Shearer, G.M. (1993). Analysis of T helper and antigen-presenting cell functions in cord blood and peripheral blood leukocytes from healthy children of different ages. *J. Clin. Invest.* **91**, 2829–2836.

Cresswell, P., Ackerman, A.L., Giodini, A., Peaper, D.R., and Wearsch, P.A. (2005). Mechanisms of MHC class I-restricted antigen processing and cross-presentation. *Immunol. Rev.* **207**, 145–157.

Dabbagh, K., Dahl, M.E., Stepick-Biek, P., and Lewis, D.B. (2002). Toll-like receptor 4 is required for optimal development of Th2 immune responses: role of dendritic cells. *J. Immunol.* **168**, 4524–4530.

Ehlers, S., and Smith, K.A. (1991). Differentiation of T cell lymphokine gene expression: the in vitro acquisition of T cell memory. *J. Exp. Med.* **173**, 25–36.

Etzioni, A., and Ochs, H.D. (2004). The hyper IgM syndrome – an evolving story. *Pediatr. Res.* **56**, 519–525.

Gandhi, M.K., and Khanna, R. (2004). Human cytomegalovirus: clinical aspects, immune regulation, and emerging treatments. *Lancet Infect. Dis.* **4**, 725–738.

Gasparoni, A., Ciardelli, L., Avanzini, A., Castellazzi, A.M., Carini, R., Rondini, G., and Chirico, G. (2003). Age-related changes in intracellular TH1/TH2 cytokine production, immunoproliferative T lymphocyte response and natural killer cell activity in newborns, children and adults. *Biol. Neonate* **84**, 297–303.

Goriely, S., Vincart, B., Stordeur, P., Vekemans, J., Willems, F., Goldman, M., and De Wit, D. (2001). Deficient IL-12(p35) gene expression by dendritic cells derived from neonatal monocytes. *J. Immunol.* **166**, 2141–2146.

Guerina, N.G., Hsu, H.W., Meissner, H.C., Maguire, J.H., Lynfield, R., Stechenberg, B., Abroms, I., Pasternack, M.S., Hoff, R., and Eaton, R.B., et al. (1994). Neonatal serologic screening and early treatment for congenital Toxoplasma gondii infection. The New England Regional Toxoplasma Working Group. *N. Engl. J. Med.* **330**, 1858–1863.

Harari, A., Zimmerli, S.C., and Pantaleo, G. (2004). Cytomegalovirus (CMV)-specific cellular immune responses. *Hum. Immunol.* **65**, 500–506.

Herget, G.W., Riede, U.N., Schmitt-Graff, A., Lubbert, M., Neumann-Haefelin, D., and Kohler, G. (2005). Generalized herpes simplex virus infection in an immunocompromised patient–report of a case and review of the literature. *Pathol. Res. Pract.* **201**, 123–129.

Hermann, E., Truyens, C., Alonso-Vega, C., Even, J., Rodriguez, P., and Berthe, A., et al. (2002). Human fetuses are able to mount an adult-like CD8 T-cell response. *Blood* **100**, 2153–2158.

Hunter, C.A. (2005). New IL-12 family members: IL-23 and IL-27, cytokines with divergent functions. *Nat. Rev. Immunol.* **5**, 521–531.

Jain, A., Atkinson, T.P., Lipsky, P.E., Slater, J.E., Nelson, D.L., and Strober, W. (1999). Defects of T-cell effector function and post-thymic maturation in X-linked hyper-IgM syndrome. *J. Clin. Invest.* **103**, 1151–1158.

Jullien, P., Cron, R.Q., Dabbagh, K., Cleary, A., Chen, L., and Tran, P., et al. (2003). Decreased CD154 expression by neonatal CD4+ T cells is due to limitations in both proximal and distal events of T cell activation. *Int. Immunol.* **15**, 1461–1472.

Kimberlin, D.W. (2004). Neonatal herpes simplex infection. *Clin. Microbiol. Rev.* **17**, 1–13.

Langrish, C.L., Buddle, J.C., Thrasher, A.J., and Goldblatt, D. (2002). Neonatal dendritic cells are intrinsically biased against Th-1 immune responses. *Clin. Exp. Immunol.* **128**, 118–123.

Liu, Y.J. (2001). Dendritic cell subsets and lineages, and their functions in innate and adaptive immunity. *Cell* **106**, 259–262.

Lucin, P., Pavic, I., Polic, S., Jonjic, S., and Kosinowski, U.H. (1992). Gamma-interferon-dependent clearance of cytomegalovirus infection in salivary glands. *J. Virol.* **66**, 1977–1984.

Marchant, A., Appay, V., Van Der Sande, M., Dulphy, N., Liesnard, C., and Kidd, M., et al. (2003). Mature CD8(+) T lymphocyte response to viral infection during fetal life. *J. Clin. Invest.* **111**, 1747–1755.

Mellman, I., and Steinman, R.M. (2001). Dendritic cells: specialized and regulated antigen processing machines. *Cell* **106**, 255–258.

Melvin, A.J., McGurn, M.E., Bort, S.J., Gibson, C., and Lewis, D.B. (1995). Hypomethylation of the interferon-gamma gene correlates with its expression by primary T-lineage cells. *Eur. J. Immunol.* **25**, 426–430.

Montoya, J.G., and Liesenfeld, O. (2004). Toxoplasmosis. *Lancet* **363**, 1965–1976.

Nonoyama, S., Penix, L.A., Edwards, C.P., Lewis, D.B., Ito, S., and Aruffo, A., et al. (1995). Diminished expression of CD40 ligand by activated neonatal T cells. *J. Clin. Invest.* **95**, 66–75.

O'Sullivan, B., and Thomas, R. (2003). CD40 and dendritic cell function. *Crit. Rev. Immunol.* **23**, 83–107.

Peggs, K.S., Verfuerth, S., Pizzey, A., Khan, N., Moss, P., and Goldstone, A.H., et al. (2003). Reconstitution of T-cell repertoire after autologous stem cell transplantation: influence of CD34 selection and cytomegalovirus infection. *Biol. Blood Marrow Transplant.* **9**, 198–205.

Picard, C., and Casanova, J.L. (2004) Inherited disorders of cytokines. *Curr. Opin. Pediatr.* **16**, 648–658.

Polic, B., Hengel, H., Krmpotic, A., Trgovcich, J., Pavic, P., and Luccaronin, P., et al. (1998) Hierarchical and redundant lymphocyte subset control precludes cytomegalovirus replication during latent infection. *J. Exp. Med.* **188**, 1047–1054.

Pirenne-Ansart, H., Paillard, F., De Groote, D., Eljaafari, A., Le Gac, S., Blot, P., Franchimont, P., Vaquero, C., and Sterkers, G. (1995). Defective cytokine expression but adult-type T-cell receptor, CD8, and p56lck modulation in CD3- or CD2-activated T cells from neonates. *Pediatr. Res.* **37**, 64–69.

Quezada, S.A., Jarvinen, L.Z., Lind, E.F., and Noelle, R.J. (2004). CD40/CD154 interactions at the interface of tolerance and immunity. *Annu. Rev. Immunol.* **22**, 307–328.

Schwartz, R.H. (2005). Natural regulatory T cells and self-tolerance. *Nat. Immunol.* **6**, 327–330.

Smith, S., Jacobs, R.F., and Wilson, C.B. (1997). Immunobiology of childhood tuberculosis: a window on the ontogeny of cellular immunity. *J. Pediatr.* **131**, 16–26.

Stagno, S., Pass, R.F., Dworsky, M.E., and Alford, C.A. (1983). Congenital and perinatal cytomegalovirus infections. *Semin. Perinatol.* **7**, 31–42.

Subauste, C.S., Wessendarp, M., Sorensen, R.U., and Leiva, L.E. (1999). CD40-CD40 ligand interaction is central to cell-mediated immunity against Toxoplasma gondii: patients with hyper IgM syndrome have a defective type 1 immune response that can be restored by soluble CD40 ligand trimer. *J. Immunol.* **162**, 6690–6700.

Sullender, W.M., Miller, J.L., Yasukawa, L.L., Bradley, J.S., Black, S.B., Yeager, A.S., and Arvin, A.M. (1987). Humoral and cell-mediated immunity in neonates with herpes simplex virus infection. *J. Infect. Dis.* **155**, 28–37.

Takeda, K., and Akira, S. (2005). Toll-like receptors and innate immunity. *Int. Immunol.* **17**, 1–14.

Trivedi, H.N., HayGlass, K.T., Gangur, V., Allardice, J.G., Embree, J.E., and Plummer, F.A. (1997). Analysis of neonatal T cell and antigen presenting cell functions. *Hum. Immunol.* **57**, 69–79.

Tu, W., Chen, S., Sharp, M., Dekker, C., Manganello, A.M., and Tongson, E.C., et al. (2004). Persistent and selective deficiency of CD4+ T cell immunity to cytomegalovirus in immunocompetent young children. *J. Immunol.* **172**, 3260–3267.

Upham, J.W., Lee, P.T., Holt, B.J., Heaton, T., Prescott, S.L., and Sharp, M.J., et al. (2002). Development of interleukin-12-producing capacity throughout childhood. *Infect. Immun.* **70**, 6583–6588.

Walter, E.A., Greenberg, P.D., Gilbert, M.J., Finch, R.J., Watanabe, K.S., Thomas, E.D., and Riddell, S.R. (1995). Reconstitution of cellular immunity against cytomegalovirus in recipients of allogeneic bone marrow by transfer of T-cell clones from the donor. *N. Engl. J. Med.* **333**, 1038–1044.

Watts, C. (2004). The exogenous pathway for antigen presentation on major histocompatibility complex class II and CD1 molecules. *Nat. Immunol.* **5**, 685–692.

White, G.P., Watt, P.M., Holt, B.J., and Holt, P.G. (2002). Differential patterns of methylation of the IFN-gamma promoter at CpG and non-CpG sites underlie differences in IFN-gamma gene expression between human neonatal and adult CD45RO- T cells. *J. Immunol.* **168**, 2820–2827.

Wilson, C.B., Remington, J.S., Stagno, S., and Reynolds, D.W. (1980). Development of adverse sequelae in children born with subclinical congenital Toxoplasma infection. *Pediatrics* **66**, 767–774.

Wu, C.Y., Demeure, C., Kiniwa, M., Gately, M., and Delespesse, G. (1993). IL-12 induces the production of IFN-gamma by neonatal human CD4 T cells. *J. Immunol.* **151**, 1938–1949.

7

Controversies in Neonatal Sepsis: Immunomodulation in the Treatment and Prevention of Neonatal Sepsis

Dr Samantha J. Moss and Dr Andrew R. Gennery

1. Introduction

Neonatal sepsis remains an important cause of morbidity and mortality, particularly in preterm or low birth weight infants. Recognition of infection is difficult because of the non specific nature of presenting signs in most infected neonates. Intravenous antimicrobial therapy should be initiated as soon as possible in neonates with suspected sepsis. Empiric therapy is normally given pending the results of bacterial cultures commonly penicillin and an aminoglycoside. In view of the relative immaturity of the neonatal immune system, immunotherapies have been considered as adjunctive treatments for neonatal sepsis but their use is still controversial. This chapter will outline the rationale for their use and summarise results of trials performed to date.

2. Background to Neonatal Infection

2.1. Difficulties in Interpreting Blood Culture Results

It is difficult to accurately gauge how many neonates develop sepsis. This will depend partly on the definition of infection. Many infants are treated for suspected infection but subsequent microbial cultures are negative. There may be a number of reasons for this including pre treatment with antibiotics and false negative results of blood cultures due to an inadequate volume of blood (Schelonka et al., 1996). Conversely not all positive blood cultures indicate infection. A positive blood culture may be due to contamination of the culture medium with an organism introduced either from the neonate's skin or indeed in the laboratory. Difficulty in interpretation of these results is compounded by the fact that many infections may be caused by

Hot Topics in Infection and Immunity in Children, edited by Andrew J. Pollard and Adam Finn.
Springer, New York, 2006

coagulase negative Staphylococci: it can be difficult to ascertain whether a positive blood culture result is truly due to infection with the organism that has been isolated or whether the isolated organism has contaminated the culture medium. Furthermore the signs of infection are often nonspecific. An infant who looks unwell, is commenced on empiric antibiotic treatment, and subsequently is found to have a positive blood culture may in fact have some other problem, for example a patent ductus arteriosus, which has coincidently been associated with a false positive blood culture result.

2.2. Epidemiology

The incidence of infection varies according to geography. A study examining the incidence of early onset group B Streptococcal disease within the United Kingdom and Republic of Ireland found a regional variation of between 0.27–1.38 cases per 1000 live births (Heath et al., 2004).

Gestation clearly has an effect on the incidence of infection, with low birth weight and preterm infants being more likely to develop sepsis and also to die from infection (Heath et al., 2004; Ronnestad et al., 2005). Low birth weight infants' who have been infected, have a mortality of 17% compared to 7% in non-infected infants (Stoll et al., 1996). Additionally, septic infants who survive infection have a worse neurodevelopmental outcome with poorer mental and psychomotor development (Stoll et al., 2004).

2.3. Factors Pre Disposing to Neonatal Infection

Why should infants be particularly at risk of sepsis, compared to other paediatric populations? A number of pre disposing factors that increase the risk of infection can be identified. Broadly, they can be classified as those due to extrinsic and those due to intrinsic factors.

2.3.1. Extrinsic Factors

Infants on a neonatal unit are subject to frequent breaches of skin and mucous membranes as they regularly undergo cannulation of the vessels, venesection, intubation and nasopharyngeal suction. Fragile skin is also more likely to be broken leaving a portal for infection to enter. Resuscitation and indwelling intravenous catheters are risk factors for culture proven neonatal sepsis (Haque et al., 2004) and there is an increased risk of positive blood culture in very low birth weight infants if they have a central venous catheter, umbilical arterial catheter or are receiving total parental nutrition for more than 7 days (Stoll et al., 2002).

Neonatal units are relatively high-risk environments for nosocomial infection. In a study examining the role of transmission of nosocomial infection from cot based toys, infection rates decreased from 4.6 to 1.99 per 1000 patient days over a 6 month evaluation period when toys were removed from the neonatal micro environment (Hanrahan and Lofgren, 2004). Finally, with frequent antibiotic use, drug resistant organisms are found more commonly in the hospital environment and in units where there is high rate of antibiotic usage drug resistant organisms are more frequent (Golan et al., 2005; Regev-Yochay et al., 2005; Linkin et al., 2004).

2.3.2. Intrinsic Factors

Neonates and particularly preterm neonates have a relatively immature immune response compared to older children. Deficiencies can be found in both the innate and the acquired immune response.

2.3.2a. The Innate Immune Response

Neutrophils are the first line of defence once mucosal and skin barriers are breached. They exhibit a non specific response to pathogens and make use of complement and immunoglobulin which opsonise extracellular organisms and bind to complement and immunoglobulin FC receptors on the neutrophil surface. There are neutrophil precursors in fetal bone marrow by 10–11 weeks of gestation and mature neutrophils can be found by 14–16 weeks gestation. There are abundant neutrophil precursors in neonatal bone marrow. However, the rate of proliferation of circulating neutrophil precursors is close to the maximum in the neonate (Ohls et al., 1995) and so in times of stress the increase in response to infection is limited. Severe sepsis is associated with neutropenia particularly in the preterm infant and this is at least partly due to depletion of the neutrophil storage pool (Christensen and Rothstein, 1980).

Growth factors such as granulocyte colony stimulating factor (G-CSF) and granulocyte macrophage colony stimulating factor (GM-CSF) promote neutrophil production and survival and optimise function. GM-CSF results in a quantitative and qualitative enhancement of monocytes and neutrophils. However it is also pro-inflammatory and augments interferon gamma and IL12 production. G-CSF enhances neutrophil numbers and function and down-regulates the pro-inflammatory response by increasing serum TNF receptor and IL1 receptor and decreasing TNF alpha, interferon gamma and IL1 beta production (Gorgen et al., 1992). Preterm infants produce less G-CSF and GM-CSF than term infants after invitro stimulation and low growth factor concentrations may contribute to the neutropenia associated with the septic preterm infant.

Neonatal neutrophils have reduced opsonisation ability compared to older infants, partly because there is less circulating complement and immunoglobulin in the serum (Fujiwara et al., 1997) and partly because there are fewer cell surface immunoglobulin and complement receptors (Carr and Davies, 1990; Smith et al., 1990). Mild phagocytic defects can also be demonstrated in preterm and septic infants.

2.3.3b. Immunoglobulin Production and B Cell Function

IgM and IgG synthesis can be demonstrated by 12 weeks gestation although preterm infants have limited ability to generate effective quantitative and qualitative antibody responses as a result of limitation in T cell help and intrinsic B cell immaturity. Neonatal B cells are able to undergo immunoglobulin class switch recombination (Punnonen et al., 1992) and also somatic hypermutation (Mortari et al., 1993). However immunoglobulin production is limited compared to that of antigen naïve adult B cells. Production of different immunoglobulin isotypes is impaired and polyclonal activators fail to cause class switching of neonatal B cells resulting in mainly IgM responses. Neonates demonstrate a restricted use of the variable, diversity and joining gene repertoire compared to adults (Mortari et al., 1993; Sanz, 1991), although it is unlikely that this limits the neonatal humoral immune response.

There is possibly impaired presentation of antigen by neonatal dendritic cells as there are less dendritic cells in cord blood than in adult peripheral blood and cord blood accessory cells express less co-stimulatory molecules than adult cells (Hunt et al., 1994). There is impaired T and B cell interaction as there is less CD40 ligand expression on neonatal T cells (Nonoyama et al., 1995), necessary to interact with CD40 on B cells and initiate immunoglobulin class switching.

The majority of neonatal IgG is maternally derived and is mainly transferred actively across the placenta from the mother to the foetus after 32 weeks gestation. There is little maternal IgG2 transferred and there is a relatively slow onset of neonatal immunoglobulin synthesis. Thus preterm infants miss out on maternal IgG transfer and will therefore be relatively hypogammaglobulinaemic compared to their term counterparts. After birth maternal IgG decays and there may be a nadir from 2–4 months of age until neonatal intrinsic IgG production has commenced. However, neither preterm nor term infants are able to produce any anti polysaccharide antibody of the IgG2 subclass, thus rendering them particularly susceptible to infection with encapsulated bacteria.

Because the neonatal immune system is relatively deficient compared to that of older children or adults, the concept of treating neonatal sepsis by replacing some of the deficiencies is attractive. However, the use of adjunct immunomodulation in the treatment or prophylaxis of neonatal infection remains controversial, and is not yet recommended for routine use. The rest of this chapter will review the evidence for the use of these experimental therapies in the treatment or prevention of neonatal sepsis.

3. Use of Granulocyte Infusions in the Treatment of Neonatal Sepsis

Because neutropenia is associated with poor outcome in neonatal sepsis, and the neutrophil pool is depleted in such patients, the use of granulocyte infusions to treat these patients has been considered for more than twenty years (Christensen et al., 1982a). A number of small studies have looked at the use of granulocyte infusion for the treatment of neonatal sepsis. A meta-analysis of some of these have been published by the Cochrane Review (Mohan and Brocklehurst, 2003). Three small randomised trials were analysed, of 9, 10 and 25 infants respectively with either proven or suspected neonatal sepsis and a neutropenia (defined as <1700 cells/µl [1 trial] or as <1500 cells/µl [2 trials]) and receiving antibiotic treatment

(Baley et al., 1987; Christensen et al., 1982b, Wheeler et al., 1987). Infants were randomised to receive either granulocyte infusions or placebo of fresh frozen plasma or red blood cell infusions (Wheeler et al., 1987) or granulocyte infusions with no placebo control in the other 2 trials. A further randomised trial of 35 infants who were septic and neutropenic (<1700 cells/μl) and on antibiotic treatment and randomised to receive either granulocyte infusions or placebo of intravenous immunoglobulin was also analysed (Cairo et al., 1992).

The meta-analysis demonstrated no significant difference in mortality of all infants, or any subgroups of infants including preterm infants, infants with early onset sepsis and infants with confirmed sepsis, treated with granulocyte infusions compared to those receiving placebo or no infusion. Four infants from 2 trials developed pulmonary complications and 2 died of pre-existing severe respiratory disease. There was no information available on the neurodevelopmental outcome at 1 year, the length of hospital stay or the immunological outcome. In the study of 35 infants who received either granulocyte infusions or intravenous immunoglobulin there was a borderline statistically significant reduction in all cause mortality (p = 0.05) in those infants receiving granulocyte infusions. No adverse events were reported and no other outcomes were reported. Three non-randomised studies comprising of 42 infants were excluded from the meta-analysis and 1 study that enrolled infants with non neutropenic sepsis was also excluded although 20 of the 35 infants enrolled were neutropenic.

In conclusion there is insufficient evidence from randomised control trials to support or refute the use of granulocyte transfusions in neonatal septic neutropenia to reduce overall mortality or morbidity. Granulocyte concentrates transfused for the treatment of sepsis have potential side effects including transmission of blood borne infection, fluid overload and Graft versus Host disease, particularly in the immuno-incompetent infant. Preparation of granulocytes needs technical expertise which is not universally available and there are substantial cost implications. There may well be a delay in the procurement and transfusion of granulocytes following a decision to use this product and the shelf life of the blood product is short. As correction of the neutrophil storage pool deficiency is possible through the use of colony stimulating factors which are more easily available and easier to give with fewer side effects there is no place for the routine infusion of donor granulocytes in the treatment of neonatal sepsis.

4. Use of Growth Factors in Neonatal Sepsis

Granulocyte – macrophage colony stimulating factor (GM-CSF) and granulocyte colony stimulating factor (G-CSF) are naturally occurring cytokines used routinely to accelerate neutrophil recovery during the aplastic stage of bone marrow transplantation. When used in this clinical setting there are no adverse side effects. Their use as a treatment to increase circulating neutrophils in septic neutropenic infants may be of benefit because the neutrophil precursor storage pool should readily respond to increased concentrations of these growth factors.

Table 7.1. Randomised controlled trials using growth factors to treat infants with sepsis

Study Author	Type of Study	Number of Infants Clinical Definitions	Growth Factor Dosage Regimen µg/kg
Ahmad 2002	Multicentre RC DB placebo	28 infants EO or LO sepsis N < 1 or culture +	G-CSF 5bd (10) GM-CSF 4bd (10) Placebo (8)
Bedford-Russell 2001	Multicentre RC DB placebo	28 neonates clinical sepsis N < 5	G-CSF 10od (13) Placebo (15)
Bilgin 2001	Single centre [R] C No placebo	60 infants clinical & culture + sepsis, N < 1.5	GM-CSF 5od (30) Control (30)
Drossou-Agakidou 1998	Single centre RC No placebo	35 infants clinical & culture + sepsis, N < 5	G-CSF 10od (19) Control (16)
Gillan 1994	Multicentre RC placebo	42 infants clinical sepsis, N < 5	G-CSF 1od (9), 5od (9) 10od (9), 5bd (3), 10bd (3) Control (9)
Miura 2001	Single centre RC placebo	44 infants clinical sepsis, N not criteria	G-CSF 10od (22) Placebo (22)
Schibler 1998	Multicentre RC DB placebo	20 infants, Manroe EO clinical sepsis	G-CSF 10od (10) Placebo (10)

RC, randomised controlled; DB, double blind; [R], quasi-randomised (alternate allocation); EO, early onset; LO, late onset; N, neutrophils; Manroe, Manroe criteria (Manroe et al., 1979); G-CSF, granulocyte colony stimulating factor; GM-CSF, granulocyte-macrophage colony stimulating factor.

4.1. Growth Factors To Treat Neonatal Sepsis

A meta-analysis on 7 randomised controlled trials using growth factors to treat infants with suspected or proven bacterial sepsis and neutropenia (Table 7.1) has been published as a Cochrane Review (Carr et al., 2003). Four different definitions of neutropenia were used between the studies and the definition of sepsis varied between a clinical or clinical and culture proven episode. Five studies used G-CSF as treatment, one used GM-CSF and one used either GM-CSF or GCSF. Six different dosage regimens for G-CSF and 2 different dosage regimens for GM-CSF were employed between the studies. Meta-analysis showed no significant reduction in immediate (by day 14) all cause mortality when either G-CSF or GM-CSF were given in addition to antibiotics in preterm infants. A retrospective analysis of those infants who were subsequently proven to have septic neutropenia (97 infants in 3 studies) showed a significant reduction in mortality to day 14 ($p = 0.03$), although 67% of infants in this sub-analysis were contributed by 1 study (Bilgin et al., 2001). However, these findings confirm those of a previously published meta-analysis (Bernstein et al., 2001) that examined 5 studies (including 2 not reported in the Cochrane Review) and found a significant reduction in mortality in neutropenic infants.

The problem with the studies to date is that infants included in the studies have had non-neutropenic sepsis and there has been a low incidence of culture proven confirmed sepsis. However, it would appear that the use of growth factors as an adjunct to antibiotics can improve the outcome in septic, neutropenic infants. Practically though, when confronted with a sick preterm or term infant, a decision needs to be taken at the time as to whether or not the child is septic, and whether or not to initiate treatment with growth factors in addition to antibiotics.

4.2. Prophylactic Growth Factors for the Prevention of Neonatal Sepsis

In order to resolve the clinical conundrums posed above, it may be possible to prevent sepsis by the use of prophylactic treatment with growth factors, rather than treating possible septic episodes. Three studies have been analysed and reported in a Cochrane Review looking at a total of 359 infants who received prophylactic growth factors (Carr et al., 2003). All infants received GM-CSF although 2 different preparations were used and there were 4 different dosages used over 2 dosage schedules (Table 7.2). There was no significant reduction in the incidence of sepsis although 1 study of 75 infants showed a significant difference in the prevention of neutropenia in the group receiving GM-CSF (Carr et al., 1999). In the same study there was a reduction in the incidence of systemic infection in neutropenic preterm infants that did not reach statistical significance. Overall, there was insufficient evidence to support the routine use of G-CSF or GM-CSF as prophylaxis to prevent systemic infection in high-risk neonates. The main problem with the studies to date is that low risk babies have been recruited, thus diluting the effects of any potential benefit to high-risk infants. There is currently a large UK based multicentre single blind randomised control trial (prophylactic granulocyte macrophage colony stimulating factor (GM-CSF) to reduce sepsis in preterm neonates – PROGRAMS) (http://www.npeu.ox.ac.uk/programs/) to investigate whether GM-CSF given prophylactically to low birth weight preterm infants less than 72 hours old can reduce

Table 7.2. Randomised controlled trials using growth factors as prophylactic treatment in infants with high-risk of sepsis

Study Author Sepsis Definition	Study Design	Study Subjects	Growth Factor Dosage Regimen μg/kg (no. of infants)
Cairo 1995 Clinical or culture	Multicentre RC placebo	20 infants Bw 500–1500 g <34 w, <72 hr	GM-CSF 5od 7/7(5) 5bd 7/7(5) 10od 7/7 (5), Placebo (5)
Cairo 1999 Clinical or culture	Multicentre RC DB placebo	264 infants Bw 501–1000 g AGA,<72 hr	8od 7/7 then alt day for 21 day (134) Placebo (130)
Carr 1999 Culture	Multicentre RC	75 infants Bw 480–760 g <32 w, <72 hr	GM-CSF 10od 5/7 (36) Control (39)

RC, randomised, controlled; DB, double blind, Bw, birth weight; w, weeks; hr, hours; AGA, appropriate gestational age; GM-CSF, granulocyte-macrophage colony stimulating factor.

the incidence of systemic infection or mortality. At time of preparation of this manuscript, 249 babies had been recruited into the study, but no results are available yet.

5. Use of Pooled Human Immunoglobulin in Neonatal Sepsis

As discussed above, preterm infants are relatively hypo-gammaglobulinaemic and have missed out on transfer of maternal IgG across the placenta. Immunoglobulin activates complement and also has intrinsic opsonic activity and binds to cell surface receptors. The concept of replacing the physiological deficit to treat or prevent neonatal sepsis is therefore attractive. Pooled human immunoglobulin contains donations from between 1000–2000 donors, and undergoes a variety of viral inactivation steps during manufacture. However, there is a risk of anaphylactic reactions, particularly in the septic patient. Furthermore, although all commercially available immunoglobulin is sourced from out with the UK, the theoretical risk of transmitting blood borne viral infection or prion disease should be borne in mind before using any blood product.

5.1. Role of Intravenous Immunoglobulin Infusions in the Treatment of Suspected or Proven Neonatal Sepsis

A Cochrane Review has examined 9 studies comprising 553 infants who received polyclonal human intravenous immunoglobulin in addition to systemic antibiotics for the treatment of suspected or microbiological proven systemic bacterial infection (Ohlsson and Lacy, 2004a). Within the 9 studies, 5 different commercially available intravenous immunoglobulin preparations were used. Four different dose regimens were employed. Four of the studies used a single dose of intravenous immunoglobulin in addition to antibiotics (Christensen et al., 1991; Mancilla-Ramirez et al., 1992; Weisman et al., 1992; Chen, 1996), 5 studies used multiple doses of intravenous immunoglobulin (Sidiropoulos et al., 1986; Haque et al., 1988; Erdem et al., 1993; Samatha et al., 1997; Shenoi et al., 1999). In 4 of the studies the definition of sepsis was a positive blood culture prior to entry to the study.

Six studies (318 infants) reported on the prevention of mortality for suspected infection and there was a borderline statistically significant reduction in mortality ($p = 0.05$). In 7 studies, a sub-analysis of infants who had proven infection showed a statistically significant reduction in mortality ($p = 0.04$) but there was no statistically significant reduction in the length of hospital stay in preterm infants.

There is currently insufficient evidence to support the routine use of polyclonal human intravenous immunoglobulin to reduce mortality from suspected or subsequently proven neonatal infection. The borderline statistical significance for outcome of mortality in neonates with suspected infection and reduced mortality in neonates with subsequently proven infection justifies further research.

5.2. Role of Prophylactic Intravenous Immunoglobulin Infusions in the Prevention of Neonatal Sepsis

Would the prophylactic administration of pooled human immunoglobulin help to preterm or low birth-weight infants help prevent systemic infection? A Cochrane Review has analysed 19 trials that randomised 5000 preterm or low birth weight infants to prophylactic intravenous immunoglobulin infusions or placebo (Ohlsson and Lacy, 2004b). Eight different IVIG preparations were used and there were 7 dosage variables. In 4 trials, infants received a single dose of intravenous immunoglobulin, and in the remaining 15 multiple doses were given. The meta-analysis of 10 of the trials comprising 3975 infants, showed a significant decrease in clinical and culture proven septic episodes in those infants who received immunoglobulin (p = 0.02). The prevention of 1 or more episodes of serious infection comprising sepsis, meningitis or urinary tract infection was studied in 16 studies comprising of 4986 infants and a combined analysis showed a significant decrease in serious infection in infants receiving immunoglobulin (p = 0.0005). Of 15 studies that reported on all cause mortality due to infection, intraventricular haemorrhage, necrotising enterocolitis or chronic lung disease, the combined analysis showed no significant decrease in mortality (4125 infants) and in ten studies of 1690 infants looking at prevention of infectious mortality, a combined analysis showed no significant decrease in mortality. Prophylactic intravenous immunoglobulin administration results in a 3% reduction in sepsis and 4% reduction in serious infection but no associated reduction in mortality or other co-morbidities.

Coagulase negative staphylococci are amongst the most common organisms causing infection in neonatal units (Urrea et al., 2003). Pooled polyclonal human immunoglobulin has high opsonic activity against group B streptococcus and pneumococcus but only low opsonic activity against coagulase negative Staphylococci, and opsonic activity varies between batches and preparations (Weisman et al., 1994; Fischer et al., 1994). Thus, routine use of commercially available pooled human immunoglobulin preparations is unlikely to significantly reduce the incidence of neonatal sepsis that is most commonly due to coagulase negative staphylococci, although it is likely to be most effective against organisms that are associated with more severe sepsis and mortality. The use of a preparation that is effective against coagulase negative staphylococci is more likely to reduce the overall incidence on sepsis, although it may have little effect on morbidity. BSYX-A110 is a monoclonal antibody that is protective for coagulase negative Staphylococci and *Staphylococcus aureus*, binding to a range of bacterial strains and is opsonic for most stains. A study involving adult healthy volunteers showed that it was safe and tolerable and antibody concentrations approached those levels that were protective in adults. Antibody concentrations were highly correlated to serum bacteriocidal activity. A randomised control trial looking at the effectiveness of BSYX-A110 to prevent episodes of neonatal sepsis is currently planned.

There is currently a multicentre randomised blinded placebo controlled trial looking at the effectiveness of pooled human immunoglobulin in the treatment of neonatal sepsis (international neonatal immunotherapy study – INIS) which has recruited 1839 babies at time of preparation of this manuscript (http://www.npeu.ox.ac.uk/inis/). Eligibility for the study requires an infant to be receiving antibiotics from proven or suspected infection and less than 1500 grams or ventilated or have evidence of infection in blood, CSF or another sterile site.

6. Conclusion

Whilst the concept of replacing the neonatal immunological deficit to prevent or treat systemic sepsis is theoretically attractive, the results from clinical trials performed to date are perhaps less dramatic than expected. The reasons for this are likely to be multifactorial, but include the recruitment of low risk infants who are not septic or neutropenic, and use of inappropriate immunoglobulin preparations that contains little or no antibody effective against pathogens causing neonatal sepsis. There may well be cheaper, more effective, but less glamorous, methods of preventing infection in the preterm infant. There is accumulating evidence regarding the role of prophylactic broad spectrum antibiotics on neonatal gut flora and the influence of early enteral feeding with a correct milk formula can have on encouraging "friendly" gut bacteria. The role of growth factors and immunoglobulin preparations in the treatment and prophylaxis of neonatal sepsis has not yet been established and these products should only be used in the context of large randomised controlled clinical trials until there is clear evidence of benefit (or absence of benefit) in their use.

Acknowledgements

We thank Alison Bedford-Russell for helpful discussions.

References

Ahmad, A., Laborada, G., Bussel, J., and Nesin, M. (2002). Comparison of recombinant granulocyte colony-stimulating factor, recombinant human granulocyte-macrophage colony-stimulating factor and placebo for treatment of septic preterm infants. *Pediatr Infect Dis J.* 21:1061–1065.

Baley, J.E., Stork, E.K., Warkentin, P.I., and Shurin, S.B. (1987). Buffy coat transfusions in neutropenic neonates with presumed sepsis: a prospective, randomized trial. *Pediatrics.* 80:712–720.

Bedford Russell, A.R., Emmerson, A.J., Wilkinson, N., Chant, T., Sweet, D.G., Halliday, H.L., Holland, B., and Davies, E.G. (2001). A trial of recombinant human granulocyte colony stimulating factor for the treatment of very low birthweight infants with presumed sepsis and neutropenia. *Arch Dis Child Fetal Neonatal Ed.* 84:F172–F176.

Bernstein, H.M., Pollock, B.H., Calhoun, D.A., and Christensen, R.D. (2001). Administration of recombinant granulocyte colony-stimulating factor to neonates with septicemia: A meta-analysis. *J Pediatr.* 138:917–920.

Bilgin, K., Yaramis, A., Haspolat, K., Tas, M.A., Gunbey, S., and Derman, O. (2001). A randomized trial of granulocyte-macrophage colony-stimulating factor in neonates with sepsis and neutropenia. *Pediatrics.* 107:36–41.

Cairo, M.S., Worcester, C.C., Rucker, R.W., Hanten, S., Amlie, R.N., Sender, L., and Hicks, D.A. (1992). Randomized trial of granulocyte transfusions versus intravenous immune globulin therapy for neonatal neutropenia and sepsis. *J Pediatr.* 120:281–285.

Cairo, M.S., Christensen, R., Sender, L.S., Ellis, R., Rosenthal, J., van de Ven, C., Worcester, C., and Agosti, J.M. (1995). Results of a phase I/II trial of recombinant human granulocyte-macrophage colony-stimulating factor in very low birthweight neonates: significant induction of circulatory neutrophils, monocytes, platelets, and bone marrow neutrophils. *Blood.* 86:2509–2515.

Cairo, M.S., Agosti, J., Ellis, R., Laver, J.J., Puppala, B., deLemos, R., Givner, L., Nesin, M., Wheeler, J.G., Seth, T., van de Ven, C., and Fanaroff, A. (1999). A randomized, double-blind, placebo-controlled trial of prophylactic recombinant human granulocyte-macrophage colony-stimulating factor to reduce nosocomial infections in very low birth weight neonates. *J Pediatr.* 134:64–70.

Carr, R., and Davies, J.M. (1990). Abnormal FcRIII expression by neutrophils from very preterm neonates. *Blood.* 76:607–611.

Carr, R., Modi, N., Dore, C.J., El-Rifai, R., and Lindo, D. (1999). A randomized, controlled trial of prophylactic granulocyte-macrophage colony-stimulating factor in human newborns less than 32 weeks gestation. *Pediatrics.* 103:796–802.

Carr, R., Modi, N., and Dore, C. (2003). G-CSF and GM-CSF for treating or preventing neonatal infections. *Cochrane Database Syst Rev.* 3:CD003066.

Chen, J.Y. (1996). Intravenous immunoglobulin in the treatment of full-term and premature newborns with sepsis. *J Formos Med Assoc.* 95:839–844.

Christensen, R.D., and Rothstein, G. (1980). Exhaustion of mature marrow neutrophils in neonates with sepsis. *J Pediatr.* 96:316–318.

Christensen, R.D., Anstall, H., and Rothstein, G. (1982a). Neutrophil transfusion on septic neutropenic neonates. *Transfusion.* 22:151–153.

Christensen, R.D., Rothstein, G., Anstall, H.B., and Bybee, B. (1982b). Granulocyte transfusions in neonates with bacterial infection, neutropenia, and depletion of mature marrow neutrophils. *Pediatrics.* 70:1–6.

Christensen, R.D., Brown, M.S., Hall, D.C., Lassiter, H.A., and Hill, H.R. (1991). Effect on neutrophil kinetics and serum opsonic capacity of intravenous administration of immune globulin to neonates with clinical signs of early-onset sepsis. *J Pediatr.* 118:606–614.

Drossou-Agakidou, V., Kanakoudi-Tsakalidou, F., Sarafidis, K., Taparkou, A., Tzimouli, V., Tsandali, H., and Kremenopoulos, G. (1998). Administration of recombinant human granulocyte-colony stimulating factor to septic neonates induces neutrophilia and enhances the neutrophil respiratory burst and beta2 integrin expression. Results of a randomized controlled trial. *Eur J Pediatr.* 157:583–588.

Erdem, G., Yurdakok, M., Tekinalp, G., and Ersoy, F. (1993). The use of IgM-enriched intravenous immunoglobulin for the treatment of neonatal sepsis in preterm infants. *Turk J Pediatr.* 35:277–281.

Fischer, G.W., Cieslak, T.J., Wilson, S.R., Weisman, L.E., and Hemming, V.G. (1994). Opsonic antibodies to Staphylococcus epidermidis: in vitro and in vivo studies using human intravenous immune globulin. *J Infect Dis.* 169:324–329.

Fujiwara, T., Kobayashi, T., Takaya, J., Taniuchi, S., and Kobayashi, Y. (1997). Plasma effects on phagocytic activity and hydrogen peroxide production by polymorphonuclear leukocytes in neonates. *Clin Immunol Immunopathol.* 85:67–72.

Gillan, E.R., Christensen, R.D., Suen, Y., Ellis, R., van de Ven, C., and Cairo, M.S. (1994). A randomized, placebo-controlled trial of recombinant human granulocyte colony-stimulating factor administration in newborn infants with presumed sepsis: significant induction of peripheral and bone marrow neutrophilia. *Blood.* 84:1427–1433.

Golan, Y., Doron, S., Sullivan, B., and Snydman, D.R. (2005). Transmission of vancomycin-resistant enterococcus in a neonatal intensive care unit. *Pediatr Infect Dis J.* 24:566–567.

Gorgen, I., Hartung, T., Leist, M., Niehorster, M., Tiegs, G., Uhlig, S., Weitzel, F., and Wendel, A. (1992). Granulocyte colony-stimulating factor treatment protects rodents against lipopolysaccharide-induced toxicity via suppression of systemic tumor necrosis factor-alpha. *J Immunol.* 149:918–924.

Hanrahan, K.S., and Lofgren, M. (2004) Evidence-based practice: examining the risk of toys in the microenvironment of infants in the neonatal intensive care unit. *Adv Neonatal Care.* 4:184–201.

Haque, K.N., Zaidi, M.H., and Bahakim, H. (1988). IgM-enriched intravenous immunoglobulin therapy in neonatal sepsis. *Am J Dis Child.* 142:1293–1296.

Haque, K.N., Khan, M.A., Kerry, S., Stephenson, J., and Woods, G. (2004). Pattern of culture-proven neonatal sepsis in a district general hospital in the United Kingdom. *Infect Control Hosp Epidemiol.* 25:759–764.

Heath, P.T., Balfour, G., Weisner, A.M., Efstratiou, A., Lamagni, T.L., Tighe, H., O'Connell, L.A., Cafferkey, M., Verlander, N.Q., Nicoll, A., and McCartney, A.C.; PHLS Group B Streptococcus Working Group. (2004). Group B streptococcal disease in UK and Irish infants younger than 90 days. *Lancet.* 363:292–294.

Hunt, D.W., Huppertz, H.I., Jiang, H.J., and Petty, R.E. (1994). Studies of human cord blood dendritic cells: evidence for functional immaturity. *Blood.* 84:4333–4343.

Linkin, D.R., Fishman, N.O., Patel, J.B., Merrill, J.D., and Lautenbach, E. (2004). Risk factors for extended-spectrum beta-lactamase-producing Enterobacteriaceae in a neonatal intensive care unit. *Infect Control Hosp Epidemiol.* 25:781–783.

Mancilla-Ramirez, J., Gonzalez-Yunes, R., Castellanos-Cruz ,C., Garcia-Roca, P., and Santos-Preciado, J.I. (1992). [Intravenous immunoglobulin in the treatment of neonatal septicemia] *Bol Med Hosp Infant Mex.* 49:4–11.

Manroe, B.L., Weinberg, A.G., Rosenfeld, C.R., and Browne, R. (1979). The neonatal blood count in health and disease.I. Reference values for neutrophilic cells. *J Pediatr.* 95:89–98.

Miura, E., Procianoy, R.S., Bittar, C., Miura, C.S., Miura, M.S., Mello, C., and Christensen, R.D. (2001). A randomized, double-masked, placebo-controlled trial of recombinant granulocyte colony-stimulating factor administration to preterm infants with the clinical diagnosis of early-onset sepsis. *Pediatrics.* 107:30–35.

Mohan, P., and Brocklehurst, P. (2003). Granulocyte transfusions for neonates with confirmed or suspected sepsis and neutropaenia. *Cochrane Database Syst Rev.* 4:CD003956.

Mortari, F., Wang, J.Y., and Schroeder, H.W. Jr. (1993). Human cord blood antibody repertoire. Mixed population of VH gene segments and CDR3 distribution in the expressed C alpha and C gamma repertoires. *J Immunol.* 150:1348–1357.

Nonoyama, S., Penix, L.A., Edwards, C.P., Lewis, D.B., Ito, S., Aruffo, A., Wilson, C.B., and Ochs, H.D. (1995). Diminished expression of CD40 ligand by activated neonatal T cells. *J Clin Invest.* 95:66–75.

Ohls, R.K., Li, Y., Abdel-Mageed, A., Buchanan, G., Jr., Mandell, L., and Christensen, R.D. (1995). Neutrophil pool sizes and granulocyte colony-stimulating factor production in human mid-trimester fetuses. *Pediatr Res.* 37:806–811.

Ohlsson, A., and Lacy, J.B. (2004a). Intravenous immunoglobulin for suspected or subsequently proven infection in neonates. *Cochrane Database Syst Rev.* 1:CD001239.

Ohlsson, A., and Lacy, J.B. (2004b). Intravenous immunoglobulin for preventing infection in preterm and/or low-birth-weight infants. *Cochrane Database Syst Rev.* 1:CD000361.

Punnonen, J., Aversa, G.G., Vandekerckhove, B., Roncarolo, M.G., and de Vries, J.E. (1992). Induction of isotype switching and Ig production by CD5+ and CD10+ human fetal B cells. *J Immunol.* 148:3398–1404.

Regev-Yochay, G., Rubinstein, E., Barzilai, A., Carmeli, Y., Kuint, J., Etienne, J., Blech, M., Smollen, G., Maayan-Metzger, A., Leavitt, A., Rahav, G., and Keller, N. (2005). Methicillin-resistant Staphylococcus aureus in neonatal intensive care unit. *Emerg Infect Dis.* 11:453–456.

Ronnestad, A., Abrahamsen, T.G., Medbo, S., Reigstad, H., Lossius, K., Kaaresen, P.I., Engelund, I.E., Irgens, L.M., and Markestad, T. (2005). Septicemia in the first week of life in a Norwegian national cohort of extremely premature infants. *Pediatrics.* 115: e262–e268.

Samatha, S., Jalalu, M.P., Hegde, R.K., Vishwanath, D., and Maiya, P.P. (1997). Role of IgM-enriched intravenous immunoglobulin as an adjuvant to antibiotics in neonatal sepsis. *Karnataka Pediatr J.* 11:1–6.

Sanz, I. (1991). Multiple mechanisms participate in the generation of diversity of human H chain CDR3 regions. *J Immunol.* 147:1720–1729.

Schelonka, R.L., Chai, M.K., Yoder, B.A., Hensley, D., Brockett, R.M., and Ascher, D.P. (1996). Volume of blood required to detect common neonatal pathogens. *J Pediatr.* 129:275–278.

Schibler, K.R., Osborne, K.A., Leung, L.Y., Le, T.V., Baker, S.I., and Thompson, D.D. (1998). A randomized, placebo-controlled trial of granulocyte colony-stimulating factor administration to newborn infants with neutropenia and clinical signs of early-onset sepsis. *Pediatrics.* 102:6–13.

Shenoi, A., Nagesh, N.K., Maiya, P.P., Bhat, S.R., and Subba Rao, S.D. (1999). Multicenter randomized placebo controlled trial of therapy with intravenous immunoglobulin in decreasing mortality due to neonatal sepsis. *Indian Pediatr.* 36:1113–1118.

Sidiropoulos, D., Boehme, U., Von Muralt, G., Morell, A., and Barandun, S. (1986). Immunoglobulin supplementation in prevention or treatment of neonatal sepsis. *Pediatr Infect Dis.* 5:S193–194.

Smith, J.B., Campbell, D.E., Ludomirsky, A., Polin, R.A., Douglas, S.D., Garty, B.Z., and Harris, M.C. (1990). Expression of the complement receptors CR1 and CR3 and the type III Fc gamma receptor on neutrophils from newborn infants and from fetuses with Rh disease. *Pediatr Res.* 28:120–126.

Stoll, B.J., Gordon, T., Korones, S.B., Shankaran, S., Tyson, J.E., Bauer, C.R., Fanaroff, A.A., Lemons, J.A., Donovan, E.F., Oh.W., Stevenson, D.K., Ehrenkranz, R.A., Papile, L.A., Verter, J., and Wright, L.L. (1996). Late-onset sepsis in very low birth weight neonates: a report from the National Institute of Child Health and Human Development Neonatal Research Network. *J Pediatr.* 129:63–71.

Stoll, B.J., Hansen, N., Fanaroff, A.A., Wright, L.L., Carlo, W.A., Ehrenkranz, R.A., Lemons, J.A., Donovan, E.F., Stark, A.R., Tyson, J.E., Oh, W., Bauer, C.R., Korones, S.B., Shankaran, S., Laptook, A.R., Stevenson, D.K., Papile, L.A., and Poole, W.K. (2002). Late-onset sepsis in very low birth weight neonates: the experience of the NICHD Neonatal Research Network. *Pediatrics.* 110:285–291.

Stoll, B.J., Hansen, N.I., Adams-Chapman, I., Fanaroff, A.A., Hintz, S.R., Vohr, B., and Higgins, R.D.; National Institute of Child Health and Human Development Neonatal Research Network. (2004). Neurodevelopmental and growth impairment among extremely low-birth-weight infants with neonatal infection. *JAMA.* 292:2357–2565.

Urrea, M., Iriondo, M., Thio, M., Krauel, X., Serra, M., LaTorre, C., and Jimenez, R. (2003). A prospective incidence study of nosocomial infections in a neonatal care unit. *Am J Infect Control.* 8:505–507.

Weisman, L.E., Stoll, B.J., Kueser, T.J., Rubio, T.T., Frank, C.G., Heiman, H.S., Subramanian, K.N., Hankins, C.T., Anthony, B.F., and Cruess, D.F. (1992). Intravenous immune globulin therapy for early-onset sepsis in premature neonates. *J Pediatr.* 121:434–443.

Weisman, L.E., Cruess, D.F., and Fischer, G.W. (1994). Opsonic activity of commercially available standard intravenous immunoglobulin preparations. *Pediatr Infect Dis J.* 13:1122–1125.

Wheeler, J.G., Chauvenet, A.R., Johnson, C.A., Block, S.M., Dillard, R., and Abramson, J.S. (1987). Buffy coat transfusions in neonates with sepsis and neutrophil storage pool depletion. *Pediatrics.* 79:422–425.

8

Chlamydia trachomatis Genital Infection in Adolescents and Young Adults

Toni Darville

1. Introduction

Chlamydia trachomatis is an obligate, intracellular, nonmotile, Gram-negative bacterium recognized as one of the most common sexually transmitted agents in the world. Chlamydial genital infection primarily affects sexually active adolescents and young adults. Large-scale screening programs routinely detect infection rates of 5–10 percent in young adults (19–25 years of age) (Miller et al., 2004) (LaMontagne et al., 2004), and 20–25 percent or greater in sexually active adolescents 15–19 years of age (Blythe et al., 1988; Burstein et al., 1998; Ford et al., 2005). Most infected persons do not have symptoms, thus they do not seek medical care, and their infections go undetected. Consequently, screening is necessary to identify and treat this infection. The large reservoir of unrecognized infected individuals helps sustain transmission of this organism. Among men, urethritis is the most common illness resulting from *C. trachomatis* infection. Complications (e.g., epididymitis) affect a minority of infected men and rarely result in sequelae. Women bear the brunt of disease due to infection, for if left untreated, up to 40% of women with untreated *C. trachomatis* infections experience pelvic inflammatory disease (PID)(Stamm et al., 1984a). Of those with PID, up to 20% will become infertile; 18% will experience debilitating, chronic pelvic pain; and 9% will have a life-threatening tubal pregnancy (Westrom et al., 1992). In addition, an infected pregnant woman can transmit the organism to her newborn at the time of delivery, potentially resulting in neonatal conjunctivitis and/or afebrile pneumonia.

The "silent epidemic" of *C. trachomatis* threatens to cause reproductive damage and infertility in many of the 50 million women who acquire it each year. This chapter reviews the microbiology, pathophysiology, epidemiology, and clinical manifestations of genital *C. trachomatis* infections in adolescents and young adults.

Hot Topics in Infection and Immunity in Children, edited by Andrew J. Pollard and Adam Finn. Springer, New York, 2006

Diagnostic methods useful for screening and data on effectiveness of screening and treatment programs are also discussed.

2. The Pathogen

Chlamydiae are obligate intracellular parasites that have been classified under the order Chlamydiales with their own family and genus (Chlamydiaceae, *Chlamydia*). Molecular sequencing analysis of their ribosomal RNA has shown that chlamydiae are a unique class of bacteria that are closely related to each other and have little relation to other eubacteria (Weisburg et al., 1986). Like other Gram-negative bacteria, chlamydiae have an outer membrane that contains lipopolysaccharide (LPS) and membrane proteins, but their outer membrane contains no detectable peptidoglycan, despite the presence of genes encoding proteins for its synthesis (Griffiths and Gupta, 2002). This recent genomic finding is the basis for the so-called chlamydial peptidoglycan paradox, for it has been known for years that chlamydial development is inhibited by beta-lactam antibiotics. Although chlamydiae contain DNA, RNA, and ribosomes, during growth and replication they obtain high-energy phosphate compounds from the host cell. Consequently, they are considered energy parasites. The chlamydial genome size is only 660 KDa which is smaller than that of any other prokaryote except *Mycoplasma spp.*. *Chlamydia trachomatis* encodes an abundant protein called the major outer membrane protein (MOMP or OmpA) that is surface exposed and is the major determinant of serologic classification.

2.1. Chlamydial Developmental Cycle

The biphasic developmental cycle of chlamydiae is unique among microorganisms and involves two highly specialized morphologic forms. The extracellular form or elementary body (EB) contains extensive disulfide cross-links both within and between outer membrane proteins giving it an almost sporelike structure that is stable outside of the cell. The small (350 nm in diameter) infectious EB is inactive metabolically. The developmental cycle is initiated when an EB attaches to a susceptible epithelial cell. The process of EB internalization is very efficient, suggesting that EBs trigger their own internalization by cells that are not considered professional phagocytes. A number of candidate adhesions have been proposed, but their identity and that of associated epithelial cell receptors remain uncertain. Intriguing data indicate chlamydial EBs interact with human endometrial host cell membrane protein disulfide isomerase (PDI) during attachment. Productive entry into and infectivity of EB in host cells is dependent on reduction of EB cross-linked outer membrane proteins. Protein disulfide isomerase carries out thiol-disulfide exchange reactions at the host cell surface, making it a feasible receptor candidate. Interestingly, PDI is a member of the estrogen receptor complex, and EB attachment is enhanced in estrogen-dominant cells (Davis et al., 2002).

Once inside the epithelial cell, surface antigens of the EB appear to prevent fusion of the endosome with lysosomes, protecting itself from enzymatic destruction. Hiding from host attack by antibody or cell mediated defenses; it reorganizes into the replicative form, the reticulate body (RB). Reticulate bodies successfully parasitize the host cell and divide and multiply. As the RB divides by binary fission it fills the endosome, now a cytoplasmic inclusion, with its progeny. After 48 to 72 hours, multiplication ceases and nucleoid condensation occurs as the reticulate bodies transform to new infectious elementary bodies. The elementary bodies are then released from the cell, allowing for infection of new host cells to occur.

The biphasic and relatively prolonged developmental cycle of chlamydiae are survival advantages. Antibiotic treatment or the host immune response must be able to kill both extracellular non-replicating infectious EBs as well as intracellular replicating RBs hidden within their protective vacuole if they are to rid the host of infection. Thus, antibiotic treatment requires multiple-dose regimens for 7–14 days. Single-dose azithromycin treats genital *C. trachomatis* infection effectively because it has a half-life in host cells of 5 to 7 days. The ability to cause prolonged, often subclinical infection is a major characteristic of chlamydiae.

2.2. Classification

Microimmunofluorescence and monoclonal antibody testing have shown that there are at least 18 serovars of *C. trachomatis* with several distinctive clinical patterns of disease. For example, ocular trachoma is caused by serovars A, B, Ba, and C; oculogenital disease by serovars B, Da, Ia, and D-K; and genitourinary disease, or lymphogranuloma venereum (LGV) is due to serovars L1, L2, L2a, and L3. LGV infections are more invasive, as these serovars can replicate in macrophages, whereas replication of the other serovars of *C. trachomatis* is confined to mucosal epithelial cells.

3. Pathogenesis

3.1. Immunopathogenesis

Most individuals with chlamydial infection spontaneously heal without disease sequelae. Individuals with severe forms of chlamydial disease often display immune responses to a common chlamydial heat shock protein 60 (hsp60) antigen. Because the protein shares nearly 50% sequence identity with the human homolog, it is speculated that molecular mimicry may result in autoimmune inflammatory damage that in turn causes chlamydial disease sequelae. Because hsp60 immune responses are genetically determined, susceptibility genes for chlamydial disease may also exist. Although high serum antibody titers generally correlate with complications of chlamydial disease, they may simply reflect increased overall exposure. However, in a study that controlled for level of exposure, serum antibodies to chlamydial heat shock protein 10 (hsp10), but not chlamydial hsp60 or MOMP were present in higher levels in women with tubal factor infertility (TFI), and the degree of the serologic response correlated with severity of infertility (LaVerda et al., 2000). These findings support the hypothesis that the serological response to *C. trachomatis* heat shock proteins is associated with the severity of disease and identifies hsp10 as an antigen recognized by a significant proportion of women with TFI.

In female patients with clinically suspected acute PID pathological features that correlate both with upper genital tract infection and tubal salpingitis include: presence of any neutrophils in the endometrial surface epithelium; neutrophils within gland lumens; dense subepithelial stromal lymphocytic infiltration; any stromal plasma cells; and germinal centers containing transformed lymphocytes (Kiviat et al., 1990). Predominant neutrophilic, lymphocytic and plasma cell infiltrates have been described in human endocervical, endometrial and fallopian tube specimens from infected patients (Kiviat et al., 1990; Paavonen et al., 1987; Paukku et al., 1999). In female patients with ectopic pregnancy who were seropositive for *C. trachomatis*, fallopian tube biopsy specimens revealed extensive subepithelial plasma cell infiltration (Brunham et al., 1986). All patients denied any history of genital tract symptoms indicating subclinical infection of the oviduct can result in tubal disease.

HLA class II alleles may present different chlamydial peptides that evoke damaging, protective, or regulatory immune responses by CD4$^+$ T cells, and cytokine polymorphisms may alter the risk of disease (Cohen et al., 2000; Cohen et al., 2003). For example, HLA class II alleles DQA1*0102 and DQB1*0602 together with a specific IL-10 promoter polymorphism (IL-10-1082AA) were found more frequently in female patients with tubal infertility than in controls (Kinnunen et al., 2002).

3.2. Immunoprotection

A major variant surface protein, major outer membrane protein (MOMP) is the principal target of neutralizing antibodies and may be the target of protective immunity. The detailed genetic and immunochemical knowledge of MOMP has stimulated multiple attempts to design an oligopeptide vaccine. Success has been limited in part because of the antigenic variation that the protein exhibits and in part because of the absence of knowledge regarding the three-dimensional structure of the protein.

The natural course of *C. trachomatis* infection was recently described in a study of Columbian women followed for a 5-year period (Molano et al., 2005). Eighty-two women found to be positive for *C. trachomatis* at the start of the study were studied at 6-month intervals. Most of the women (57.3%) were >30 years of age (70.7% were >25 years of age). Infection was classified as persistent if the same serotype was found at follow-up visits. Women who had taken antibiotics effective against *C. trachomatis* while infected were excluded. All study women reported 1–2 lifetime sex partners (82.9% reported a single lifelong sex partner), thus the potential for repeated infection from an untreated male sex partner was high. Approximately 46% of the infections were persistent at 1 year, 18% at 2 years, and 6% at 4 years of follow-up as determined by PCR of cervical scrape samples. Thus, in nearly half of this female cohort, an adaptive immune response effective in eradicating their infection or in preventing repeat infection did not develop for up to one year. Rates of clearance were faster for women who had ever used oral contraceptives and for women who had their first sexual intercourse at >20 years of age suggesting hormonal effects may affect the immune response to *C. trachomatis* infection.

There is a strong inverse relationship between age and susceptibility to chlamydial infection even when corrected for frequency of sexual contact, suggesting effective adaptive immunity eventually develops. Lymphoproliferative responses, but not serum antibody titers increase with age (Arno et al., 1994). Data from humans (Brunham et al., 1996) point to MHC Class II – restricted $CD4^+$ T cells of the Th1 phenotype as being critical to recovery from chlamydial infection as well as having a role in protection from disease (Kimani et al., 1996). In a cohort of female commercial sex workers with HIV, susceptibility to chlamydial PID increased as numbers of $CD4^+$ T cells decreased (Kimani et al., 1996).

Debattista et al. (Debattista et al., 2002) reported that PBMCs from women with chlamydial PID or a history of repeated *C. trachomatis* infection produced less IFN-gamma in response to chlamydial hsp60 than did women with a single episode of *C. trachomatis* infection, again suggesting an important role for host factors in protection from disease. In a prospective cohort study of commercial sex workers in Nairobi at high risk of exposure, production of IFN-gamma by PBMCs stimulated with chlamydial hsp60 strongly correlated with protection against incident *C. trachomatis* infection. In contrast, levels of chlamydial EB or hsp60-specific IgA and IgG detected in endocervical mucus and plasma were not significantly associated with an altered risk of infection (Cohen et al., 2005).

Although antibody may not play a primary role in protection from reinfection, studies suggest it may help control the shedding of organisms and protect against upper tract disease. One study reported the prevalence of mucosal IgA antibodies was inversely related to the quantity of *C. trachomatis* shed from the human endocervix (Brunham et al., 1983), and another found the presence of serum IgA and IgG antibodies reduced the risk for ascending infection among women undergoing therapeutic abortion (Brunham et al., 1987).

4. Epidemiology

Chlamydia trachomatis is the most common bacterial sexually transmitted infection, with an estimated 92 million cases occurring globally each year, including more than four million in sexually active adolescents and adults in the United States(Geneva: WHO, 2001). Many men and most women infected with *C. trachomatis* are either asymptomatic or minimally symptomatic and presentation for diagnosis is a result of screening or a contact being symptomatic. Regional estimates are hampered by under diagnosis and underreporting of cases. Because symptoms are absent or minimal in most women and many men, a large reservoir of asymptomatic infection is present that can sustain the pathogen within a community.

In the U.S., substantial racial/ethnic disparities are present in the prevalence of both chlamydial and gonococcal infections. One large study of U.S. female military recruits found a chlamydial prevalence of 9% that was maintained over 4 consecutive years (Gaydos et al., 2003). Young age, black race, home-of-record from the south, more than one sex partner, a new sex partner, lack of condom use, and a history of having a sexually transmitted disease were correlates of chlamydia infection.

A cross-sectional analyses of a prospective cohort study of a nationally representative sample of 14,322 young adults aged 18 to 26 years conducted in 2001–2002 revealed the overall prevalence of chlamydial infection to be 4.2 percent (Miller et al., 2004). Women (4.7%) were more likely to be infected than men (3.7%; prevalence ratio, 1.29). The prevalence of chlamydial infection was highest among black women (14%) and black men (11%). Lowest prevalences were among Asian men (1.1%), white men (1.4%) and white women (2.5%). Prevalence of chlamydial infection was highest in the south (5.4%); and lowest in the northeast (2.4%). Overall prevalence of gonorrhea was 0.4%, with prevalence among black men and women being 2.1% compared to 0.1% among white young adults. These results provide compelling evidence that nationwide disparities in chlamydial and gonococcal infections across racial/ethnic groups are real rather than the result of biased estimates (Miller et al., 2004).

Urine screening for chlamydial infection in Louisiana public schools revealed the overall prevalence of *C. trachomatis* was 6.5%, with rates among girls more than twice that of boys (9.7% vs. 4.0%). The highest prevalence for boys occurred in 12th grade (8.9%), whereas the highest prevalence for girls occurred in 10th grade (15.8%) (Cohen et al., 1998). The high prevalence rates in this cohort are in contrast to a rate of 0.9% for a cohort of 1114 patients aged 15 to 24 years in two pediatric private practices in suburban North Carolina. In sexually active participants, prevalence was 2.1%; in sexually active females, 2.7%; and in sexually active males, 0.9%. One case of gonococcal infection was found. Most participants were female (63%), white (87%), and from highly educated families (64% of their mothers graduated from college) (Best et al., 2001).

In a study of adolescents, use of depot-medroxyprogesterone acetate injections (adjusted OR, 5.44; 95% CI, 1.25–23.6) but not oral contraceptives (adjusted OR, 0.92, 95% CI, 0.10–8.78) were found to increase the risk for chlamydial infection compared to non-hormone users; whereas, oral contraceptives did not (Jacobson et al., 2000). Both of these hormonal contraceptives were associated with an increased risk of chlamydial infection in HIV-infected women (Lavreys et al., 2004).

Identification of persons with the highest risk of infection should enhance cost-effectiveness of screening and treatment programs. However, in a recent study among 3202 sexually active adolescent females attending middle school health centers in Baltimore, MD chlamydial infection was found in 771 first visits (24.1%) and 299 repeat visits (13.9%); 29.1% had at least 1 positive test result (Burstein et al., 1998). Females who were 14 years old had the highest age-specific prevalence rate (63 [27.5%] of 229 cases; P = 0.01). Unfortunately, independent predictors of chlamydial infection – reason for clinic visit, clinic type, prior sexually transmitted diseases, multiple or new partners, or inconsistent condom use-failed to identify a subset of adolescent females with the majority of infections.

In reports from other parts of the world, the prevalence ranges from 28.5% among female sex workers in Dakar (Sturm-Ramirez et al., 2000), to 5.7 percent among pregnant women in Thailand (Kilmarx et al., 1998), and 0.8 percent overall among women seen in private gynecology practices in Paris and 5.2 percent for those under the age of 21 years (Warszawski et al., 1999). A population based study in China estimated the prevalence of chlamydial infection at a surprisingly high rate of 2.6 per 100 population for women (Parish et al., 2003).

The rate of transmission after exposure was addressed in a report of female sexual partners of men with urethritis. Among 31 partners of men with *C. trachomatis* alone, infection occurred in 20 (65 percent) (Lin et al., 1998). The number of exposures or coinfection with gonorrhea did not affect the transmission rate.

5. Clinical Manifestations

Most studies report that 25% of men and 70% of women infected with *C. trachomatis* are asymptomatic or minimally symptomatic. The National Longitudinal Study of Adolescent Health Study collected data prospectively from 14,322 U.S adolescents and followed them into adulthood (Miller et al., 2004). Of the participants that tested positive for chlamydial infection, 95% did not report symptoms in the 24 hours preceding specimen collection. Among men with chlamydial infection, the prevalences of urethral discharge and dysuria were only 3.3% and 1.9%, respectively. Among women with chlamydial infection, the prevalences of vaginal discharge and dysuria were 0.3% and 4.2%, respectively. Among the small number of young men reporting urethral discharge (n = 17), the prevalence of chlamydial infection was high (38.5%), whereas the prevalence of chlamydial infection was only 6.0% among the women reporting dysuria (n = 232) and 0.9% among those reporting vaginal discharge (n = 98) (Miller et al., 2004).

5.1. Infections in Males

When symptomatic, males frequently complain of dysuria or note a clear or mucopurulent urethral discharge at least 7 to 14 days following contact with an infected partner (Stamm et al., 1984b). The discharge may be so slight as to be demonstrable only after penile stripping and then only in the morning. Some patients may deny the presence of discharge but may note stained underwear in the morning resulting from scant discharge overnight. The primary complications of chlamydial urethritis in men are (1) epididymitis; (2) sexually reactive arthritis, including Reiter's syndrome; and (3) transmission to women. *C. trachomatis* and *N. gonorrhoeae* are the most frequent causes of epididymitis in men under age 35; urethritis also is usually present.

Approximately 1 percent of men presenting with nongonococcal urethritis develop an acute aseptic arthritis syndrome referred to as sexually reactive arthritis (Rahman et al., 1992). It appears to be an immune-mediated inflammatory response to an infection that occurs at a site distant from the primary infection. One-third of cases have the full complex of Reiter syndrome, consisting of the triad of arthritis, nonbacterial urethritis, and conjunctivitis (Keat et al., 1980). Most patients carry the histocompatibility antigen HLA-B27.

5.2. Infections in Females

In women, chlamydial infections may cause pelvic inflammatory disease, tubal infertility, chronic pelvic pain, and ectopic pregnancy. Chlamydial infection may also be linked to cervical cancer (Koskela et al., 2000). Chlamydial and gonococcal infections may increase susceptibility to and transmission of HIV in both men and women (Plummer et al., 1991).

Symptoms in females include mild abdominal pain, intermittent bleeding, vaginal discharge, or dysuria-pyuria syndrome. The cervix can appear normal or exhibit edema, erythema, friability, or mucopurulent discharge. In prepubertal girls, vaginitis can occur secondary to infection of transitional cell epithelium by *C. trachomatis*. In contrast, the squamous epithelium of the adult vagina is not susceptible to chlamydiae, and vaginal discharge generally reflects endocervical infection.

Some women develop ascending infection of the genital tract, resulting in endometritis (infection of the uterine tissues) and salpingitis (infection of the fallopian tubes). In one study, eighteen of 109 (16.5%) infected asymptomatic adolescent women followed for 2 months or more became symptomatic, but only 2 (1.8%) developed clinical pelvic inflammatory disease (PID) (Rahm et al., 1988). However, when women infected with both *C. trachomatis* and *N. gonorrhoeae* were treated with antibiotics active only against *N. gonorrhoeae*, 6 of 20 (30 percent) developed evidence of upper genital tract infection (Stamm et al., 1984a). Why ascending infection develops in some women with cervical infections is not known.

The definition of "pelvic inflammatory disease" is a sexually transmitted infection that ascends from the vagina and cervix to involve the uterus, ovaries, and peritoneal tissues as well as the fallopian tubes. Lower abdominal pain, usually bilateral, is the most common presenting symptom. Pain may be associated with an abnormal vaginal discharge, abnormal uterine bleeding, dysuria, dyspareunia, nausea, vomiting, fever, or other constitutional symptoms. It is more commonly present in a subclinical form that lacks the typical acute symptoms, but continues to lead to the associated long-term sequelae of infertility and ectopic pregnancy (Paavonen et al., 1982). The most important causative organisms are *C. trachomatis* and *N. gonorrhoeae*; well over half of cases are caused by one or both of these agents. Other microorganisms implicated in PID include organisms found in the abnormal vaginal flora of women with bacterial vaginosis, such as bacteroides species, anaerobic cocci, *Mycoplasma hominis*, and *Ureaplasma urealyticum*. *Escherichia coli* and other enteric organisms have also been found.

The spectrum of PID associated with *C. trachomatis* infection ranges from acute, severe disease with perihepatitis and ascites (Fitz-Hugh-Curtis syndrome), to asymptomatic or "silent" disease. When women with chlamydial salpingitis are compared to women with gonococcal or with nongonococcal-nonchlamydial salpingitis, they are more likely to experience a chronic, subacute course with a longer duration of abdominal pain before seeking medical care. Yet, they have as much or more tubal inflammation at laparoscopy (Svensson et al., 1980). Routine screening of asymptomatic women for chlamydial infection and treating those identified as infected has been shown to reduce the incidence of PID in a health maintenance organization setting (Scholes et al., 1996).

6. Laboratory Diagnosis

A positive laboratory test for *C. trachomatis* can be utilized for patient education and increases both compliance with drug therapy and the likelihood of referral of sexual partners. Although the development of tissue cell culture methods in the 1960s for detecting *C. trachomatis* was a major advance, the availability of non-culture tests has dramatically increased the availability and decreased the cost of laboratory detection. Definitive diagnosis of chlamydial infection, as would be required in a medicolegal setting (i.e., suspected sexual abuse or rape), requires isolation of *C. trachomatis* in cell culture or a positive nucleic acid amplification test (NAAT) confirmed by a second NAAT that targets a different sequence (Johnson et al., 2002).

6.1. Diagnostic Specimens

Many screening tests for *C. trachomatis* require appropriately handled samples containing columnar epithelium from mucosal sites (e.g., endocervix, urethra, or conjunctiva) rather than exudate; the adequacy of specimens should be verified by periodic cytologic evaluations.

The discomfort caused by obtaining a urethral swab in males has precluded its widespread use in asymptomatic men. A dipstick test for leukocyte esterase (LE) performed on the first portion of a voided urine is a cost-effective and moderately sensitive screen (47% to 58%) for detection of chlamydial infection in asymptomatic young males (Blake et al., 2005). When feasible, urine NAAT provides a much more sensitive and equally noninvasive method of detecting *Chlamydia*.

6.2. Cell Culture

Use of chlamydial transport media containing antibiotics maximizes recovery and reduces the likelihood of culture overgrowth by other bacteria. Swabs used to obtain a specimen should have plastic or metal shafts, as soluble components from wooden shafts can have a toxic effect on cell cultures. Storage at 4°C or maintenance at −70°C is required if inoculation within 24 hours is not possible. Cycloheximide-treated McCoy or HeLa cell lines are used most frequently to isolate *C. trachomatis*. Centrifugation techniques appear to enhance absorption of chlamydiae to cells. Intracytoplasmic inclusions can be detected at 48 to 72 hours with species-specific immunofluorescent monoclonal antibodies for *C. trachomatis* and Giemsa or iodine stains.

Generally, a higher isolation rate using cell culture is found in symptomatic patients than asymptomatic ones. Nonculture methods (except for polymerase chain reaction) appear to have poorer clinical performance characteristics than culture in low-risk asymptomatic patients, when nongenital specimens (i.e., rectal) are obtained, or when specimens are obtained from young children.

6.3. Nonculture Tests for *C. trachomatis*

The previous gold standard of cell culture is being outperformed by more sensitive molecular techniques; confirmatory studies of discrepant test results frequently find true positives that were negative by cell culture. Amplification tests based on the detection of chlamydial DNA or specific chlamydial ribosomal RNA are now available. Both detect *C. trachomatis* in urine or in self-administered vaginal swab specimens, with sensitivity comparable to that obtained with urogenital swab specimens, and make noninvasive testing for chlamydial infections possible (Gaydos et al., 2004; Johnson et al., 2002).

Recent prevalence studies of *C. trachomatis* infection in sexually active adolescent girls suggest that testing for *C. trachomatis* should be offered twice per year to this population (Johnson et al., 2002). A recent cost effectiveness study in women attending family planning clinics revealed that a strategy that combined use of PCR on cervical specimens in women receiving pelvic examinations, and PCR of urine in women with no medical indication for a pelvic examination prevented the most cases of pelvic inflammatory disease and provided the highest cost savings (Howell et al., 1998).

6.4. Serology

Antibodies to *Chlamydia spp.* are best detected with a microimmunofluorescent (MIF) assay, but these assays are not widely available. Serologic screening is of very little value in uncomplicated genital infections but may be useful for population studies. Patients who have LGV infections demonstrate elevated specific IgG and IgM antibody levels compared to those with other chlamydial infections. The MIF is species-specific and sensitive but is available only at a limited number of clinical laboratories.

7. Treatment

The most widely used treatments for uncomplicated oculogenital infections caused by *C. trachomatis* in nonpregnant adolescents and adults are doxycycline (100 mg orally twice daily) for 7 days or azithromycin (1 g orally) in a single dose. In populations with poor compliance with treatment, azithromycin may be more cost-effective because it provides single-dose, directly observed therapy (Martin et al., 1992). Doxycycline costs less than azithromycin, and it has been used extensively for a longer period. Ofloxacin is similar in efficacy to doxycycline and azithromycin, but it is more expensive and offers no advantage with regard to dosage regimen. Erythromycin is less efficacious than either azithromycin and doxycycline, and gastrointestinal side effects discourage compliance (2002).

The recommended regimen for chlamydial infection during pregnancy is erythromycin base (500 mg four times daily) for 7 days; erythromycin estolate is contraindicated because of its potential hepatotoxicity. Patients unable to tolerate this regimen should be treated with either a smaller dose of erythromycin base or erythromycin ethylsuccinate for 14 days or amoxicillin for 7 days (2002). Doxycycline and ofloxacin are contraindicated in pregnant women. The safety and efficacy of azithromycin use in pregnant and lactating women have not been established although the CDC lists azithromycin as a potential alternative for treatment during pregnancy.

Sex partners should be evaluated, tested, and treated if they had sexual contact with the patient during the 60 days preceding onset of symptoms in the patient or diagnosis of chlamydia. The most recent sex partner should be treated even if the time of the last sexual contact was >60 days before diagnosis of the index case. Patients do not need to be retested for chlamydia after completing treatment with doxycycline or azithromycin unless symptoms persist or reinfection is suspected, because these therapies are highly efficacious. A test of cure may be considered 3 weeks after completion of treatment with erythromycin. Testing at <3 weeks after completion of therapy to identify patients who did not respond to therapy may not be valid (Gaydos et al., 1998).

8. Complications and Sequelae

Chlamydia trachomatis has been implicated as a pathogen in 8% to 54% of women who have pelvic inflammatory disease (PID) and has been associated with the long-term consequences of tubal infertility (17%), ectopic pregnancy (10%), or chronic pelvic pain (17%) (Cates, Jr. and Wasserheit, 1991; Stamm and Holmes, 1999; Westrom et al., 1992). Perihepatitis or Fitz-Hugh-Curtis syndrome also can occur (Eschenbach, 1984; Wang et al., 1980). Epididymitis, prostatitis, and reactive arthritis are the most common sequelae in males.

Approximately 20% to 30% of untreated pregnant women with chlamydial infection are at risk for endometritis following delivery or induced abortions (McGregor and French, 1991). Studies have examined the role of maternal chlamydial infections in adverse pregnancy outcomes such as premature rupture of membranes, prematurity, low birth weight, and stillbirth. Several studies found that adverse outcomes were more common in infected women who had Chlamydia-specific IgM antibody (Sweet et al., 1987) and in those who were not treated (Ryan, Jr. et al., 1990).

9. Prevention

Because chlamydial infections usually are not associated with overt symptoms, prevention of infection and screening of asymptomatic high risk patients is the most effective means of preventing disease and sequelae. Behavioral interventions (i.e., delaying intercourse, decreasing the number of sex partners, and use of barrier contraception) should be pursued aggressively. A decreased prevalence of *C. trachomatis* infections has been reported in regions with active chlamydial screening programs. High-risk patients who should be routinely tested for Chlamydia include women with mucopurulent cervicitis, sexually active women less than 20 years old, and older women with more than one sex partner during the last 3 months or inconsistent use of barrier contraception while in a nonmonogamous relationship (Johnson et al., 2002). Because of the frequency of repeated chlamydial infections within the first several months following treatment of an initial infection, (Burstein et al., 1998; Fortenberry et al., 1999) more frequent (e.g., every 6 months) screening of asymptomatic sexually active adolescents may be necessary. Repeat infection confers an elevated risk of PID and other complications when compared with initial infection. Therefore, recently infected women are a high priority for repeat testing for *C. trachomatis*. For these reasons, clinicians and health-care agencies should consider advising all women with chlamydial infection to be rescreened 3–4 months after treatment. Providers are also strongly encouraged to rescreen all women treated for chlamydial infection whenever they next present for care within the following 12 months, regardless of whether the patient believes that her sex partners were treated (Johnson et al., 2002).

To prevent maternal postnatal complications and chlamydial infections among infants, pregnant women should be screened for *Chlamydia* during the third trimester to permit completion of treatment before delivery. Ocular prophylaxis with topical erythromycin or tetracycline has reduced the incidence of gonococcal ophthalmia but does not appear to be effective against *C. trachomatis* (Hammerschlag et al., 1989).

10. Future Directions

An effective vaccine against *C. trachomatis* will have to activate both the antibody and cellular arms of the immune system more effectively than the body's natural response does, yet somehow limit inflammation as well. An increased knowledge of protective host antigens and better understanding of protective host responses to chlamydiae are needed. In addition, markers for protection from upper genital tract infection and/or disease in the female will be necessary if vaccine candidates are to be tested in humans. Stimulation of long-term mucosal immunity in the genital tract is a challenge; persons are susceptible to reinfection with *C. trachomatis* after a brief period of immunity because memory cells are not retained in the genital tract. It is unclear whether all genital infections could be prevented or whether only more invasive disease, such as salpingitis, might be preventable using vaccine technology.

Although current antibiotic treatment is highly successful when administered, most persons infected with *C. trachomatis* are asymptomatic and thus go undiagnosed and untreated. Therefore, more widespread screening of high-risk individuals is needed. Researchers have already proved the feasibility and benefit of employing screening of sexually active young men and women. Public health officials should pursue such strategies in parallel with the ongoing research for effective vaccines.

Acknowledgements

Toni Darville is supported by a United States National Institutes of Health grant (AI054624) and by the Horace C. Cabe Foundation and the Bates-Wheeler Foundation, Arkansas Children's Hospital Research Institute and University of Arkansas for Medical Sciences.

References

Sexually transmitted diseases treatment guidelines 2002. Centers for Disease Control and Prevention2002. *MMWR Recomm. Rep.* 51:30–36.

Arno, J.N., B.P. Katz, R. McBride, G.A. Carty, B.E. Batteiger, V.A. Caine, and R.B. Jones. 1994. "Age and clinical immunity to infections with Chlamydia trachomatis." *Sex Transm. Dis.* 21:47–52.

Best, D., C.A. Ford, and W.C. Miller. 2001. "Prevalence of Chlamydia trachomatis and Neisseria gonorrhoeae infection in pediatric private practice." *Pediatrics.* 108:E103.

Blake, D.R., C.A. Lemay, C.A. Gaydos, and T.C. Quinn. 2005. "Performance of urine leukocyte esterase in asymptomatic male youth: another look with nucleic acid amplification testing as the gold standard for Chlamydia detection." *J. Adolesc Health.* 36:337–341.

Blythe, M.J., B.P. Katz, D.P. Orr, V.A. Caine, and R.B. Jones. 1988. "Historical and clinical factors associated with Chlamydia trachomatis genitourinary infection in female adolescents." *J. Pediatr.* 112:1000–1004.

Brunham, R.C., B. Binns, J. McDowell, and M. Paraskevas. 1986. "Chlamydia trachomatis infection in women with ectopic pregnancy." *Obstetrics and Gynecology.* 67:722–726.

Brunham, R.C., J. Kimani, J. Bwayo, G. Maitha, I. Maclean, C. Yang, C. Shen, S. Roman, N.J. Nagelkerke, M. Cheang, and F.A. Plummer. 1996. "The epidemiology of Chlamydia trachomatis within a sexually transmitted diseases core group." *J. Infect. Dis.* 173:950–956.

Brunham, R.C., C.C. Kuo, L. Cles, and K.K. Holmes. 1983. "Correlation of host immune response with quantitative recovery of Chlamydia trachomatis from the human endocervix." *Infection and Immunity.* 39:1491–1494.

Brunham, R.C., R. Peeling, I. Maclean, J. McDowell, K. Persson, and S. Osser. 1987. "Postabortal Chlamydia trachomatis salpingitis: correlating risk with antigen-specific serological responses and with neutralization." *J. Infect. Dis.* 155:749–755.

Burstein, G.R., C.A. Gaydos, M. Diener-West, M.R. Howell, J.M. Zenilman, and T.C. Quinn. 1998. "Incident Chlamydia trachomatis infections among inner-city adolescent females [see comments]." *JAMA.* 280:521–526.

Cates, W., Jr. and J.N. Wasserheit. 1991. "Genital chlamydial infections: Epidemiology and reproductive sequelae." *American Journal of Obstetrics and Gynecology.* 164 Suppl.:1771–1781.

Cohen, C.R., J. Gichui, R. Rukaria, S.S. Sinei, L.K. Gaur, and R.C. Brunham. 2003. "Immunogenetic correlates for Chlamydia trachomatis-associated tubal infertility." *Obstetrics and Gynecology.* 101:438–444.

Cohen, C.R., K.M. Koochesfahani, A.S. Meier, C. Shen, K. Karunakaran, B. Ondondo, T. Kinyari, N.R. Mugo, R. Nguti, and R.C. Brunham. 2005. "Immunoepidemiologic profile of Chlamydia trachomatis infection: importance of heat-shock protein 60 and interferon-gamma." *J. Infect. Dis.* 192:591–599.

Cohen, C.R., S.S. Sinei, E.A. Bukusi, J.J. Bwayo, K.K. Holmes, and R.C. Brunham. 2000. "Human leukocyte antigen class II DQ alleles associated with Chlamydia trachomatis tubal infertility." *Obstetrics and Gynecology.* 95:72–77.

Cohen, D.A., M. Nsuami, R.B. Etame, S. Tropez-Sims, S. Abdalian, T.A. Farley, and D.H. Martin. 1998. "A School-based Chlamydia Control Program Using DNA Amplification Technology." *Pediatrics.* 101:e1.

Davis, C.H., J.E. Raulston, and P.B. Wyrick. 2002. "Protein disulfide isomerase, a component of the estrogen receptor complex, is associated with Chlamydia trachomatis serovar E attached to human endometrial epithelial cells." *Infection and Immunity.* 70:3413–3418.

Debattista, J., P. Timms, J. Allan, and J. Allan. 2002. "Reduced levels of gamma-interferon secretion in response to chlamydial 60 kDa heat shock protein amongst women with pelvic inflammatory disease and a history of repeated Chlamydia trachomatis infections." *Immunology Letters.* 81:205–210.

Eschenbach, D.A. 1984. "Acute pelvic inflammatory disease." *Urol. Clin. North. Am.* 11:65–81.

Ford, C.A., B.W. Pence, W.C. Miller, M.D. Resnick, L.H. Bearinger, S. Pettingell, and M. Cohen. 2005. "Predicting adolescents' longitudinal risk for sexually transmitted infection: results from the National Longitudinal Study of Adolescent Health." *Archives of Pediatrics Adolescent Medicine.* 159:657–664.

Fortenberry, J.D., E.J. Brizendine, B.P. Katz, K.K. Wools, M.J. Blythe, and D.P. Orr. 1999. "Subsequent sexually transmitted infections among adolescent women with genital infection due to Chlamydia trachomatis, Neisseria gonorrhoeae, or Trichomonas vaginalis." *Sex Transm. Dis.* 26:26–32.

Gaydos, C.A., K.A. Crotchfelt, M.R. Howell, S. Kralian, P. Hauptman, and T.C. Quinn. 1998. "Molecular amplification assays to detect chlamydial infections in urine specimens from high school female students and to monitor the persistence of chlamydial DNA after therapy." *Journal of Infectious Diseases.* 177:417–424.

Gaydos, C.A., M.R. Howell, T.C. Quinn, K.T. McKee, Jr., and J.C. Gaydos. 2003. "Sustained high prevalence of Chlamydia trachomatis infections in female army recruits." *Sex Transm. Dis.* 30:539–544.

Gaydos, C.A., M. Theodore, N. Dalesio, B.J. Wood, and T.C. Quinn. 2004. "Comparison of three nucleic acid amplification tests for detection of Chlamydia trachomatis in urine specimens." *J. Clin. Microbiol.* 42:3041–3045.

Geneva: WHO. World Health Organization (WHO). Global prevalence and incidence of selected curable sexually transmitted infections: overview and extimates. 2001.

Ref Type: Report

Griffiths, E. and R.S. Gupta. 2002. "Protein signatures distinctive of chlamydial species: horizontal transfers of cell wall biosynthesis genes glmU from archaea to chlamydiae and murA between chlamydiae and Streptomyces." *Microbiology.* 148:2541–2549.

Hammerschlag, M.R., C. Cummings, P.M. Roblin, T.H. Williams, and I. Delke. 1989. "Efficacy of neonatal ocular prophylaxis for the prevention of chlamydial and gonococcal conjunctivitis." *N. Eng. J. Med.* 320:769–772.

Howell, M.R., T.C. Quinn, and C.A. Gaydos. 1998. "Screening for Chlamydia trachomatis in asymptomatic women attending family planning clinics. A cost-effectiveness analysis of three strategies." *Ann. Intern. Med.* 128:277–284.

Jacobson, D.L., L. Peralta, N.M. Farmer, N.M. Graham, C. Gaydos, and J. Zenilman. 2000. "Relationship of hormonal contraception and cervical ectopy as measured by computerized planimetry to chlamydial infection in adolescents." *Sex Transm. Dis.* 27:313–319.

Johnson, R.E., W.J. Newhall, J.R. Papp, J.S. Knapp, C.M. Black, T.L. Gift, R. Steece, L.E. Markowitz, O.J. Devine, C.M. Walsh, S. Wang, D.C. Gunter, K.L. Irwin, S. Delisle, and S.M. Berman. 2002. "Screening tests to detect Chlamydia trachomatis and Neisseria gonorrhoeae infections–2002." *MMWR Recomm. Rep.* 51:1–38.

Keat, A.C., B.J. Thomas, D. Taylor-Robinson, G.D. Pegrum, R.N. Maini, and J.T. Scott. 1980. "Evidence of Chlamydia trachomatis infection in sexually acquired reactive arthritis." *Annals of the Rheumatic Diseases.* 39:431–437.

Kilmarx, P.H., C.M. Black, K. Limpakarnjanarat, N. Shaffer, S. Yanpaisarn, P. Chaisilwattana, W. Siriwasin, N.L. Young, C.E. Farshy, T.D. Mastro, and M.E. St Louis. 1998. "Rapid assessment of sexually transmitted diseases in a sentinel population in Thailand: prevalence of chlamydial infection, gonorrhoea, and syphilis among pregnant women–1996." *Sex Transm. Infect.* 74:189–193.

Kimani, J., I.W. Maclean, J.J. Bwayo, K. MacDonald, J. Oyugi, G.M. Maitha, R.W. Peeling, M. Cheang, N.J. Nagelkerke, F.A. Plummer, and R.C. Brunham. 1996. "Risk factors for Chlamydia trachomatis pelvic inflammatory disease among sex workers in Nairobi, Kenya." *J. Infect. Dis.* 173:1437–1444.

Kinnunen, A.H., H.M. Surcel, M. Lehtinen, J. Karhukorpi, A. Tiitinen, M. Halttunen, A. Bloigu, R.P. Morrison, R. Karttunen, and J. Paavonen. 2002. "HLA DQ alleles and interleukin-10 polymorphism associated with Chlamydia trachomatis-related tubal factor infertility: a case-control study." *Hum. Reprod.* 17:2073–2078.

Kiviat, N.B., P. Wolner-Hanssen, D.A. Eschenbach, J.N. Wasserheit, J.A. Paavonen, T.A. Bell, C.W. Critchlow, W.E. Stamm, D.E. Moore, and K.K. Holmes. 1990. "Endometrial histopathology in patients with culture-proved upper genital tract infection and laparoscopically diagnosed acute salpingitis." *Am. J. Surg. Pathol.* 14:167–175.

Koskela, P., T. Anttila, T. Bjorge, A. Brunsvig, J. Dillner, M. Hakama, T. Hakulinen, E. Jellum, M. Lehtinen, P. Lenner, T. Luostarinen, E. Pukkala, P. Saikku, S. Thoresen, L. Youngman, and J. Paavonen. 2000. "*Chlamydia trachomatis* infection as a risk factor for invasive cervical cancer." *International Journal of Cancer.* 85:35–39.

LaMontagne, D.S., K.A. Fenton, S. Randall, S. Anderson, and P. Carter. 2004. "Establishing the National Chlamydia Screening Programme in England: results from the first full year of screening." *Sex Transm. Infect.* 80:335–341.

LaVerda, D., L.N. Albanese, P.E. Ruther, S.G. Morrison, R.P. Morrison, K.A. Ault, and G.I. Byrne. 2000. "Seroreactivity to Chlamydia trachomatis Hsp10 correlates with severity of human genital tract disease." *Infection and Immunity.* 68:303–309.

Lavreys, L., V. Chohan, J. Overbaugh, W. Hassan, R.S. McClelland, J. Kreiss, K. Mandaliya, J. Ndinya-Achola, and J.M. Baeten. 2004. "Hormonal contraception and risk of cervical infections among HIV-1-seropositive Kenyan women." *AIDS.* 18:2179–2184.

Lin, J.-S.L., S.P. Donegan, T.C. Heeren, M. Greenberg, E.E. Flaherty, R. Haivanis, X.-H. Su, D. Dean, W.J. Newhall, J.S. Knapp, S.K. Sarafian, R.J. Rice, S.A. Morse, and P.A. Rice. 1998. "Transmission of *Chlamydia trachomatis* and *Neisseria gonorrhoeae* among men with urethritis and their female sex partners." *Journal of Infectious Diseases.* 178:1707–1712.

Martin, D.H., T.F. Mroczkowski, Z.A. Dalu, J. McCarty, R.B. Jones, S.J. Hopkins, and R.B. Johnson. 1992. "A controlled trial of a single dose of azithromycin for the treatment of chlamydial urethritis and cervicitis." *New England Journal of Medicine.* 327:921–925.

McGregor, J.A. and J.I. French. 1991. "*Chlamydia trachomatis* infection during pregnancy." *American Journal of Obstetrics and Gynecology.* 164 Suppl.:1782–1789.

Miller, W.C., C.A. Ford, M. Morris, M.S. Handcock, J.L. Schmitz, M.M. Hobbs, M.S. Cohen, K.M. Harris, and J.R. Udry. 2004. "Prevalence of chlamydial and gonococcal infections among young adults in the United States." *JAMA.* 291:2229–2236.

Molano, M., C.J. Meijer, E. Weiderpass, A. Arslan, H. Posso, S. Franceschi, M. Ronderos, N. Munoz, and A.J. van den Brule. 2005. "The natural course of Chlamydia trachomatis infection in asymptomatic Colombian women: a 5-year follow-up study." *J. Infect. Dis.* 191:907–916.

Paavonen, J., K. Teisala, P.K. Heinonen, R. Aine, S. Laine, M. Lehtinen, A. Miettinen, R. Punnonen, and P. Gronroos. 1987. "Microbiological and histopathological findings in acute pelvic inflammatory disease." *Br. J. Obstet. Gynaecol.* 94:454–460.

Paavonen, J., E. Vesterinen, and P.A. Mardh. 1982. "Infertility as a sequela of chlamydial pelvic inflammatory disease." *Scand. J. Infect. Dis. [Suppl].* 32:73–76.

Parish, W.L., E.O. Laumann, M.S. Cohen, S. Pan, H. Zheng, I. Hoffman, T. Wang, and K.H. Ng. 2003. "Population-based study of chlamydial infection in China: a hidden epidemic." *JAMA.* 289: 1265–1273.

Paukku, M., M. Puolakkainen, T. Paavonen, and J. Paavonen. 1999. "Plasma cell endometritis is associated with Chlamydia trachomatis infection." *Am. J. Clin. Pathol.* 112:211–215.

Plummer, F.A., J.N. Simonsen, D.W. Cameron, J.O. Ndinya-Achola, J.K. Kreiss, M.N. Gakinya, P. Waiyaki, M. Cheang, P. Piot, A.R. Ronald, and ••. 1991. "Cofactors in male-female sexual transmission of human immunodeficiency virus type 1." *J. Infect. Dis.* 163:233–239.

Rahm, V.-A., H. Gnarpe, and V. Odlind. 1988. "*Chlamydia trachomatis* among sexually active teenage girls. Lack of correlation between chlamydial infection, history of the patient and clinical signs of infection." *British Journal of Obstetrics and Gynaecology.* 95:916–919.

Rahman, M.U., R. Cantwell, C.C. Johnson, R.L. Hodinka, H.R. Schumacher, and A.P. Hudson. 1992. "Inapparent genital infection with *Chlamydia trachomatis* and its potential role in the genesis of Reiters syndrome." *DNA Cell Biol.* 11:215–219.

Ryan, G.M., Jr., T.N. Abdella, S.G. Mcneeley, V.S. Baselski, and D.E. Drummond. 1990. *"Chlamydia trachomatis* infection in pregnancy and effect of treatment on outcome." *American Journal of Obstetrics and Gynecology.* 162:34–39.

Scholes, D., K.K. Holmes, A. Stergachis, and W.E. Stamm. 1996. "Screening for chlamydia to prevent pelvic inflammatory disease – Reply." *New England Journal of Medicine.* 335:1532–1533.

Stamm, W.E., M.E. Guinan, C. Johnson, T. Starcher, K.K. Holmes, and W.M. McCormack. 1984a. "Effect of treatment regimens for Neisseria gonorrhoeae on simultaneous infection with Chlamydia trachomatis." *New England Journal of Medicine.* 310:545–549.

Stamm, W.E. and K.K. Holmes. 1999. *"Chlamydia trachomatis* infections of the adult." In K.K. Holmes, P.-A. Mardh, P.F. Sparling, and P.J. Wiesner, editors, *Sexually transmitted diseases.* McGraw-Hill Book company. New York. 407–422.

Stamm, W.E., L.A. Koutsky, J.K. Benedetti, J.L. Jourden, R.C. Brunham, and K.K. Holmes. 1984b. "Chlamydia trachomatis urethral infections in men. Prevalence, risk factors, and clinical manifestations." *Ann. Intern. Med.* 100:47–51.

Sturm-Ramirez, K., H. Brumblay, K. Diop, A. Gueye-Ndiaye, J.L. Sankale, I. Thior, I. N'Doye, C.C. Hsieh, S. Mboup, and P.J. Kanki. 2000. "Molecular epidemiology of genital Chlamydia trachomatis infection in high-risk women in Senegal, West Africa." *J. Clin. Microbiol.* 38:138–145.

Svensson, L., L. Westrom, K.T. Ripa, and P.A. Mardh. 1980. "Differences in some clinical and laboratory parameters in acute salpingitis related to culture and serologic findings." *American Journal of Obstetrics and Gynecology.* 138(7 Pt 2):1017–1021.

Sweet, R.L., D.V. Landers, C. Walker, and J. Schachter. 1987. "Chlamydia trachomatis infection and pregnancy outcome." *American Journal of Obstetrics and Gynecology.* 156:824–833.

Wang, S.P., D.A. Eschenbach, K.K. Holmes, G. Wager, and J.T. Grayston. 1980. "Chlamydia trachomatis infection in Fitz-Hugh-Curtis syndrome." *American Journal of Obstetrics and Gynecology.* 138(7 Pt 2):1034–1038.

Warszawski, J., L. Meyer, and P. Weber. 1999. "Criteria for selective screening of cervical Chlamydia trachomatis infection in women attending private gynecology practices." *Eur. J. Obstet. Gynecol. Reprod. Biol.* 86:5–10.

Weisburg, W.G., T.P. Hatch, and C.R. Woese. 1986. "Eubacterial origin of chlamydiae." *Journal of Bacteriology.* 167:570–574.

Westrom, L., R. Joesoef, G. Reynolds, A. Hagdu, and S.E. Thompson. 1992. "Pelvic inflammatory disease and fertility. A cohort study of 1,844 women with laparoscopically verified disease and 657 control women with normal laparoscopic results." *Sex. Transm. Dis.* 19:185–192.

9

The Role of Inflammation and Infection in the Development of Chronic Lung Disease of Prematurity

Philip L. Davies, Nicola C. Maxwell, and Sailesh Kotecha

1. Introduction

Chronic lung disease of prematurity (CLD) remains a common disorder of infants who require mechanical ventilatory support for neonatal respiratory distress syndrome (RDS). The definition is applied to infants who continue to require oxygen supplementation beyond 28 days of age or, more recently, beyond 36 weeks corrected gestational age and who have characteristic chest x-ray changes (Kotecha et al., 1999). CLD is an important cause of childhood morbidity including long term oxygen dependency, recurrent respiratory tract infections and persistent respiratory symptoms (Ghezzi et al., 1998). In addition CLD is an independent risk factor for cerebral palsy (Teberg et al., 1991) and increases mortality (Shankaran et al., 1984).

Traditionally lung injury in newborn infants is thought to occur as a result of oxygen therapy, barotrauma and volutrauma, as well as the presence of other risk factors such as a patent ductus arteriosus. Advances in neonatal care, including antenatal glucocorticoids, exogenous surfactant therapy and more gentle ventilatory strategies have decreased the incidence of CLD in more mature newborn babies. Nevertheless overall rates of CLD have not declined as increasing numbers of extremely premature infants are now surviving to develop lung complications. Eighty percent of infants with birth weights of between 500–1000 g now survive and CLD occurs in 51% of infants born with birth weights of between 501–750 g and 35% of newborn infants weighing 751–1000 g (Fanaroff et al., 1995).

Many infants present with mild or no respiratory distress at birth but progress to develop CLD. A large proportion of these children may not even need continuous

Hot Topics in Infection and Immunity in Children, edited by Andrew J. Pollard and Adam Finn.
Springer, New York, 2006.

supplemental oxygen in the first week of life and only a small number will need an oxygen concentration of greater than 40% (Bancalari et al., 2003). Five percent of preterm infants without any initial lung disease will develop CLD (Charafeddine et al., 1999). It is clear that oxygen supplementation and mechanical ventilation by themselves can not fully explain the lung injury observed in very preterm infants and there is an increasing recognition that infection and inflammation are important factors in the pathogenesis of CLD.

2. Lung Development

Classical CLD has been described histologically as having variable areas of atelectasis and hyperinflation, smooth muscle hyperplasia and extensive fibroproliferation (Northway et al., 1967). The more recent pattern that is seen in extremely preterm infants is quite different with a decreased number of large, over simplified alveoli and significantly less fibrosis than seen previously (Coalson, 2003). This is felt to represent an aberration of normal lung development.

Newborn preterm infant lungs are still in the canalicular or saccular phase of development with the process of alveolarisation not beginning until the last trimester (Kotecha, 2000). Any event that adversely alters the development programme of the fetal or newborn lung can result in dysregulated alveolar development (Jobe, 1999). Similar findings of abnormal alveolarisation have been observed in animal models of CLD with preterm lambs treated with surfactant and requiring minimal oxygen supplementation and ventilatory support developing fewer and larger alveoli than control animals (Albertine et al., 1999).

Studies using transgenic mice have shown that inflammatory cytokines are able to disrupt lung development. Transgenic mice over-expressing tumour necrosis factor alpha (TNF-α) develop lung inflammation and reduced alveolar numbers (Hardie et al., 1997). Similarly an over-expression of interleukin-6 (IL-6) causes lymphocytic infiltration of the airways with fewer and larger alveoli (DiCosmo et al., 1994). These observations are in keeping with the importance of inflammation in the development of CLD.

3. Lung Inflammation in CLD

Preterm infants who develop CLD have high numbers of inflammatory cells, predominantly neutrophils, in their airways over the first days of life compared to infants who develop and recover from their RDS and control infants who are ventilated for non-respiratory reasons (Kotecha et al., 1995). By three to four days of life, the number of neutrophils in those infants who recover from RDS starts to decline but remains high in those who progress to develop CLD. Similarly, by day four, macrophages have reached their peak in infants whose RDS recovers but remain high beyond day four of age in the CLD group (Clement et al., 1988).

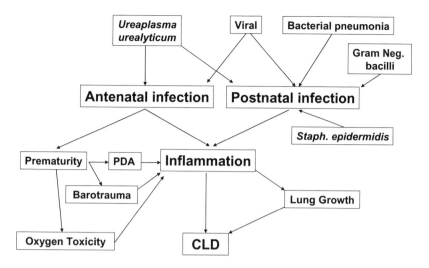

Figure 9.1. Model demonstrating the key role of pulmonary inflammation in the development of CLD.

In parallel to the cellular influx, lung lavage fluid from those infants who develop CLD is significantly more chemotactic than lavage fluid from infants who recover from RDS (Groneck et al., 1994). High levels of chemotactic cytokines such as leukotriene B4, C5a and particularly IL-8 (Kotecha et al., 1995) are observed in infants who develop CLD. In addition, proinflammatory cytokines such as TNF-α, IL-1 and IL-6 (Kotecha et al., 1996) are elevated as are soluble cell surface adhesion molecules such as intercellular adhesion molecule-1 (ICAM-1) in infants who develop CLD (Kotecha, 1996; Speer, 1999).

Thus inflammation is well established in children who develop CLD both compared to controls and those who develop and recover from RDS. Postnatal oxygen therapy and ventilation contribute to the inflammatory process. However, there is evidence to suggest that, at least in some infants, the inflammatory process begins *in utero*. The important role of inflammation in the pathogenesis of CLD is demonstrated in Figure 9.1.

4. Antenatal Inflammation

Chorioamnionitis, the histological inflammation of the placenta, is common in mothers who deliver prematurely (Goldenberg et al., 2000). The presence of chorioamnionitis has been reported to be protective against the development of RDS (Hannaford et al., 1999). Despite not developing RDS, infants born to mothers with chorioamnionitis have an increased risk of developing CLD and also of cerebral palsy (Nelson et al., 1998).

Thirty percent of preterm infants with RDS have histological evidence of chorioamnionitis compared to 82% without RDS. By contrast 63% of infants who develop CLD had chorioamnionitis compared to 21% who did not (Watterberg et al., 1996). Proinflammatory cytokines have been shown to be increased antenatally in amniotic fluid from mothers whose preterm infants subsequently develop CLD (Yoon et al., 1997).

The importance of antenatal inflammation in the development of CLD has been demonstrated using animal models. Intra-amniotic injection of E. coli endotoxin to pregnant ewes has been shown to cause chorioamnionitis and subsequently lung inflammation with elevated inflammatory cytokines in the newborn preterm animal (Kramer et al., 2002). Once delivered these newborn preterm lambs developed less RDS than controls and had significantly better lung compliance whilst surfactant proteins in the airways increased 100 fold. In spite of these early markers, histologically the lung had markedly fewer alveoli when compared to control animals. In addition, preterm lambs born to mothers exposed antenatally with intra-amniotic endotoxin generated a greater inflammatory response when mechanically ventilated compared to ventilated unexposed control animals. These findings suggest that antenatal inflammation is effective in rapidly maturing the fetal lung and in particular upregulating surfactant production, resulting in reduced incidence of RDS. However, this is at the expense of alveolarisation. Antenatal inflammation may also be responsible for upregulating the response to postnatal risk factors of CLD, including supplemental oxygen and mechanical ventilation.

5. Antenatal Infection and Preterm Delivery

Antenatal infection is a likely candidate to trigger the antenatal inflammatory process. Eighty percent of preterm deliveries occur as a result of spontaneous onset of labour or premature rupture of membranes (Tucker et al., 1991). A large body of evidence now suggests that infection plays a major part in this process. 61% of placentas from infants with preterm labour had one or more organisms isolated, compared to 21% of term placentas (Hillier et al., 1988), whilst other studies have reported infected amniotic fluid in up to 80% of preterm births (Goldenberg et al., 1998).

The use of more advanced microbial detection techniques, such as the polymerase chain reaction (PCR) to detect microbial 16s ribosomal RNA (16s rRNA) genes, means that organisms can be identified where more traditional culture techniques were negative. We demonstrated that when histological chorioamnionitis was present, microbial genes could be identified in all cases in at least one intrauterine fluid or tissue sample (Miralles et al., 2005). We also noted that 70% of deliveries following preterm prelabour rupture of membranes (pPROM) and 80% of deliveries following idiopathic spontaneous preterm labour were associated with the presence of 16s rRNA genes compared to only one of 21 infants where preterm delivery occurred without the onset of labour or as a result of a preterm twin delivery.

The most common organisms involved are *Ureaplasma urealyticum*, *Mycoplasma hominis*, *Gardnerella vaginalis* and *Bacteroides*, all of which are vaginal commensals (Goldenberg et al., 2000). These organisms ascend through the choriodecidual space and may cross the chorioamniotic membranes to colonise the amniotic fluid. The presence of bacteria stimulates the release of pro-inflammatory cytokines and causes cellular influx. Prostaglandin levels increase which trigger uterine contractions whilst inflammatory cells release proteases such as matrix metalloproteinases that can remodel the cervix and cause the chorioamniotic membranes to rupture (Goldenberg et al., 2000).

A number of studies have been performed to determine if antibiotics can significantly reduce the rate of preterm delivery (King et al., 2002; Kenyon et al., 2003). The largest of these studies were the ORACLE trials. ORACLE I compared the outcome of treatment with different regimes of antibiotics which included erythromycin, and placebo in 4826 women with pPROM and reported that antibiotic treatment significantly reduced the number of babies born within 48 hours and also seven days from treatment (Kenyon et al., 2001a). The ORACLE II trial investigated 7428 women with preterm labour and intact membranes to determine if antibiotic therapy could reduce preterm deliveries compared to placebo and reported no proven benefit (Kenyon et al., 2001b). This means that although antibiotics may eradicate the bacteria this is too late to prevent the effects of the infection on establishing labour.

Thus whilst antenatal infection remains a major cause for preterm delivery, antibiotic therapy has only been shown to be of limited benefit to those mothers with pPROM.

6. Antenatal and Postnatal Infection and CLD

Both ante- and post-natal infection have been linked to the development of CLD. A study of 119 ventilated preterm infants with a birth weight less than 1000 g, reported systemic sepsis to be a risk factor for CLD (Rojas et al., 1995). Of the 64 infants who had a positive blood culture with clinical or laboratory signs of sepsis, 35 (55%) developed CLD, compared to 9 (16%) without sepsis (odds ratio 4.4). This study also reported that a patent ductus arteriosus was a significant risk factor for CLD (odds ratio 6.2). If both these risk factors were combined the odds ratio markedly increased to 48.4.

Other studies have suggested an association between nosocomial infections such as *Staphylococcus epidermidis* and the development of CLD with 64% of infants with *Staphylococcus epidermidis* sepsis developing CLD compared to 24% of controls (Liljedahl et al., 2004). Ventilated preterm neonates frequently have endotracheal secretions colonised with Gram positive bacteria, particularly *Staphylococcus epidermidis*. In some cases Gram negative bacilli, such as *Klebsiella pneumoniae*, *Enterobacter cloacae* and *Escherichia coli* can also colonise endotracheal secretions. Such colonisation has been reported to significantly increase the incidence of CLD (Cordero et al., 1997).

Viral infections, particularly with adenovirus or cytomegalovirus (CMV) have been associated with the development of CLD. Sawyer et al. (1978) identified 32 infants born with a birth weight less than 2000 g as being infected with CMV over a five year period. Of these 24 (75%) went on to develop CLD compared to 12 of

32 (38%) matched controls. It was believed that the CMV infection was acquired postnatally. Other studies have found an association between adenovirus and the development of CLD with 27% of preterm infants who developed CLD having adenovirus detected by PCR in tracheal aspirates in the first week of life compared to 3% in the tracheal fluid from infants who did not develop CLD (Couroucli et al., 2000).

Neither the CMV nor the adenovirus finding have been widely replicated (Prösch et al., 2002) and it seems likely that for the majority of children viral infection is not the key factor in the development of CLD.

7. *Ureaplasma urealyticum*

Ureaplasma urealyticum has the potential to be extremely important in the development of CLD, since it is capable of inducing a potent inflammatory response in cell culture systems, animal models and in humans.

Cell culture studies have shown that *Ureaplasma urealyticum* is able to trigger a significant inflammatory response in neonatal fibroblasts, causing the release of significant levels of IL-6 and IL-8 (Stancombe et al., 1993). Animal models also demonstrate the ability of *Ureaplasma urealyticum* to generate an inflammatory response. *Ureaplasma urealyticum* injected into the amniotic fluid of pregnant baboons, generated a neutrophil influx and increased concentrations of the cytokines IL-6 and IL-8 in tracheal aspirates of the delivered preterm animals (Yoder et al., 2003). Histologically, pneumonitis and bronchiolitis were observed, suggesting a pulmonary inflammatory process. Similar studies in which amniotic fluid of pregnant ewes was injected with *Ureaplasma urealyticum*, showed improvements in lung compliance and increased surfactant production in the preterm offspring compared to control intubated preterm animals (Moss et al., 2005). Significant lung inflammation was also present in exposed animals when compared to control animals.

Ureaplasma urealyticum has the potential to infect the human fetus since it is present in the lower genital tract of 40–80% of pregnant women (van Waarde et al., 1997); is the most common organism grown in the amniotic fluid in chorioamnionitis and may be important in triggering preterm labour. In addition, the vertical transmission rate from mother to infant, is 46–89%, being greatest in preterm infants (Mabanta et al., 2003).

The presence of *Ureaplasma urealyticum* within the amniotic fluid is associated with an inflammatory response in the amniotic fluid, with significantly increased levels of IL-1β, TNF-α and IL-6 identified compared to culture negative fluid (Yoon et al., 1998). In human preterm infants, a significant association exists between the presence of *Ureaplasma urealyticum* in the airways and the development of CLD although its presence has not been proven to be causative. Wang et al. carried out a meta-analysis of 17 studies in 1995 and reported the relative risk of developing CLD in infants colonised with *Ureaplasma urealyticum* was 1.72 (95% confidence interval, 1.5–1.96). However, not all studies have found such an association, once prematurity had been taken into consideration (van Waarde et al., 1997).

One reason for the difficulty in proving that *Ureaplasma urealyticum* is an independent risk factor for the development of CLD is the difficulty in culturing the organism. Newer and more sensitive techniques, looking for the presence of *Ureaplasma urealyticum* genes using PCR in lung lavage fluid have suggested a link between ureaplasma and CLD. The presence of *Ureaplasma urealyticum* has also been linked with increased inflammatory cytokines in lung lavage fluid, in keeping with the role of inflammation in CLD (Kotecha et al., 2004). Using PCR, 5/6 (83%) of preterm infants, without any clinical or laboratory evidence of infection, in whom pulmonary *Ureaplasma urealyticum* was identified, developed CLD compared to 4/11 (36%) where *Ureaplasma urealyticum* was negative.

In view of the potential role of infection in CLD, it is disappointing that studies involving antibiotic therapy have so far been inconclusive. Antenatal treatment with erythromycin is able to lengthen time to delivery in preterm infants with pPROM but not if the membranes were still intact (Kenyon et al., 2001a; Kenyon et al., 2001b). The ORACLE studies also showed erythromycin improved some neonatal outcome measures such as reducing the need for oxygen or surfactant treatment after delivery and reducing the number of abnormal head ultrasound scans, if there was prolonged rupture of membranes (Kenyon et al., 2001a). If membranes were intact, however, no differences were observed in neonatal outcomes (Kenyon et al., 2001b).

A recent Cochrane review looking at postnatal antibiotic therapy to treat *Ureaplasma urealyticum* to prevent CLD only identified two small control studies, involving intubated children under 30 weeks gestation (Mabanta et al., 2003). The first study treated all 75 preterm children with erythromycin or placebo regardless of status (Lyon et al., 1998). Twenty-four children went on to develop CLD but only nine of the 75 infants had pulmonary colonisation with *Ureaplasma urealyticum* on culture. No significant difference between the two groups was observed. The second study screened 155 children for *Ureaplasma urealyticum* and treated the 28 children with *Ureaplasma* urealyticum colonisation with erythromycin or placebo once the results were available at a mean age of seven days (Jonsson et al., 1998). Again this study reported no significant difference between the treatment and placebo groups. Given the small size of these studies there is a clear need for adequately powered studies to determine if *Ureaplasma urealyticum* is causative of CLD or simply a bystander in the lungs of preterm babies.

8. Summary

CLD is a significant cause of infant morbidity and mortality. The lung injury is multifactorial in origin with supplemental oxygen and ventilatory damage being only part of the picture. Antenatal and postnatal infection and inflammation are also important in the development of CLD, although their precise role has still to be fully ascertained.

In the future, therapeutic strategies need to be considered to decrease the incidence and severity of CLD. In particular a definitive trial investigating the role of antibiotics against *Ureaplasma urealyticum* in preventing CLD needs to be performed. Increased use of newer microbiological methods will also improve our understanding of the role of infection in CLD and further guide research and clinical management.

References

Albertine, K.H., Jones, G.P., Starcher, B.C., Bohnsack, J.F., Davis, P.L., Cho, S.C., Carlton, D.P., Bland, R.D. (1999). Chronic lung injury in preterm lambs. Disordered respiratory tract development. *Am J Respir Crit Care Med.* **159**: 945–58.

Bancalari, E., Claure, N., Sosenko, I.R. (2003). Bronchopulmonary dysplasia: changes in pathogenesis, epidemiology and definition. *Semin Neonatol.* **8**: 63–71.

Charafeddine, L., D'Angio, C.T., Phelps, D.L. (1999). Atypical chronic lung disease patterns in neonates. *Pediatrics.* **103**: 759–65.

Clement, A., Chadelat, K., Sardet, A., Grimfeld, A., Tournier, G. (1988). Alveolar macrophage status in bronchopulmonary dysplasia. *Pediatr Res.* **23**: 470–3.

Coalson, J.J. (2003). Pathology of new bronchopulmonary dysplasia. *Semin Neonatol.* **8**: 73–81.

Cordero, L., Ayers, L.W., Davis, K. (1997). Neonatal airway colonization with gram-negative bacilli: association with severity of bronchopulmonary dysplasia. *Pediatr Infect Dis J.* **16**: 18–23.

Couroucli, X.I., Welty, S.E., Ramsay, P.L., Wearden, M.E., Fuentes-Garcia, F.J., Ni, J., Jacobs, T.N., Towbin, J.A., Bowles, N.E. (2000). Detection of microorganisms in the tracheal aspirates of preterm infants by polymerase chain reaction: association of adenovirus infection with bronchopulmonary dysplasia. *Pediatr Res.* **47**: 225–32.

DiCosmo, B.F., Geba, G.P., Picarella, D., Elias, J.A., Rankin, J.A., Stripp, B.R., Whitsett, J.A., Flavell, R.A. (1994). Airway epithelial cell expression of interleukin-6 in transgenic mice. Uncoupling of airway inflammation and bronchial hyperreactivity. *J Clin Invest.* **94**: 2028–35.

Fanaroff, A.A., Wright, L.L., Stevenson, D.K., Shankaran, S., Donovan, E.F., Ehrenkranz, R.A., Younes, N., Korones, S.B., Stoll, B.J., Tyson, J.E., et al. (1995). Very-low-birth-weight outcomes of the National Institute of Child Health and Human Development Neonatal Research Network, May 1991 through December 1992. *Am J Obstet Gynecol.* **173**: 1423–31.

Ghezzi, F., Gomez, R., Romero, R., Yoon, B.H., Edwin, S.S., David, C., Janisse, J., Mazor, M. (1998). Elevated interleukin-8 concentrations in amniotic fluid of mothers whose neonates subsequently develop bronchopulmonary dysplasia. *Eur J Obstet Gynecol Reprod Biol.* **78**: 5–10.

Goldenberg, R.L., Rouse, D.J. (1998). Prevention of premature birth. *N Engl J Med.* **339**: 313–20.

Goldenberg, R.L., Hauth, J.C., Andrews, W.W. (2000). Intrauterine infection and preterm delivery. *N Engl J Med.* **342**: 1500–7.

Groneck, P., Gotze-Speer, B., Oppermann, M., Eiffert, H., Speer, C.P. (1994). Association of pulmonary inflammation and increased microvascular permeability during the development of bronchopulmonary dysplasia: a sequential analysis of inflammatory mediators in respiratory fluids of high-risk preterm neonates. *Pediatrics.* **93**: 712–8.

Hannaford, K., Todd, D.A., Jeffery, H., John, E., Blyth, K., Gilbert, G.L. (1999). Role of ureaplasma urealyticum in lung disease of prematurity. *Arch Dis Child Fetal Neonatal Ed.* **81**: F162–7.

Hardie, W.D., Bruno, M.D., Huelsman, K.M., Iwamoto, H.S., Carrigan, P.E., Leikauf, G.D., Whitsett, J.A., Korfhagen, T.R. (1997). Postnatal lung function and morphology in transgenic mice expressing transforming growth factor-alpha. *Am J Pathol.* **151**: 1075–83.

Hillier, S.L., Martius, J., Krohn, M., Kiviat, N., Holmes, K.K., Eschenbach, D.A. (1988). A case-control study of chorioamniotic infection and histological chorioamnionitis in prematurity. *N Engl J Med.* **319**: 972–8.

Jobe, A.J. (1999). The new BPD: an arrest of lung development. *Pediatr Res.* **46**: 641–3.

Jonsson, B., Rylander, M., Faxelius, G. (1998). Ureaplasma urealyticum, erythromycin and respiratory morbidity in high-risk preterm neonates. *Acta Paediatr.* **87**: 1079–84.

Kenyon, S.L., Taylor, D.J., Tarnow-Mordi, W. (2001a). Broad-spectrum antibiotics for preterm, prelabour rupture of fetal membranes: the ORACLE I randomised trial. ORACLE Collaborative Group. *Lancet.* **357**: 979–88.

Kenyon, S.L., Taylor, D.J., Tarnow-Mordi, W. (2001b). Broad spectrum antibiotics for spontaneous preterm labour: the ORACLE II randomised trial. *Lancet.* **357**: 991–6.

Kenyon, S.L., Boulvain, M., Neilson, J. (2003). Antibiotics for preterm rupture of membranes. *The Cochrane Database of Systematic Reviews.* **2**: CD001058.

King, J., Fenady, V. (2002). Prophylactic antibiotics for inhibiting preterm labour with intact membranes. *Cochrane Database Syst Rev.* **4**: CD000246.

Kotecha, S. (1996). Cytokines in chronic lung disease of prematurity. *Eur J Pediatr.* **155**: S14–7.

Kotecha, S. (2000). Lung growth: implications for the newborn infant. *Arch Dis Child Fetal Neonatal Ed.* **82**: F69–74.

Kotecha, S., Chan, B., Azam, N., Silverman, M., Shaw, R.J. (1995). Increase in interleukin-8 and soluble intercellular adhesion molecule-1 in bronchoalveolar lavage fluid from premature infants who develop chronic lung disease. *Arch Dis Child Fetal Neonatal Ed.* **72**: F90–6.

Kotecha, S., Wilson, L., Wangoo, A., Silverman, M., Shaw, R.J. (1996). Increase in interleukin (IL)-1 beta and IL-6 in bronchoalveolar lavage fluid obtained from infants with chronic lung disease of prematurity. *Pediatr Res.* **40**: 250–6.

Kotecha, S., Silverman, M. (1999). Chronic Respiratory complications of neonatal disorders. *Textbook of Pediatric Respiratory Medicine.* Eds. Landau, L.I., Tausssig, L.M. Mosby. 488–521.

Kotecha, S., Hodge, R., Schaber, J.A., Miralles, R., Silverman, M., Grant, W.D. (2004). Pulmonary Ureaplasma urealyticum is associated with the development of acute lung inflammation and chronic lung disease in preterm infants. *Pediatr Res.* **55**: 61–8.

Kramer, B.W., Kramer, S., Ikegami, M., Jobe, A.H. (2002). Injury, inflammation, and remodeling in fetal sheep lung after intra-amniotic endotoxin. *Am J Physiol Lung Cell Mol Physiol.* **283**: L452–9.

Liljedahl, M., Bodin, L., Schollin, J. (2004). Coagulase-negative staphylococcal sepsis as a predictor of bronchopulmonary dysplasia. *Acta Paediatr.* **93**: 211–5.

Lyon, A.J., McColm, J., Middlemist, L., Fergusson, S., McIntosh, N., Ross, P.W. (1998). Randomised trial of erythromycin on the development of chronic lung disease in preterm infants. *Arch Dis Child Fetal Neonatal Ed.* **78**: F10–4.

Mabanta, C.G., Pryhuber, G.S., Weinberg, G.A., Phelps, D.L. (2003). Erythromycin for the prevention of chronic lung disease in intubated preterm infants at risk for, or colonized or infected with Ureaplasma urealyticum. *Cochrane Database Syst Rev.* **4**: CD003744.

Miralles, R., Hodge, R., McParland, P.C., Field, D.J., Bell, S.C., Taylor, D.J., Grant, W.D., Kotecha, S. (2005). Relationship between antenatal inflammation and antenatal infection identified by detection of microbial genes by polymerase chain reaction. *Pediatr Res.* **57**: 570–7.

Moss, T.J., Nitsos, I., Ikegami, M., Jobe, A.H., Newnham, J.P. (2005). Experimental intrauterine Ureaplasma infection in sheep. *Am J Obstet Gynecol.* **192**: 1179–86.

Nelson, K.B., Dambrosia, J.M., Grether, J.K., Phillips, T.M. (1998). Neonatal cytokines and coagulation factors in children with cerebral palsy. *Ann Neurol.* **44**: 665–75.

Northway, W.H., Rosan, R.C., Porter, D.V. (1967). Pulmonary disease following respirator therapy of hyaline membrane disease: bronchopulmonary dysplasia. *N Engl J Med.* **276**: 357–68.

Prösch, S., Lienicke, U., Priemer, C., Flunker, G., Seidel, W.F., Kruger, D.H., Wauer, R.R. Human adenovirus and human cytomegalovirus infections in preterm newborns: no association with bronchopulmonary dysplasia. *Pediatr Res.* **52**: 219–24.

Rojas, M.A., Gonzalez, A., Bancalari, E., Claure, N., Poole, C., Silva-Neto, G. (1995). Changing trends in the epidemiology and pathogenesis of neonatal chronic lung disease. *J Pediatr.* **126**: 605–10.

Sawyer, M.H., Edwards, D.K., Spector, S.A. (1987). Cytomegalovirus infection and bronchopulmonary dysplasia in premature infants. *Am J Dis Child.* **141**: 303–5.

Shankaran, S., Szego, E., Eizert, D., Siegel, P. (1984). Severe bronchopulmonary dysplasia. Predictors of survival and outcome. *Chest.* **86**: 607–10.

Speer, C.P. (1999). Inflammatory mechanisms in neonatal chronic lung disease. *Eur J Pediatr.* **158**: S18–22.

Stancombe, B.B., Walsh, W.F., Derdak, S., Dixon, P., Hensley, D. (1993). Induction of human neonatal pulmonary fibroblast cytokines by hyperoxia and Ureaplasma urealyticum. *Clin Infect Dis.* **17**: S154–7.

Tucker, J.M., Goldenberg, R.L., Davis, R.O., Copper, R.L., Winkler, C.L., Hauth, J.C. (1991). Etiologies of preterm birth in an indigent population: is prevention a logical expectation? *Obstet Gynecol.* **77**: 343–7.

van Waarde, W.M., Brus, F., Okken, A., Kimpen, J.L. (1997). Ureaplasma urealyticum colonization, prematurity and bronchopulmonary dysplasia. *Eur Respir J.* **10**: 886–90.

Teberg, A.J., Pena, I., Finello, K., Aguilar, T., Hodgman, J.E. (1991). Prediction of neurodevelopmental outcome in infants with and without bronchopulmonary dysplasia. *Am J Med Sci.* **301**: 369–74.

Wang, E.E., Ohlsson, A., Kellner, J.D. (1995). Association of Ureaplasma urealyticum colonization with chronic lung disease of prematurity: results of a metaanalysis. *J Pediatr.* **127**: 640–4.

Watterberg, K.L., Demers, L.M., Scott, S.M., Murphy, S. (1996). Chorioamnionitis and early lung inflammation in infants in whom bronchopulmonary dysplasia develops. *Pediatrics.* **97**: 210–5.

Yoder, B.A., Coalson, J.J., Winter, V.T., Siler-Khodr, T., Duffy, L.B., Cassell, G.H. (2003). Effects of antenatal colonization with ureaplasma urealyticum on pulmonary disease in the immature baboon. *Pediatr Res.* **54**: 797–807.

Yoon, B.H., Romero, R., Jun, J.K., Park, K.H., Park, J.D., Ghezzi, F., Kim, B.I. (1997). Amniotic fluid cytokines (interleukin-6, tumor necrosis factor-alpha, interleukin-1 beta, and interleukin-8) and the risk for the development of bronchopulmonary dysplasia. *Am J Obstet Gynecol.* **177**: 825–30.

Yoon, B.H., Romero, R., Park, J.S., Chang, J.W., Kim, Y.A., Kim, J.C., Kim, K.S. (1998). Microbial invasion of the amniotic cavity with Ureaplasma urealyticum is associated with a robust host response in fetal, amniotic, and maternal compartments. *Am J Obstet Gynecol.* **179**: 1254–60.

10

Streptococcus pneumoniae: Infection, Inflammation and Disease

Tim J. Mitchell

1. Introduction

Streptococcus pneumoniae (the pneumococcus) is a major human pathogen, causing diseases such as pneumonia, meningitis and otitis media. The organism is also carried asymptomatically in a large proportion of the population. The primary niche of the pneumococcus is the human nasopharynx where it exists asymptomatically as a commensal. Colonisation can occur within hours of birth and by twelve days post-birth the carriage rates are similar to that of the babies' mothers (Gray et al., 1980). Colonisation is usually transient lasting several weeks to months with up to four different capsular serotypes present at one time (Gillespie, 1989). Carriage rates are highest in young children reaching up to 60%. The most common diseases caused by the pneumococcus are pneumonia, otitis media, bacteraemia and meningitis. In the USA the pneumococcus causes several million cases of otitis media, over 500,000 cases of pneumonia, 50,000 cases of bacteraemia and 3000 cases of meningitis per year, representing a substantial health care burden (Obaro, 2000). Additionally in developing countries, pneumococcal pneumonia is estimated to cause one million deaths in children under the age of five each year (Denny and Loda, 1986). However, the true disease burden caused by the pneumococcus is uncertain because these ailments can be caused by a variety of different organisms and are often treated without bacteriological confirmation of the cause. This is particularly true in the developing word where the disease burden may be highest but diagnostic capacity limited. Improved diagnostics or a highly efficacious vaccine is suggested to provide a truer reflection of pneumococcal disease burden and to show current values to be underestimates (Obaro, 2000). Understanding pneumococcal virulence may assist in the development of new treatment or vaccination strategies.

Hot Topics in Infection and Immunity in Children, edited by Andrew J. Pollard and Adam Finn. Springer, New York, 2006.

2. Virulence Factors Produced by the Pneumococcus

2.1. The Pneumococcal Cell Surface

One of the main virulence factors of the pneumococcus is the polysaccharide capsule, which is believed to be anti-phagocytic (Jonsson et al., 1985). Over 90 different serotypes of the pneumococcus have been identified based on the antigenically distinct polysaccharide capsule (Kalin, 1998).

As well as the capsule the pneumococcus produces a range of other molecules on its cell surface that are associated with pathogenesis including components of the cell wall itself and cell-surface proteins (choline-binding proteins and LPXTG-anchored proteins). The cell wall of the pneumococcus is important in mediating attachment of the bacterium to activated lung cells (Cundell et al., 1995). The phosphorylcholine of the cell wall binds to the receptor for platelet-activating factor (PAF). The PAF receptor is up-regulated during the inflammatory response and during viral infection, which might explain the increased occurrence of pneumococcal pneumonia following viral infection. The cell wall also plays a role in initiating the inflammation associated with pneumococcal infection (Tuomanen et al., 1985).

Anchored to the pneumococcal cell surface are a range of proteins. The surface proteins of pneumococcus can be broadly divided into three families depending on how they are linked to the cell surface.

2.1.1. Choline-Binding Proteins

The choline-binding proteins are anchored to the cell surface by a non-covalent interaction of a repeat region at the carboxy-terminal end of the protein with the phosphorylcholine of the pneumococcal cell wall. This family contains several proteins known to be important in pneumococcal virulence including the autolysin, pneumococcal surface protein A (PspA) and pneumococal surface protein C (PspC). PspA is a highly variable protein expressed by all clinically important pneumococcal serotypes (Crain et al., 1990). Mutant bacteria unable to produce PspA are more easily cleared from the bloodstream and are therefore less virulent in animal models of infection (Mcdaniel et al., 1987). PspA inhibits complement activation and reduces the effectiveness of complement receptor-mediated clearance mechanisms (Neeleman et al., 1999) and might act by preventing the deposition of C3b on the cell surface or may inhibit the formation of the alternate pathway C3 convertase (Tu et al., 1999). PspA binds lactoferrin and therefore could be involved in iron acquisition (Hammerschmidt et al., 1999). PspC (also known as CbpA or SpsA) is involved in bacterial adhesion to the cells in the nasopharynx. PspC-deficient mutants colonize the nasopharynx of rats less well and show reduced binding to human cells (Rosenow et al., 1997). PspC has several activities which might be important in the disease process, including binding to complement component C3 (Smith and Hostetter, 2000). Some forms of PspC can also bind to the complement-control protein factor H (Dave et al., 2001; Janulczyk et al., 2000). PspC also stimulates the production of IL-8 from pulmonary epithelial cells and might therefore be involved in immune cell recruitment and chemotaxis (Madsen et al., 2000). It also mediates attachment to the polymeric immunoglobulin receptor (pIgR) and translocation of pneumococci across human nasopharyngeal epithelial cells (Zhang et al., 2000). Other choline-binding proteins known to be involved in virulence include the major autolysin LytA (Berry et al., 1989a), which is believed to mediate its effect through the release of other virulence factors and cell wall components. CbpD, CbpE, CbpG, LytB and LytC play a role in the colonization of the nasopharynx (Gosink et al., 2000) and mutations in CbpE and CbpG reduce adherence to human cells. CbpG, which might be a serine protease, also plays a role in sepsis (Gosink et al., 2000). CbpE is known to be a phosphorylcholine esterase (Holtje and Tomasz, 1974) whose molecular structure has recently been solved.

2.1.2. LPXTG-Anchored Proteins

This family of pneumococcal surface proteins are anchored to the cell wall by covalent linkage to peptidoglycan via a carboxy-terminal motif, LPXTG. This motif is recognized by a sortase enzyme, which links the threonine residue of the motif to the cell wall. Analysis of the pneumococal genome sequence (Tettelin et al., 2001) reveals a family of these proteins including hyaluronidase and neur-aminidase enzymes. Hyaluronidase breaks down the hyaluronic acid component of mammalian connective tissue and extracellular matrix and is secreted by 99% of clinical isolates of pneumococcus (Humphrey, 1948). Deletion of the hyaluronidase gene alone does not affect virulence in a mouse model of infection but deletion of hyaluronidase in a pneumolysin-negative background reduces the virulence of the pneumolysin-negative mutant (Berry and Paton, 2000). Neuramindase cleaves N-acetylneuraminic acid from glycolipids, lipoproteins and oligosaccharides on cell surfaces and in body fluids (Camara et al., 1994). The pneumococcus has genes for the production of at least three neuraminidases and the role of neuraminidase A in infection has been investigated. It plays a role in nasopharyngeal colonization and development of otitis media in the chinchilla model (Tong et al., 2000) but does not play a role in meningitis-associated deafness (Winter et al., 1997). Analysis of the pneumococcal genome sequence (Tettelin et al., 2001) reveals the presence of four putative sortases. The major sortase (sortase A) has been deleted (Kharat and Tomasz, 2003) and shown to release neuraminidase into the growth medium rather than anchoring it to the cell surface. The *srtA*⁻ mutant was not deficient in virulence in mice but was reduced in its ability to bind to pharyngeal cells (Kharat and Tomasz, 2003). This suggests that anchoring of LPXTG proteins to the cell surface might play a role in attachment and colonization rather than in gross virulence of the pneumococcus. Our recent data show that sortase A is a pneumococcal fitness factor in experimental models of pneumonia and bacteraemia and that it contributes to nasopharyngeal colonization in vivo (Paterson and Mitchell, 2006). Sortase A plays a role in the adherence of pneumococci to human cells, but only in the absence of capsule.

2.1.3. Lipoproteins

Several lipoproteins have been shown to be important in the adhesion of the pneumococcus to cells and the virulence of the organism. Pneumococcal surface antigen A (PsaA) is part of an ABC transporter system that transports manganese (Lawrence et al., 1998). Mutants in PsaA have been shown to have reduced binding to cells, reduced virulence and increased susceptibility to oxidative damage (Tseng et al., 2002).

2.1.4. Other Surface Proteins

There are several proteins present on the surface of the pneumococcus that have no obvious secretion signal sequence or mechanism of anchoring. The proteins include glyceraldhehyde-3-phosphate dehydrogenase (Bergmann et al., 2004), enolase (Bergmann et al., 2003) and PavA (Holmes et al., 2001). The surface bound enzymes can bind plaminogen which then becomes activated to the serine protease plasmin (Bergmann et al., 2003; Bergmann et al., 2004). PavA is present on the surface of the pneumococcus and has high sequence similarity to an FBP from other streptococci. Inactivation of the *pavA* gene in the pneumococcus reduces the binding of bacteria to fibronectin and dramatically reduces the virulence of the organism (Holmes et al., 2001).

2.2. Pneumolysin

2.2.1. Biological Properties of Pneumolysin

The haemolysin produced by the pneumococcus (pneumolysin) is a pore-forming protein belonging to the family of cholesterol dependent cytolysins (for review see (Tweten, 2005)). This toxin does not have a typical secretion signal, is released by the action of the cell-bound autolysin and has been shown to play a role in pathogenesis in several animal models of disease (Alcantara et al., 1999; Benton et al., 1995; Berry et al., 1989b; Braun et al., 2002; Canvin et al., 1995; Comis et al., 1993; Johnson, 1979). The importance of pneumolysin to pathogenesis can be demonstrated by studies of pneumolysin deletion mutants in animal models of infection (Figure 10.1). Pneumolysin has a range of activities that play a role in these virulence models, including the ability to stimulate production of inflammatory mediators including TNFα, IL-1β (Houldsworth et al., 1994), nitric oxide (Braun et al., 1999), IL-8 (Cockeran et al., 2002) and prostaglandins and leukotrienes (Cockeran et al., 2001). The toxin also activates phospholipases in endothelial cells (Rubins et al., 1994) and is toxic to pulmonary endothelial and epithelial cells (Rubins et al., 1992; Rubins et al., 1993). Additionally, pneumolysin can inhibit non-specific defences such as respiratory cilial beat (Steinfort et al., 1989) and plays a role in evasion of the immune system as it can inhibit phagocyte function (Paton and Ferrante, 1983), lymphocyte function (Ferrante et al., 1984) and interfere with the complement pathway (Paton et al., 1984).

Pneumolysin has at least two major activities important in its role in pathogenesis: the ability to form pores and the ability to activate the complement pathway. The functional regions for these activities have been located within the molecule and both activities have been shown to be important in the causation of disease (Rubins et al., 1996b). The toxin is important in the pathogenesis of meningitis as it can cause damage to the ependymal cilia of the brain (Hirst et al., 2000) and can

Pneumonia model:Survival proportions

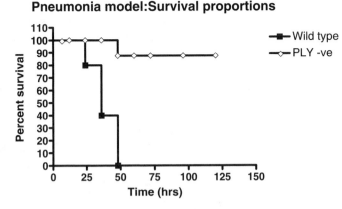

Figure 10.1. The effect of pneumolysin on mortality in a murine model of pneumonia. Mice were challenged intranasally with 10^6 *Streptococcus pneumoniae*. Mice were challenged with either wild-type organisms (squares) or organisms in which the gene for pneumolysin has been deleted (diamonds). Deletion of the pneumolysin gene attenuates the organism.

induce apoptosis of brain cells (Braun et al., 2002). The neurotoxicity of pneumolysin has been shown to be due to alterations in calcium flux into cells and signalling via activation of the p38 MAP kinase (Braun et al., 2002; Stringaris et al., 2002). Pneumolysin has been shown to bind to Toll-like receptor 4 (TLR-4) (Malley et al., 2003). This interaction with TLR-4 was essential for the protection of mice against invasive disease caused by the pneumococcus. In the absence of functional TLR-4, mice were more heavily colonized and much more likely to develop invasive disease. While the inflammatory response to pneumolysin may contribute to this protection, it has also been shown that pneumolysin-TLR4 signalling can induce host cell apoptosis in vitro and in vivo (Srivastava et al., 2005). Pneumolysin therefore plays a diverse and important role in the pathogenesis of pneumoccocal infections.

2.2.2. Variation in Pneumolysin Sequence and Activity in Serotype 1 Pneumococci

Although there is an overlap in strains of pneumococcus that cause disease and those associated with carriage, some serotypes are more likely to be recovered from invasive disease (serotypes 1, 4, 14 and 18C) and others are more commonly isolated from nasopharyngeal swabbing of healthy individuals (serotype 3 ST180, 26B, 19F and 23F) (Brueggemann et al., 2003). Serotype 1 *S. pneumoniae* has remained one of the most prevalent invasive serotypes with reports of increases in serotype 1 invasive pneumococcal disease (IPD) in Scotland (McChlery et al., 2005), Sweden (Henriques Normark et al., 2001) and Denmark (Konradsen and Kaltoft, 2002) and a high prevalence of serotype 1 disease throughout Europe, South America, Africa and Asia (Hausdorff et al., 2000; Hausdorff WP, 2000). Serotype 1 is associated with complicated pneumonia, pulmonary empyema, peritonitis and salpingitis (Eltringham et al., 2003; McFarlane et al., 1979; Sirotnak et al., 1996; Tan et al., 2002; Westh et al., 1990) and has been directly linked to mortality; irrespective of factors such as age, environment and leukocyte count of patients (Martens et al., 2004). Along with serotypes 5 and 7, serotype 1 is also associated with higher ratio of hospitalization versus ambulatory care compared with pneumococcal infections from other serotypes (Alpern et al., 2001). Serotype 1 is one of the few serotypes linked with non-hospital outbreaks of pneumococcal disease, such outbreaks occur in over-crowded institutions and/or where alcoholism is a problem (Dagan et al., 2000; DeMaria et al., 1980; Gratten et al., 1993; Mercat et al., 1991). Nasopharyngeal swabbing of healthy patients rarely results in the isolation of serotype 1 pneumococci, because of this and a global association with IPD, serotype 1 is said to have a high attack rate (Brueggemann and Spratt, 2003).

Generally, the amino acid sequence of pneumolysin is thought to be highly conserved throughout all pneumococcal serotypes with little variance over time and geographic distance (Mitchell et al., 1990). Serotypes 7 and 8 have been reported to possess a threonine to isoleucine substitution at amino acid position 172 that reduces the specific activity of the toxin (Lock et al., 1996) but this is the only reported naturally occurring mutation within pneumolysin that affects function. It is important to note that residual hemolytic activity of pneumolysin is adequate for full virulence of the pneumococcus as demonstrated by the chromosomal replacement of pneumolysin with pneumolysin carrying a mutation that reduces the specific activity of the toxin to 0.1% of wild type pneumolysin (Berry et al., 1995a).

We have identified a number of clinical isolates with mutations in their pneumolysin gene (*ply*). The mutations were predominantly in the *ply* gene of serotype 1, 7 and 8. The serotype 1 isolates had additional mutations in the *ply* gene to those previously described for serotype 7 and 8 (Lock et al., 1996) and were chosen for further investigation due to their high attack rate and recent reports of an increase in serotype 1 disease. From an initial study of 250 Scottish clinical pneumococcal isolates (Jefferies et al., 2004) four were serotype 1 and of these, two had mutations within the *ply* gene. Further analysis of an additional 28 serotype 1 isolates revealed that more than half had mutations within the *ply* gene, which resulted in the abrogation of the toxin's hemolytic activity. Multi Locus Sequence Typing (MLST) of the serotype 1 isolates revealed a correlation between mutations in the *ply* gene and sequence type. All the serotype 1 isolates examined belonged to either sequence type (ST) 227 or 306. ST227 and ST306 are from the same lineage and clonal group (Brueggemann and Spratt, 2003) and are almost exclusive to serotype 1 pneumococci, although ST227 has been shown to switch capsule (Jefferies et al., 2004). ST306 IPD is predominant in continental Europe whereas ST227 is the most prevalent serotype 1 ST in England, North America and Canada (Brueggemann and Spratt, 2003). All of the ST306 isolates were shown to have mutations in their *ply* gene that resulted in the expression of non-hemolytic pneumolysin, yet these clinical isolates were from patients with IPD. Studies with defined point mutations in the *ply* gene showed that the hemolytic activity of pneumolysin was important during sepsis (Berry et al., 1995b). The hemolytic property of pneumolysin has also been shown to be involved in the initial stages of pneumonia including invasion of the lung tissue and neutrophil recruitment, however hemolytic activity is not important in facilitating pneumococcal growth in the alveoli (Rubins et al., 1996a; Rubins et al., 1995). As serotype 1 disease is associated with pneumonia rather than meningitis in Europe and the United States (Hausdorff et al., 2000; Ispahani et al., 2004), it may be that hemolytically active pneumolysin is not essential for the pathogenesis of certain serotype 1 sequence types, such as ST306. This may mean that there will be less selective pressure to conserve the nucleotide sequences that encode the hemolytic activity of pneumolysin, and may allow the mutations we have identified to arise. Recently, an increase in pneumococcal meningitis in Ghana has been attributed to serotype 1 clonal complex ST217 (Leimkugel et al., 2005), which the authors propose may be better equipped to cause meningitis. ST227 and ST306 were not found in the Ghana study but it would be interesting to analyze the *ply* gene from serotype 1 isolates that have caused meningitis, in particular ST217, to see if the Ply227 allele is required. This highlights the need for closer investigation of not only serotype specific virulence factors but the possibility of variation within sero- types and underlines the importance of MLST surveillance.

2.2.3. Pneumolysin as a Vaccine Candidate

The heptavalent polysaccharide conjugate vaccine (7PCV) is currently the most effective conjugate vaccine available for protection against *S. pneumoniae* infection (Black et al., 2000; Obaro, 2002). The protein used for conjugation of the seven capsular polysaccharides is a diphtheria toxoid, CRM_{197}, conferring an increased immune response to the capsule polysaccharides in infants (Black et al., 2000; Pelton et al., 2003). However, as only seven out of a possible ninety (Henrichsen, 1995) pneumococcal serotypes are covered in 7PCV (serotypes 4, 6B, 9V, 14, 18C, 19F and 23F), protection from this vaccine is somewhat limited and varies globally with pneumococcal serotype prevalence (Hausdorff WP, 2000; Spratt and Greenwood, 2000). For example, the serotypes included in 7PCV cover almost 90% of invasive pneumococcal disease (IPD) causing serotypes in North America and Canada but cover <60% of the predominant serotypes in Asia where serotypes 1 and 5 are the predominant cause of IPD (Hausdorff WP, 2000).

Studies of children vaccinated with 7PCV in Finland have revealed a 57% decrease of otitis media caused by the seven vaccine serotypes. However, this was found to coincide with a 34% increase in otitis media from non-vaccine serotypes (Kipli et al., 2000). This promoted selection for non-vaccine serotypes is a major shortcoming of current pneumococcal vaccines, with an increase in serotype replace- ment reported by countries where the 7PCV has been introduced (Lipsitch, 1999). Alternatives to serotype-specific vaccination are currently being investigated with efforts focused on using a protective species-common pneumococcal protein such as pneumolysin.

Immunogenic pneumolysin mutants (Alexander et al., 1994; Paton et al., 1991) with reduced cytotoxicity have previously been constructed (Baba et al., 2001; Boulnois et al., 1990), however, these mutants retain the ability to damage host cell membranes. One pneumolysin mutant extensively researched and commonly referred to as the Pdb toxoid has a tryptophan to phenylalanine substitution as position 433. This mutant forms large pores in host cell membranes (Korchev et al., 1998) and retains 0.1 to 1% hemolytic activity compared with WT pneumolysin (Korchev et al., 1998; Paton, 1996). Another pneumolysin mutant with negligible activity was reported by Michon et al. (Michon et al., 1998), but the location of the mutation was not identified and protection against pneumococcal disease was not proven. We have constructed a series of mutations in pneumolysin in a region implied to have a role in oligomerization (de los Toyos et al., 1996). The resulting mutants had no hemolytic activity. All mutants were recognized by monoclonal antibody Ply 4, an antibody that recognizes an important antigenic region of pneumolysin and can block oligomerization (de los Toyos et al., 1996). One single amino acid deletion pneumolysin mutant, ΔA146 Ply, was further characterized and demonstrated to be unable to form pores in cell membranes or stimulate the in vivo inflammatory effects associated with native pneumolysin treatment. Mice vaccinated with ΔA146 Ply + Alum had high titers of neutralizing anti-pneumolysin IgG and were protected from *S. pneumoniae* infection significantly longer than mice given Alum alone. These pneumolysin toxoids may be suitable candidates for inclusion into the next generation of pneumococcal vaccines.

2.3. Other Virulence Factors

At the time of writing, three signature-tagged mutagenesis screens have been carried out on the pneumococcus which have identified extensive lists of putative virulence factors (Hava and Camilli, 2002; Lau et al., 2001; Polissi et al., 1998). Other proven and potential virulence factors produced by the pneumococcus include a surface-bound enolase that can bind plasmin(ogen) (Bergmann et al., 2003), a superoxide dismutase (Yesilkaya et al., 2000), NADH oxidase (Auzat et al., 1999), hydrogen peroxide (Duane et al., 1993) and iron-uptake systems (Brown et al., 2001).

3. Conclusions

Streptococcus pneumoniae remains an important cause of morbidity and mortality in humans, especially young children. With the availability of bacterial genome sequences and other molecular biology approaches, understanding the interaction of this organism with its host is progressing at an increasing rate. The use of the information provided by these studies will make it easier to design new strategies for the treatment and prevention of the diseases caused by this important human pathogen.

References

Alcantara, R.B., Preheim, L.C., and Gentry, M.J. (1999) Role of pneumolysin's complement-activating activity during pneumococcal bacteremia in cirrhotic rats. *Infection and Immunity* **67**: 2862–2866.
Alexander, J.E., Lock, R.A., Peeters, C.C., Poolman, J.T., Andrew, P.W., Mitchell, T.J., Hansman, D., and Paton, J.C. (1994) Immunization of mice with pneumolysin toxoid confers a significant degree

of protection against at least nine serotypes of *Streptococcus pneumoniae*. *Infect Immun* **62**: 5683–5688.

Alpern, E.R., Alessandrini, E.A., McGowan, K.L., Bell, L.M., and Shaw, K.N. (2001) Serotype prevalence of occult pneumococcal bacteremia. *Pediatrics* **108**: E23.

Auzat, I., Chapuy-Regaud, S., Le Bras, G., Dos Santos, D., Ogunniyi, A.D., Le Thomas, I., Garel, J.R., Paton, J.C., and Trombe, M.C. (1999) The NADH oxidase of Streptococcus pneumoniae: its involvement in competence and virulence. *Molecular Microbiology* **34**: 1018–1028.

Baba, H., Kawamura, I., Kohda, C., Nomura, T., Ito, Y., Kimoto, T., Watanabe, I., Ichiyama, S., and Mitsuyama, M. (2001) Essential role of domain 4 of pneumolysin from *Streptococcus pneumoniae* in cytolytic activity as determined by truncated proteins. *Biochemical and Biophysical Research Communications* **281**: 37–44.

Benton, K.A., Everson, M.P., and Briles, D.E. (1995) A pneumolysin-negative mutant of Streptococcus pneumoniae causes chronic bacteremia rather than acute sepsis in mice. *Infection and Immunity* **63**: 448–455.

Bergmann, S., Wild, D., Diekmann, O., Frank, R., Bracht, D., Chhatwal, G.S., and Hammerschmidt, S. (2003) Identification of a novel plasmin(ogen)-binding motif in surface displayed alpha-enolase of Streptococcus pneumoniae. *Mol Microbiol* **49**: 411–423.

Bergmann, S., Rohde, M., and Hammerschmidt, S. (2004) Glyceraldehyde-3-Phosphate Dehydrogenase of Streptococcus pneumoniae Is a Surface-Displayed Plasminogen-Binding Protein. *Infect. Immun.* **72**: 2416–2419.

Berry, A., Alexander, J., Mitchell, T., Andrew, P., Hansman, D., and Paton, J. (1995a) Effect of defined point mutations in the pneumolysin gene on the virulence of Streptococcus pneumoniae. *Infect. Immun.* **63**: 1969–1974.

Berry, A.M., Lock, R.A., Hansman, D., and Paton, J.C. (1989a) Contribution of autolysin to virulence of *Streptococcus pneumoniae*. *Infect. Immun.* **57**: 2324–2330.

Berry, A.M., Yother, J., Briles, D.E., Hansman, D., and Paton, J.C. (1989b) Reduced virulence of a defined pneumolysin-negative mutant of *Streptococcus pneumoniae*. *Infect. Immun.* **57**: 2037–2042.

Berry, A.M., Alexander, J.E., Mitchell, T.J., Andrew, P.W., Hansman, D., and Paton, J.C. (1995b) Effect of defined point mutations in the pneumolysin gene on the virulence of Streptococcus pneumoniae. *Infect Immun* **63**: 1969–1974.

Berry, A.M., and Paton, J.C. (2000) Additive Attenuation of Virulence of Streptococcus pneumoniae by Mutation of the Genes Encoding Pneumolysin and Other Putative Pneumococcal Virulence Proteins. *Infect. Immun.* **68**: 133–140.

Black, S., Shinefield, H., Fireman, B., Lewis, E., Ray, P., Hansen, J.R., Elvin, L., Ensor, K.M., Hackell, J., Siber, G., Malinoski, F., Madore, D., Chang, I., Kohberger, R., Watson, W., Austrian, R., and Edwards, K. (2000) Efficacy, safety and immunogenicity of heptavalent pneumococcal conjugate vaccine in children. Northern California Kaiser Permanente Vaccine Study Center Group. *Pediatr Infect Dis J* **19**: 187–195.

Boulnois, G.J., Mitchell, T.J., Saunders, F.K., Mendez, F.J., and Andrew, P.W. (1990) Structure and function of pneumolysin, the thiol-activated toxin of *Streptococcus pneumoniae*. In *Bacterial Protein Toxins*. Rapuoli, R.e.a. (ed): Stuttgart, pp. pp. 43–51.

Braun, J.S., Novak, P., Gao, G.L., Murray, P.J., and Shenep, J.L. (1999) Pneumolysin, a protein toxin of Streptococcus pneumoniae, induces nitric oxide production from macrophages. *Infection and Immunity* **67**: 3750–3756.

Braun, J.S., Sublett, J.E., Freyer, D., Mitchell, T.J., Cleveland, J.L., Tuomanen, E.I., and Weber, J.R. (2002) Pneumococcal pneumolysin and H2O2 mediate brain cell apoptosis during meningitis. *Journal of Clinical Investigation* **109**: 19–27.

Brown, J.S., Gilliland, S.M., and Holden, D.W. (2001) A Streptococcus pneumoniae pathogenicity island encoding an ABC transporter involved in iron uptake and virulence. *Mol Microbiol* **40**: 572–585.

Brueggemann, A.B., Griffiths, D.T., Meats, E., Peto, T., Crook, D.W., and Spratt, B.G. (2003) Clonal relationships between invasive and carriage Streptococcus pneumoniae and serotype- and clone-specific differences in invasive disease potential. *J Infect Dis* **187**: 1424–1432.

Brueggemann, A.B., and Spratt, B.G. (2003) Geographic Distribution and Clonal Diversity of Streptococcus pneumoniae Serotype 1 Isolates. *J. Clin. Microbiol.* **41**: 4966–4970.

Camara, M., Boulnois, G.J., Andrew, P.W., and Mitchell, T.J. (1994) A neuraminidase from *Streptococcus pneumoniae* has the features of a surface protein. *Infect. Immun.* **62**: 3688–3695.

Canvin, J.R., Marvin, A.P., Sivakumaran, M., Paton, J.C., Boulnois, G.J., Andrew, P.W., and Mitchell, T.J. (1995) The Role of Pneumolysin and Autolysin in the Pathology of Pneumonia and Septicemia in Mice Infected With a Type-2 Pneumococcus. *Journal of Infectious Diseases* **172**: 119–123.

Cockeran, R., Steel, H.C., Mitchell, T.J., Feldman, C., and Anderson, R. (2001) Pneumolysin potentiates production of prostaglandin E-2 and leukotriene B-4 by human neutrophils. *Infection and Immunity* **69**: 3494–3496.

Cockeran, R., Durandt, C., Feldman, C., Mitchell, T.J., and Anderson, R. (2002) Pneumolysin activates the synthesis and release of interleukin-8 by human neutrophils in vitro. *J Infect Dis* **186**: 562–565.

Comis, S.D., Osborne, M.P., Stephen, J., Tarlow, M.J., Hayward, T.L., Mitchell, T.J., Andrew, P.W., and Boulnois, G.J. (1993) Cytotoxic effects on hair cells of guinea pig cochlea produced by pneumolysin, the thiol activated toxin of Streptococcus pneumoniae. *Acta Otolaryngol* **113**: 152–159.

Crain, M.J., II, W.D.W., Turner, J.S., Yother, J., Talkington, D.F., McDaniel, L.S., Gray, B.M., and Briles, D.E. (1990) Pneumococcal surface protein A (PspA) is serologically highly variable and is expressed by all clinically important capsular serotypes of *Streptococcus pneumoniae*. *Infect Immun* **58**: 3293–3299.

Cundell, D.R., Gerard, N.P., Gerard, C., Idanpaan-Heikkila, I., and Tuomanen, E.I. (1995) *Streptococcus pneumoniae* anchor to activated human cells by the receptor for platelet-activating factor. *Nature* **377**: 435–438.

Dagan, R., Gradstein, S., Belmaker, I., Porat, N., Siton, Y., Weber, G., Janco, J., and Yagupsky, P. (2000) An outbreak of Streptococcus pneumoniae serotype 1 in a closed community in southern Israel. *Clin Infect Dis* **30**: 319–321.

Dave, S., Brooks-Walter, A., Pangburn, M.K., and McDaniel, L.S. (2001) PspC, a Pneumococcal Surface Protein, Binds Human Factor H. *Infect. Immun.* **69**: 3435–3437.

de los Toyos, J.R., Mendez, F.J., Aparicio, J.F., Vazquez, F., Del Mar Garcia Suarez, M., Fleites, A., Hardisson, C., Morgan, P.J., Andrew, P.W., and Mitchell, T.J. (1996) Functional analysis of pneumolysin by use of monoclonal antibodies. *Infect Immun* **64**: 480–484.

DeMaria, A., Browne, K., Berk, S.L., Sherwood, E.J., and McCabe, W.R. (1980) An outbreak of type 1 pneumococcal pneumonia in a men's shelter. *J. American Medical Association* **244**: 1446–1449.

Denny, F.W., and Loda, F.A. (1986) Acute respiratory infections are the leading cause of death in children in developing countries. *Am J Trop Med Hyg* **35**: 1–2.

Duane, P.G., Rubins, J.B., Weisel, H.R., and Janoff, E.N. (1993) Identification of hydrogen peroxide as a Streptococcus pneumoniae toxin for rat alveolar epithelial cells. *Infection and Immunity*: 4392–4397.

Eltringham, G., Kearns, A., Freeman, R., Clark, J., Spencer, D., Eastham, K., Harwood, J., and Leeming, J. (2003) Culture-Negative Childhood Empyema Is Usually Due to Penicillin-Sensitive Streptococcus pneumoniae Capsular Serotype 1. *J. Clin. Microbiol.* **41**: 521–522.

Ferrante, A., RowanKelly, B., and Paton, J.C. (1984) Inhibition of in vitro human lymphocyte response by the pneumococcal toxin pneumolysin. *Infect. Immun.* **46**: 585–589.

Gillespie, S.H. (1989) Aspects of pneumococcal infection including bacterial virulence, host response and vaccination. *J. Med. Microbiol.* **28**: 237–248.

Gosink, K.K., Mann, E.R., Guglielmo, C., Tuomanen, E.I., and Masure, H.R. (2000) Role of novel choline binding proteins in virulence of Streptococcus pneumoniae. *Infection and Immunity* **68**: 5690–5695.

Gratten, M., Morey, F., Dixon, J., Manning, K., Torzillo, P., Matters, R., Erlich, J., Hanna, J., Asche, V., and Riley, I. (1993) An outbreak of serotype 1 Streptococcus pneumoniae infection in central Australia. *Med J Aust* **158**: 340–342.

Gray, B.M., Converse, G.M.I., and Dillon, H.C.J. (1980) Epidemiologic Studies of *Streptococcus pneumoniae* in Infants: Acquisition, Carriage, and Infection During the First 24 Months of Life. *The Journal of Infectious Disease* **142**: 923–933.

Hammerschmidt, S., Bethe, G., H. Remane, P., and Chhatwal, G.S. (1999) Identification of Pneumococcal Surface Protein A as a Lactoferrin-Binding Protein of Streptococcus pneumoniae. *Infect. Immun.* **67**: 1683–1687.

Hausdorff, W.P., Bryant, J., Kloek, C., Paradiso, P.R., and Siber, G.R. (2000) The contribution of specific pneumococcal serogroups to different disease manifestations: implications for conjugate vaccine formulation and use, part II. *Clin Infect Dis* **30**: 122–140.

Hausdorff WP, B.J., Paradiso P.R., Siber G.R. (2000) Which pneumococcal serogroups cause the most invasive disease: implications for conjugate vaccine formulation and use, part I. *Clin Infect Dis* **30**: 100–121.

Hava, D., and Camilli, A. (2002) Large-scale identification of serotype 4 Streptococcus pneumoniae virulence factors. *Mol Microbiol* **45**: 1389–1406.

Henrichsen, J. (1995) Six newly recognized types of *Streptococcus pneumoniae*. *J. Clin. Microbiol.* **33**: 2759–2762.

Henriques Normark, B., Kalin, M., Ortqvist, A., Akerlund, T., Liljequist, B.O., Hedlund, J., Svenson, S.B., Zhou, J., Spratt, B.G., Normark, S., and Kallenius, G. (2001) Dynamics of penicillin-susceptible clones in invasive pneumococcal disease. *J Infect Dis* **184**: 861–869.

Hirst, R.A., Sikand, K.S., Rutman, A., Mitchell, T.J., Andrew, P.W., and Ocallaghan, C. (2000) Relative roles of pneumolysin and hydrogen peroxide from Streptococcus pneumoniae in inhibition of ependymal ciliary beat frequency. *Infection and Immunity* **68**: 1557–1562.

Holmes, A.R., McNab, R., Millsap, K.W., Rohde, M., Hammerschmidt, S., Mawdsley, J.L., and Jenkinson, H.F. (2001) The pavA gene of Streptococcus pneumoniae encodes a fibronectin-binding protein that is essential for virulence. *Mol Microbiol* **41**: 1395–1408.

Holtje, J.-V., and Tomasz, A. (1974) Teichoic Acid Phosphorylcholine Esterase. A Novel Enzyme Acivity in Pneumococcus. *J. Biol. Chem.* **249**: 7032–7034.

Houldsworth, S., Andrew, P.W., and Mitchell, T.J. (1994) Pneumolysin Stimulates Production of Tumor Necrosis Factor Alpha and Interleukin-1β by Human Mononuclear Phagocytes. *Infection and Immunity* **62**: 1501–1503.

Humphrey, J.H. (1948) Hyaluronidase Production by Pneumococci. *Journal of Pathology and Bacteriology* **55**: 273–275.

Ispahani, P., Slack, R.C., Donald, F.E., Weston, V.C., and Rutter, N. (2004) Twenty year surveillance of invasive pneumococcal disease in Nottingham: serogroups responsible and implications for immunisation. *Arch Dis Child* **89**: 757–762.

Janulczyk, R., Iannelli, F., Sjoholm, A.G., Pozzi, G., and Bjorck, L. (2000) Hic, a Novel Surface Protein of Streptococcus pneumoniae That Interferes with Complement Function. *J. Biol. Chem.* **275**: 37257–37263.

Jefferies, J.M.C., Smith, A., Clarke, S.C., Dowson, C., and Mitchell, T.J. (2004) Genetic Analysis of Diverse Disease-Causing Pneumococci Indicates High Levels of Diversity within Serotypes and Capsule Switching. *J. Clin. Microbiol.* **42**: 5681–5688.

Johnson, M.K. (1979) The role of pneumolysin in ocular infections with Streptococcus pneumoniae. *Curr. Eye Res.* **9**: 1107–1114.

Jonsson, S., Musher, D.M., Chapman, A., Goree, A., and Lawrence, E.C. (1985) Phagoocytosis and Killing of Common Bacterial pathogens of the Lung by Human Alveolar Macrophages. *The Journal of Infectious Diseases* **152**: 4–13.

Kalin, M. (1998) Pneumococcal serotypes and their clinical relevance. *Thorax* **53**: 159–162.

Kharat, A.S., and Tomasz, A. (2003) Inactivation of the srtA Gene Affects Localization of Surface Proteins and Decreases Adhesion of Streptococcus pneumoniae to Human Pharyngeal Cells In Vitro. *Infect. Immun.* **71**: 2758–2765.

Kipli, T., Jokinen, J., Herva, E., Palmu, A., Lockhart, S., Siber, G., Eskola, J., and Group, t.F.O.M.S. (2000) Effect of a heptavalent pneumococcal vaccine (PNCCRM) on pneumococcal acute otitis media (AOM) by serotype. In *Second International Symposium on Pneumococci and Pneumococcal Disease, Sun City, South Africa, March 19–23, abstract O20.*

Konradsen, H.B., and Kaltoft, M.S. (2002) Invasive Pneumococcal Infections in Denmark from 1995 to 1999: Epidemiology, Serotypes, and Resistance. *Clin. Diagn. Lab. Immunol.* **9**: 358–365.

Korchev, Y.E., Bashford, C.L., Pederzolli, C., Pasternak, C.A., Morgan, P.J., Andrew, P.W., and Mitchell, T.J. (1998) A conserved tryptophan in pneumolysin is a determinant of the characteristics of channels formed by pneumolysin in cells and planar lipid bilayers. *Biochem J* **329 (Pt 3)**: 571–577.

Lau, G.W., Haataja, S., Lonetto, M., Kensit, S.E., Marra, A., Bryant, A.P., McDevitt, D., Morrison, D.A., and Holden, D.W. (2001) A functional genomic analysis of type 3 Streptococcus pneumoniae virulence. *Mol Microbiol* **40**: 555–571.

Lawrence, M.C., Pilling, P.A., Epa, V.C., Berry, A.M., Ogunniyi, A.D., and Paton, J.C. (1998) The crytal structure of pneumococcal surface antigen PsaA reveals a metal-binding site and a novel structure for a putative ABC-type binding protein. *Structure* **6**: 1553–1561.

Leimkugel, J., Adams Forgor, A., Gagneux, S., Pfluger, V., Flierl, C., Awine, E., Naegeli, M., Dangy, J.P., Smith, T., Hodgson, A., and Pluschke, G. (2005) An Outbreak of Serotype 1 Streptococcus pneumoniae Meningitis in Northern Ghana with Features That Are Characteristic of Neisseria meningitidis Meningitis Epidemics. *J Infect Dis* **192**: 192–199.

Lipsitch, M. (1999) Bacterial vaccines and serotype replacement: lessons from *Haemophilus influenzae* and prospects for *Streptococus pneumoniae*. *Em. Inf. Dis.* **5**: 336–345.

Lock, R.A., Zhang, Q.Y., Berry, A.M., and Paton, J.C. (1996) Sequence variation in the Streptococcus pneumoniae pneumolysin gene affecting haemolytic activity and electrophoretic mobility of the toxin. *Microb Pathog* **21**: 71–83.

Madsen, M., Lebenthal, Y., Cheng, Q., Smith, B.L., and Hostetter, M.K. (2000) A pneumococcal protein that elicits interleukin-8 from pulmonary epithelial cells. *Journal of Infectious Diseases* **181**: 1330–1336.

Malley, R., Henneke, P., Morse, S.C., Cieslewicz, M.J., Lipsitch, M., Thompson, C.M., Kurt-Jones, E., Paton, J.C., Wessels, M.R., and Golenbock, D.T. (2003) Recognition of pneumolysin by Toll-like receptor 4 confers resistance to pneumococcal infection. *PNAS*: 0435928100.

Martens, P., Worm, S.W., Lundgren, B., Konradsen, H.B., and Benfield, T. (2004) Serotype-specific mortality from invasive Streptococcus pneumoniae disease revisited. *BMC Infect Dis* **4**: 21.

McChlery, S.M., Scott, K.J., and Clarke, S.C. (2005) Clonal analysis of invasive pneumococcal isolates in Scotland and coverage of serotypes by the licensed conjugate polysaccharide pneumococcal vaccine: possible implications for UK vaccine policy. *Eur J Clin Microbiol Infect Dis.*

Mcdaniel, L.S., Yother, J., Vijayakumar, M., Mcgarry, L., Guild, W.R., and Briles, D.E. (1987) Use of insertional inactivation to facilitate studies of biological properties of pneumococcal surface protein-a (pspa). *Journal Of Experimental Medicine* **165**: 381–394.

McFarlane, A.C., Hamra, L.K., Reiss-Levy, E., and Hansman, D. (1979) Pneumococcal peritonitis in adolescent girls. *Med J Aust* **1**: 100–101.

Mercat, A., Nguyen, J., and Dautzenberg, B. (1991) An outbreak of pneumococcal pneumonia in two men's shelters. *Chest* **99**: 147–151.

Michon, F., Fusco, P.C., Minetti, C.A.S.A., Laude-Sharp, M., Uitz, C., Huang, C.-H., D'Ambra, A.J., Moore, S., Remeta, D.P., Heron, I., and Blake, M.S. (1998) Multivalent pneumococcal capsular polysaccharide conjugate vaccines employing genetically detoxified pneumolysin as a carrier protein. *Vaccine* **16**: 1732–1741.

Mitchell, T.J., Mendez, F., Paton, J.C., Andrew, P.W., and Boulnois, G.J. (1990) Comparison of pneumolysin genes and proteins from Streptococcus pneumoniae types 1 and 2. *Nucleic Acids Res* **18**: 4010.

Neeleman, C., Geelen, S.P.M., Aerts, P.C., Daha, M.R., Mollnes, T.E., Roord, J.J., Posthuma, G., vanDijk, H., and Fleer, A. (1999) Resistance to both complement activation and phagocytosis in type 3 pneumococci is mediated by the binding of complement regulatory protein factor H. *Infection and Immunity* **67**: 4517–4524.

Obaro, S.K. (2000) Confronting the pneumococcus: a target shift or bullet change? *Vaccine* **19**: 1211–1217.

Obaro, S.K. (2002) The new pneumococcal vaccine. *Clin Microbial Infect* **8**: 623–633.

Paterson, G.K., and Mitchell, T.J. (2006) The role of Streptococcus pneumoniae sortase A in colonisation and pathogenesis. *Microbes and Infection* **8**: 145–153.

Paton, J.C., and Ferrante, A. (1983) Inhibition of human polymorphonuclear leukocyte respiratory burst, bactericidal activity, and migration by pneumolysin. *Infect. Immun.* **41**: 1212–1216.

Paton, J.C., Rowan-Kelly, B., and Ferrante, A. (1984) Activation of human complement by the pneumococcal toxin pneumolysin. *Infect. Immun.* **43**: 1085–1087.

Paton, J.C., Lock, R.A., Lee, C.J., Li, J.P., Berry, A.M., Mitchell, T.J., Andrew, P.W., Hansman, D., and Boulnois, G.J. (1991) Purification and immunogenicity of genetically obtained pneumolysin toxoids and their conjugation to Streptococcus pneumoniae type 19F polysaccharide. *Infect Immun* **59**: 2297–2304.

Paton, J.C. (1996) The contribution of pneumolysin to the pathogenicity of Streptococcus pneumoniae. *Trends Microbiol* **4**: 103–106.

Pelton, S.I., Dagan, R., Gaines, B.M., Klugman, K.P., Laufer, D., O'Brien, K., and Schmitt, H.J. (2003) Pneumococcal conjugate vaccines: proceedings from an interactive symposium at the 41st Interscience Conference on Antimicrobial Agents and Chemotherapy. *Vaccine* **21**: 1562–1571.

Polissi, A., Pontiggia, A., Feger, G., Altieri, M., Mottl, H., Ferrari, L., and Simon, D. (1998) Large-scale identification of virulence genes from Streptococcus pneumoniae. *Infection and Immunity* **66**: 5620–5629.

Rosenow, C., Ryan, P., Weiser, J., Johnson, S., Fontan, P., Ortqvist, A., and Masure, H. (1997) Contribution of a Novel Choline Binding Protein to Adherence, Colonization, and Immunogenicity of *Streptococcus pneumoniae*. *Molecular Microbiology* **25**: 819–829.

Rubins, J., Charboneau, D., Fasching, C., Berry, A., Paton, J., Alexander, J., Andrew, P., Mitchell, T., and Janoff, E. (1996a) Distinct roles for pneumolysin's cytotoxic and complement activities in the pathogenesis of pneumococcal pneumonia. *Am. J. Respir. Crit. Care Med.* **153**: 1339–1346.

Rubins, J.B., Duane, P.G., Charboneau, D., and Janoff, E.N. (1992) Toxicity of Pneumolysin to Pulmonary Endothelial Cells In Vitro. *Infection and Immunity* **60**: 1740–1746.

Rubins, J.B., Duane, P.G., Clawson, D., Charboneau, D., Young, J., and Niewoehner, D.E. (1993) Toxicity of Pneumolysin to Pulmonary Alveolar Epithelial Cells. *Infection and Immunity* **61**: 1352–1358.

Rubins, J.B., Mitchell, T.J., Andrew, P.W., and Niewoehner, D.E. (1994) Pneumolysin activates phospholipase A in pulmonary artery endothelial cells. *Infection and Immunity* **62**: 3829–3836.

Rubins, J.B., Charboneau, D., Paton, J.C., Mitchell, T.J., Andrew, P.W., and Janoff, E.N. (1995) Dual function of pneumolysin in the early pathogenesis of murine pneumococcal pneumonia. *J Clin Invest* **95**: 142–150.

Rubins, J.B., Charboneau, D., Fasching, C., Berry, A.M., Paton, J.C., Alexander, J.E., Andrew, P.W., Mitchell, T.J., and Janoff, E.N. (1996b) Distinct role for pneumolysin's cytotoxic and complement activities in the pathogenesis of pneumococcal pneumonia. *Am. J. Respir. and Critical Care Medicine* **153**: 1339–1346.

Sirotnak, A.P., Eppes, S.C., and Klein, J.D. (1996) Tuboovarian abscess and peritonitis caused by Streptococcus pneumoniae serotype 1 in young girls. *Clin Infect Dis* **22**: 993–996.

Smith, B.L., and Hostetter, M.K. (2000) C3 as substrate for adhesion of Streptococcus pneumoniae. *J Infect Dis* **182**: 497–508.

Spratt, B.G., and Greenwood, B.M. (2000) Prevention of pneumococcal disease by vaccination: does serotype replacement matter? *The Lancet* **356**: 1210–1211.

Srivastava, A., Henneke, P., Visintin, A., Morse, S.C., Martin, V., Watkins, C., Paton, J.C., Wessels, M.R., Golenbock, D.T., and Malley, R. (2005) The Apoptotic Response to Pneumolysin Is Toll-Like Receptor 4 Dependent and Protects against Pneumococcal Disease. *Infect. Immun.* **73**: 6479–6487.

Steinfort, C., Wilson, R., Mitchell, T., Feldman, C., Rutman, A., Todd, H., Sykes, D., Walker, J., Saunders, K., Andrew, P.W., Boulnois, G.J., and Cole, P.J. (1989) Effect of Streptococcus pneumoniae on human respiratory epithelium in vitro. *Infection and Immunity* **57**: 2006–2013.

Stringaris, A.K., Geisenhainer, J., Bergmann, F., Balshusemann, C., Lee, U., Zysk, G., Mitchell, T.J., Keller, B.U., Kuhnt, U., Gerber, J., Spreer, A., Bahr, M., Michel, U., and Nau, R. (2002) Neurotoxicity of Pneumolysin, a Major Pneumococcal Virulence Factor, Involves Calcium Influx and Depends on Activation of p38 Mitogen-Activated Protein Kinase. *Neurobiol Dis* **11**: 355–368.

Tan, T.Q., Mason, E.O., Jr, Wald, E.R., Barson, W.J., Schutze, G.E., Bradley, J.S., Givner, L.B., Yogev, R., Kim, K.S., and Kaplan, S.L. (2002) Clinical Characteristics of Children With Complicated Pneumonia Caused by Streptococcus pneumoniae. *Pediatrics* **110**: 1–6.

Tettelin, H., Nelson, K.E., Paulsen, I.T., Eisen, J.A., Read, T.D., Peterson, S., Heidelberg, J., DeBoy, R.T., Haft, D.H., Dodson, R.J., Durkin, A.S., Gwinn, M., Kolonay, J.F., Nelson, W.C., Peterson, J.D., Umayam, L.A., White, O., Salzberg, S.L., Lewis, M.R., Radune, D., Holtzapple, E., Khouri, H., Wolf, A.M., Utterback, T.R., Hansen, C.L., McDonald, L.A., Feldblyum, T.V., Angiuoli, S., Dickinson, T., Hickey, E.K., Holt, I.E., Loftus, B.J., Yang, F., Smith, H.O., Venter, J.C., Dougherty,

B.A., Morrison, D.A., Hollingshead, S.K., and Fraser, C.M. (2001) Complete genome sequence of a virulent isolate of Streptococcus pneumoniae. *Science* **293**: 498–506.

Tong, H.H., Blue, L.E., James, M.A., and DeMaria, T.F. (2000) Evaluation of the virulence of a Streptococcus pneumoniae neuraminidase-deficient mutant in nasopharyngeal colonization and development of otitis media in the chinchilla model. *Infection and Immunity* **68**: 921–924.

Tseng, H.-J., McEwan, A.G., Paton, J.C., and Jennings, M.P. (2002) Virulence of Streptococcus pneumoniae: PsaA Mutants Are Hypersensitive to Oxidative Stress. *Infect. Immun.* **70**: 1635–1639.

Tu, A.T., Fulgham, R.L., McCrory, M.A., Briles, D.E., and Szalai, A.J. (1999) Pneumococcal Surface Protein A Inhibits Complement Activation by Streptococcus pneumoniae. *Infection and Immunity* **67**: 4720–4724.

Tuomanen, E., Liu, H., Hengstler, B., Zak, O., and Tomasz, A. (1985) The induction of meningeal inflammation by components of the pneumococcal cell wall. *J. Infect. Dis.* **151**: 859–868.

Tweten, R.K. (2005) Cholesterol-Dependent Cytolysins, a Family of Versatile Pore-Forming Toxins. *Infect. Immun.* **73**: 6199–6209.

Westh, H., Skibsted, L., and Korner, B. (1990) Streptococcus pneumoniae infections of the female genital tract and in the newborn child. *Rev Infect Dis* **12**: 416–422.

Winter, A.J., Comis, S.D., Osborne, M.P., Tarlow, M.J., Stephen, J., Andrew, P.W., Hill, J., and Mitchell, T.J. (1997) A role for pneumolysin but not neuraminidase in the hearing loss and cochlear damage induced by experimental pneumococcal meningitis in guinea pigs. *Infect. Immun.* **65**: 4411–4418.

Yesilkaya, H., Kadioglu, A., Gingles, N., Alexander, J.E., Mitchell, T.J., and Andrew, P.W. (2000) Role of manganese-containing superoxide dismutase in oxidative stress and virulence of Streptococcus pneumoniae. *Infect Immun* **68**: 2819–2826.

Zhang, J.R., Mostov, K.E., Lamm, M.E., Nanno, M., Shimida, S., Ohwaki, M., and Tuomanen, E. (2000) The polymeric immunoglobulin receptor translocates pneumococci across human nasopharyngeal epithelial cells. *Cell* **102**: 827–837.

11

Impact of Antimicrobial Resistance on Therapy of Bacterial Pneumonia in Children

B. Keith English and Steven C. Buckingham

1. Introduction

Since the development and widespread use of cefotaxime and ceftriaxone in pediatric medicine, the potential for increasing/emerging resistance to these agents has been a concern. However, more than 20 years after their development, these potent antibiotics remain active against all clinical isolates of the respiratory pathogens *Haemophilus influenzae*, *Moraxella catarrhallis*, *Neisseria meningitidis*, and *Streptococcus pyogenes*. On the other hand, two emerging developments threaten to undermine the remarkable success rates of these third-generation cephalosporins in the treatment of certain serious pediatric infections, including pneumonia: (A) The marked increase in clinically relevant resistance to beta-lactam antibiotics (including cefotaxime and ceftriaxone) in *Streptococcus pneumoniae* (Bradley and Connor, 1991; Sloas et al., 1992; McCracken, 1995; Abbasi et al., 1996; Cunha, 2002; Adam, 2002; File, 2002; Applebaum, 2002), and (B) The striking increase in community-acquired infections caused by methicillin-resistant strains of *Staphylococcus aureus* (MRSA) in children, adolescents and adults (Herold et al., 1998; Hunt et al., 1999; Gorak et al., 1999; Frank et al., 2002; Martinez-Aguilar et al., 2003; Buckingham et al., 2003; Buckingham et al., 2004; Martinez-Aguilar et al., 2004; Alfaro et al., 2005; Miller et al., 2005; Adem et al., 2005; Francis et al., 2005; Gonzalez et al., 2005a; Gonzalez et al., 2005b).

Increases in pneumococcal resistance to beta-lactam antibiotics first received widespread attention when cefotaxime and ceftriaxone treatment failures were reported in children with bacterial meningitis caused by pneumococci resistant to these agents (Bradley and Connor, 1991; Sloas et al., 1992), leading to changes in the recommended empiric therapy of probable or definite bacterial meningitis in U.S. children (American Academy of Pediatrics, 1997). Since that time, many

Hot Topics in Infection and Immunity in Children, edited by Andrew J. Pollard and Adam Finn.
Springer, New York, 2006.

investigators have sought to determine whether resistance to these agents is also associated with treatment failure in the setting of non-meningeal infections caused by the pneumococcus.

Barrett and colleagues first reported hospital-associated infections caused by methicillin-resistant strains of *Staphylococcus aureus* (MRSA) in the United States in 1968 (Barrett et al., 1968). For the next three decades, serious infections caused by MRSA were largely confined to the nosocomial setting (Martin, 1994; Lowy, 1998), though community-acquired infections caused by MRSA have been reported since the 1970s. Recently, a dramatic increase in the number of infections caused by community-acquired MRSA (CA-MRSA) has been reported in children and adults in many parts of the United States and other parts of the world. This was first documented by Herold and colleagues in Chicago, who reported an alarming 25-fold increase in methicillin resistance amongst community-acquired *Staphylococcus aureus* isolates from Chicago-area children between 1990 and 1995 (Herold et al., 1998). As would be expected, treatment failures routinely resulted when infections caused by MRSA were empirically treated with beta-lactam antibiotics, and this problem came to national attention in the United States in the summer of 1999 when the CDC reported four deaths resulting from CA-MRSA infections of children living in Minnesota and North Dakota (Hunt et al., 1999). Subsequently, we (Figure 11.1) (Buckingham et al., 2004) and others (Gorak et al., 1999; Frank et al., 2002: Martinez-Aguilar et al., 2003; Martinez-Aguilar et al., 2004; Gonzalez et al., 2005a; Gonzalez et al., 2005b; Alfaro et al., 2005; Miller et al., 2005; Adem et al., 2005; Francis et al., 2005) have also documented increases in the number of serious infections caused by CA-MRSA strains in children and adults in different regions of the United States. Relative to the current discussion, a number of recent reports indicate that MRSA is emerging as an important cause of severe or complicated pneumonia in otherwise healthy children (Herold et al., 1998; Gorak et al., 1999; Buckingham et al., 2003; Buckingham et al., 2004; Alfaro et al., 2005; Gonzalez et al., 2005b).

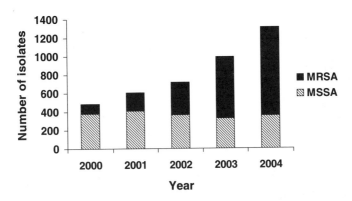

Figure 11.1. Increasing numbers of infections caused by MRSA in Memphis children. Annual numbers of *Staphylococcus aureus* isolates recovered in the clinical microbiology laboratory at Le Bonheur Children's Medical Center, Memphis, TN, demonstrating rising incidence of methicillin resistance from 2000 to 2004.

In this review, we will focus on the implications of these two trends in infections caused by antibiotic-resistant bacteria on the empiric therapy of severe bacterial pneumonia in hospitalized children. We will refer to our recent experience at Le Bonheur Children's Medical Center, a 190-bed pediatric hospital located in Memphis, Tennessee, as an example. We will argue that the first trend (increasing beta-lactam resistance in pneumococci) has had little impact on the efficacy of cefotaxime or ceftriaxone therapy of severe bacterial pneumonia in children, but that the second trend (emergence of MRSA as a cause of severe bacterial pneumonia in otherwise healthy children) has made it impossible to rely on cefotaxime or ceftriaxone monotherapy in the treatment of severe community-acquired pediatric pneumonia in areas where CA-MRSA infections are being recognized frequently. Like other beta-lactam antibiotics, these cephalosporins lack activity against the "atypical" respiratory pathogens *Mycoplasma pneumoniae* and *Chlamydia pneumoniae* (necessitating the addition of additional antibiotic, usually a macrolide, to treat pneumonia caused by these organisms). However, pneumonia caused by *Mycoplasma* or *Chlamydia* is generally less severe than that caused by *Streptococcus pneumoniae* or *Staphylococcus aureus* and is not the focus of this discussion.

2. Epidemiology of Bacterial Pneumonia in Children

Many episodes of pediatric pneumonia are caused by respiratory viruses, *Mycoplasma pneumoniae*, or *Chlamydia pneumoniae* (McIntosh, 2002). More severe or complicated cases of pneumonia are usually caused by pyogenic bacteria, including streptococci and staphylococci, though many studies indicate that clinical or radiological criteria do not reliably differentiate bacterial versus viral versus "atypical" pneumonias (McIntosh, 2002; Michelow et al., 2004). Lung puncture studies performed in children in the developing world 30–40 years ago identified *Streptococcus pneumoniae*, *Haemophilus influenzae*, and *Staphylococcus aureus* as causes of severe bacterial pneumonia in children (McIntosh, 2002). Group A streptococci and *Moraxella catarrhalis* were also implicated in some series.

Most studies of bacterial pneumonia in children do not include the use of an invasive procedure like lung biopsy, and except for the small number of pediatric pneumonia patients with associated bacteremia it is often difficult to identify the specific etiology of presumed bacterial pneumonia in children. Recent use of molecular diagnostic techniques confirm the important role of *Streptococcus pneumoniae* as a cause of pneumonia in children (Michelow et al., 2004) and indicate that co-infection (usually, but not always, involving a virus and a bacterial pathogen) is common in children with pneumonia (Michelow et al., 2004; Tsolia et al., 2004). In addition, the demonstrated reductions in the rates of pneumonia in children receiving the 7-valent conjugated pneumococcal vaccine (PCV-7) (Black et al., 2002) provide additional support for the prominent role of the pneumococcus as a cause of pneumonia in children.

The first pneumococcus resistant to penicillin was isolated from a patient in Australia in 1967. Beginning in the late 1980s, increasing numbers of pneumococcal isolates with reduced susceptibility to both penicillins and cephalosporins were recognized in the United States and around the world (McCracken, 1995; Adam, 2002). The global spread of beta-lactam resistant pneumococci was largely due to the spread of a handful of highly resistant clones (Klugman, 2002). Today, the rates of beta-lactam resistance in pneumococci remain high, though they have declined in some areas – possibly due to reduced antimicrobial use (Guillemot et al., 2005) and to the impact of PCV-7 vaccine (Whitney et al., 2003; Whitney and Klugman, 2004) (several of the most highly resistant clones are members of vaccine serotypes).

Staphylococcus aureus is a well-known and greatly-feared cause of community-acquired pneumonia in children, particularly in young infants (Rebhan and Edwards, 1960; Chartrand and McCracken, 1982), but has been a relatively uncommon pathogen in most series of pediatric pneumonia. Staphylococcal pneumonia was first reported as a complication of influenza during the 1918 Spanish Flu pandemic (Chickering and Park, 1919). Two large pediatric studies of patients with staphylococcal pneumonia have been published: both reports found that most childhood cases of staphylococcal pneumonia occurred in young infants. In the series of 329 patients from Toronto published in 1960, 68% of the children with staphylococcal pneumonia were less than 1 year of age (Rebhan and Edwards, 1960). Similarly, in the experience published by McCracken and colleagues in Dallas, the median age of 79 children with staphylococcal pneumonia was 6 months (Chartrand and McCracken, 1982). *Staphylococcus aureus* is recognized as particularly virulent cause of pneumonia, and is associated with a high rate (as high as 55–80%) of complications, including empyema, pneumatocole and pyopneumo-thorax. Until recently, MRSA were rarely associated with community-acquired pneumonia in children or adults, but this changed with the emergence and spread of CA-MRSA in the United States and other parts of the world. Unlike the previous experience with pneumonia caused by methicillin-susceptible isolates, the marked recent increase in pneumonia caused by CA-MRSA strains has not been limited to young infants, instead affecting children (Herold et al., 1998; Gorak et al., 1999; Buckingham et al., 2003; Buckingham et al., 2004; Alfaro et al., 2005; Gonzalez et al., 2005b), adolescents (Alfaro et al., 2005; Gonzalez et al., 2005b) and adults (Gorak et al., 1999; Francis et al., 2005).

Recently, four reports (analyzing the experience at eleven different pediatric hospitals in the United States) documented increasing rates of parapneumonic empyema in U.S. children in the 1990s (Hardie et al., 1996; Byington et al., 2002; Tan et al., 2002; Buckingham et al., 2003). In the earlier years of these studies, most of the pediatric empyemas were caused by *Streptococcus pneumoniae*, but there was no evidence that beta-lactam resistant strains were more likely to be associated with this complication. In the most recent report from our children's hospital in Memphis, we documented declining rates of pneumococcal empyema after the introduction of PCV-7 (in 2000) and increasing rates of empyema caused by MRSA in 2000 and 2001 (Figure 11.2) (Buckingham et al., 2003). During the past five years, several other children's hospitals have also reported increasing numbers of episodes of pneumonia caused by CA-MRSA (Alfaro et al., 2005; Gonzalez et al., 2005b) as well as declines in pneumonia and other invasive infections caused by pneumococci after the introduction of PCV-7 (Whitney et al., 2003; Whitney and

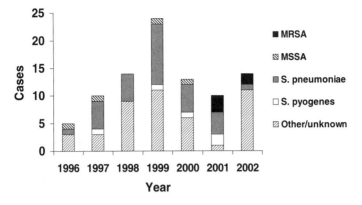

Figure 11.2. Bacterial causes of complicated parapneumonic effusions in Memphis children. Annual cases of parapneumonic empyema with breakdown by culture-confirmed etiologic agents at Le Bonheur Children's Medical Center, Memphis, TN, 1996–2002.

Klugman, 2004). However, the efficacy of PCV-7 in preventing complicated pneumonia in children will likely be limited by the lack of inclusion of certain serotypes (e.g., serotypes 1, 3) that are highly associated with empyema and pleural effusions in children (Tan et al., 2002; Buckingham et al., 2003).

3. Impact of Discordant Empiric Therapy on Outcome in Pediatric Pneumonia

3.1. Beta-Lactam Resistance in *Streptococcus pneumoniae*

Penicillin and cephalosporin resistance in the pneumococcus was first clearly associated with cefotaxime/ceftriaxone treatment failure in children with bacterial meningitis caused by these strains (Bradley and Connor, 1991; Sloas et al., 1992), leading to the routine use of combination antimicrobial therapy (vancomycin plus cefotaxime or ceftriaxone) for the therapy of definite or probable pneumococcal meningitis in U.S. children. In Memphis, we also reported two cefotaxime treatment failures in children with bone and joint infections caused by highly-resistant pneumococci (Abbasi et al., 1996). Documented failures of cefotaxime/ceftriaxone therapy for other non-meningeal infections caused by pneumococci are very uncommon.

Specifically, the impact of penicillin or cephalosporin resistance on the outcome of beta-lactam therapy of patients with pneumococcal pneumonia remains uncertain (Bradley, 2002; File, 2004; Yu and Baddour, 2004). Several studies of children or adults with pneumococcal pneumonia have revealed no impact of penicillin or cephalosporin resistance and/or discordant therapy on the outcome of the infection (Pallares et al., 1995; Deeks et al., 1999; Tan et al., 2002; Yu et al., 2003; Aspa et al., 2004). However, all of these studies included relatively few patients with pneumonia caused by more highly resistant pneumococci (e.g., penicillin MICs of 4–8 µg/ml or greater; cefotaxime/ceftriaxone MICs of 2–4 µg/ml or greater). Two case reports suggest that empiric therapy with second (cefuroxime) or third generation cephalosporins may occasionally result in treatment failure when pneumonia is caused by more highly-resistant pneumococci (Buckingham et al., 1998; Dowell et al., 1999). Furthermore, a retrospective review of 5837 cases of pneumococcal pneumonia (93% of the patients were adults) found that pneumonia caused by pneumococci with higher beta-lactam MICs was associated with markedly increased late mortality (Feikin et al., 2000). Pneumonia caused by pneumococcal strains with penicillin MICs of 4 µg/ml or greater was associated with a 7.1 fold increase in mortality after the 4th day of hospitalization; similarly, pneumonia caused by pneumococci with cefotaxime/ceftriaxone MICs of 2 µg/ml or greater was associated with a 5.9 fold increase in late mortality (Feikin et al., 2000). This study is limited by its retrospective nature and by the lack of information regarding discordant versus concordant therapy for the patients in this cohort. However, the observation of an impact on late but not early mortality is concerning and has biological plausibility. The landmark study published by Austrian and Gold in 1964 demonstrated that penicillin therapy of adults with bacteremic pneumococcal pneumonia reduced overall mortality from 85 per cent to 17 per cent – but penicillin treatment had little or no effect on mortality during the first 5 days of hospitalization (Austrian and Gold, 1964).

3.2. Emergence of CA-MRSA as a Cause of Pneumonia in Otherwise Healthy Children

The emergence of CA-MRSA as a cause of pneumonia in otherwise healthy children (and adults) has now been reported in many parts of the United States (Herold et al., 1998; Gorak et al., 1999; Buckingham et al., 2003; Buckingham et al., 2004; Alfaro et al., 2005; Gonzalez et al., 2005b), adolescents (Alfaro et al., 2005; Gonzalez et al., 2005b) and adults (Gorak et al., 1999; Francis et al., 2005). Pneumonia caused by CA-MRSA has been associated with severe local (pleural effusion, empyema, pneumatocele) and systemic (metastatic foci, toxic shock syndrome, necrotizing fasciiitis, sepsis, Waterhouse-Friedrichsen syndrome) complications. And, unlike the controversy regarding the effects of discordant therapy on the outcome of pneumonia caused by penicillin/cefotaxime-resistant pneumococci, there is no doubt that the available beta-lactam antibiotics (including cefotaxime and ceftriaxone) will fail to successfully treat pneumonia and other serious infections caused by CA-MRSA.

Thus the impact of CA-MRSA infections on the outcome of empiric cefotaxime/ceftriaxone therapy of pediatric pneumonia in a particular geographic area can be predicted on the basis of local epidemiology. Once CA-MRSA are recognized as frequent causes of skin and soft tissue infections in a community, increased vigilance is necessary to determine whether CA-MRSA is also a cause of pneumonia in that community. Once pneumonia caused by CA-MRSA is documented in that area, empiric therapy of severe or complicated pneumonia should include an agent active against CA-MRSA (in addition to cefotaxime or ceftriaxone in most cases).

The optimal therapy of pneumonia caused by CA-MRSA is not known. The time-honored therapy for serious infections caused by MRSA is vancomycin, but many reports indicate that this agent is suboptimal for therapy of infections caused by either methicillin-susceptible or methicillin-resistant strains of *S. aureus* (Karchmer, 1991; Levine et al., 1991; Sakoulas et al., 2004). Recently, new agents such as linezolid have been reported to be at least as effective as (Stevens et al., 2002; Kaplan et al., 2003) and perhaps superior to (Wunderink et al., 2003; Kollef et al., 2004; Weigelt et al., 2005) vancomycin for treatment of certain MRSA infections, including ventilator-associated pneumonia caused by hospital-associated strains of MRSA (Wunderink et al., 2003; Kollef et al., 2004). Whether linezolid is equivalent to or superior to vancomycin or other agents in the treatment of pneumonia caused by CA-MRSA strains (in children or adults) is not known.

Published reports indicate that clindamycin is effective for the treatment of pneumonia and other infections caused by susceptible strains of CA-MRSA (Frank et al., 2002; Martinez-Aguilar et al., 2003; Alfaro et al., 2005). However, clindamycin must be used with caution in the treatment of serious CA-MRSA infections because MRSA (and MSSA) isolates may harbor either constitutive or inducible resistance to clindamycin. All MRSA isolates found to be susceptible to clindamycin but resistant to erythromycin on initial testing should be examined for the presence of inducible resistance to clindamycin by performance of the double-disk diffusion "D" test (Weisblum, 1995). Again, local epidemiology is critical: rates of clindamycin resistance amongst CA-MRSA isolates vary widely in different regions (Frank et al., 2002; Martinez-Aguilar et al., 2003; Buckingham et al., 2004; Alfaro et al., 2005; Kaplan et al., 2005; Braun et al., 2005; Chavez-Bueno et al., 2005) and may be increasing in some areas (Kaplan et al., 2005; Braun et al., 2005).

4. Summary, Conclusions, and Recommendations

Two trends threaten the efficacy of empiric therapy of pediatric pneumonia with cefotaxime or ceftriaxone alone.

(A) Increasing penicillin and cephalosporin resistance amongst pneumococci may result in cefotaxime/ceftriaxone treatment failure of pneumococcal pneumonia, but this appears to be rare and may only apply to infections caused by isolates with unusually high cefotaxime/ceftriaxone MICs. Less than one percent of pneumococci isolated from U.S. patients with invasive disease have very-high-level penicillin resistance (MIC 8 µg/mL or higher), but the rates of high-level resistance to penicillin are higher in some regions (e.g., 3.8% in Tennessee) (Schrag et al., 2004).

(B) On the other hand, the recognition of CA-MRSA as a cause of pneumonia in a community augurs poorly for the continued success of cephalosporin monotherapy of pneumonia. In Memphis, Tennessee, the emergence of CA-MRSA infections (including pneumonia) in our community (Buckingham et al., 2003; Buckingham et al., 2004) led us to recommend the addition of clindamycin or vancomycin (to cefotaxime or ceftriaxone) for the empiric therapy of severe or complicated bacterial pneumonia in children in our community. Given the relatively low prevalence of clindamycin-resistant CA-MRSA in our area, we recommend the addition of this agent for the empiric treatment of most children with pneumonia complicated by pleural effusion or empyema. However, in more severe or potentially life-threatening cases of pneumonia (with or without the presence of pleural effusion/empyema), we recommend the addition of empiric vancomycin (+/− clindamycin). We agree with Bradley (Bradley, 2002) that vancomycin should be part of the empiric therapy for children with life-threatening pulmonary infections in communities where MRSA is a potential pathogen, but we remain concerned that vancomycin may represent suboptimal therapy of serious staphylococcal infections. More studies are needed to assess the efficacy of linezolid and other antimicrobials in the therapy of pneumonia and other severe infections caused by CA-MRSA in children and adults.

Finally, knowledge of the local epidemiology of pneumonia in a community will be required to determine the optimal empiric antimicrobial therapy for presumed bacterial pneumonia in hospitalized children. Because MRSA pneumonia is likely to be associated with pleural effusions/empyema, obtaining pleural fluid for microbiologic studies may play an important role in the detection of CA-MRSA as a cause of pneumonia in a community and in individual patients. Furthermore, monitoring of antibiotic susceptibility testing data of CA-MRSA isolates will provide essential information for the clinician faced with the difficult choice of both empiric and definitive therapy for these patients.

References

Abbasi, S., Orlicek, S.L., Almohsen, I., Luedtke, G., and English, B.K. (1996). Septic arthritis and osteomyelitis caused by penicillin and cephalosporin-resistant *Streptococcus pneumoniae* in a children's hospital. *Pediatr. Infect. Dis. J.* **15**, 78–83.

Adam, D. (2002) Global antibiotic resistance in *Streptococcus pneumoniae*. *J. Antimicrob. Chemother.* **50**, Suppl: 1–5.

Adem, P.V., Montgomery, C.P., Husain, A.N., Koogler, T.K., Arangelovich, V., Humilier, M., Boyle-Vavra, S., and Daum, R.S. (2005). Brief Report: *Staphylococcus aureus* Sepsis and the Waterhouse–Friderichsen Syndrome in Children. *N. Engl. J. Med.* **353**, 1245–51.

Alfaro, C., Fergie, J., and Purcell, K. (2005). Emergence of community-acquired methicillin-resistant *Staphylococcus aureus* in complicated parapneumonic effusions. *Pediatr. Infect. Dis. J.* **24**, 274–6.

American Academy of Pediatrics, Committee on Infectious Diseases (1997). Therapy for children with invasive pneumococcal infections. *Pediatrics* **99**, 289–99.

Applebaum, P.C. (2002) Resistance among *Streptococcus pneumoniae*: implications for drug selection. *Clin. Infect. Dis.* **34**, 1613–20.

Aspa, J., Rajas, O., Rodriguez de Castro, F., Blanquer, J., Zalacain R., Fenoll, A., de Celis, R., Vargas, A., Rodriguez Salvanes, F., Espana, P.P., Rello, J., and Torres, A. (2004). Pneumococcal Pneumonia in Spain Study Group. Drug-resistant pneumococcal pneumonia: clinical relevance and related factors. *Clin. Infect. Dis.* **38**, 787–98.

Austrian, R., and Gold, J. (1964) Pneumococcal bacteremia with especial reference to bacteremic pneumococcal pneumonia. *Ann. Intern. Med* **60**, 759–70.

Barrett, F.F., McGehee, R.F., Jr., and Finland, M. (1968). Methicillin-resistant *Staphylococcus aureus* at Boston City Hospital. Bacteriologic and epidemiologic observations. *N. Engl. J. Med.* **279**, 441–8.

Black, S.B., Shinefield, H.R., Ling, S., Hansen, J., Fireman, B., Spring, D., Noyes, J., Lewis, E., Ray, P., Lee, J., and Hackell, J. (2002). Effectiveness of heptavalent pneumococcal conjugate vaccine in children younger than five years of age for prevention of pneumonia. *Pediatr. Infect. Dis. J.* **21**, 810–5.

Bradley, J.S., and Connor, J.D. (1991). Ceftriaxone failure in meningitis caused by *Streptococcus pneumoniae* with reduced susceptibility to beta-lactam antibiotics. *Pediatr. Infect. Dis. J.* **10**, 871–3.

Bradley, J.S. (2002). Management of community-acquired pediatric pneumonia in an era of increasing antibiotic resistance and conjugate vaccines. *Pediatr. Infect. Dis. J.* **21**, 592–8.

Braun, L., Craft, D., Williams, R., Tuamokumo, F., and Ottolini, M. (2005). Increasing clindamycin resistance among methicillin-resistant *Staphylococcus aureus* in 57 northeast United States military treatment facilities. *Pediatr. Infect. Dis. J.* **24**, 622–6.

Buckingham, S.C., Brown, S.P., and Joaquin, V.H. (1998). Breakthrough bacteremia and meningitis during treatment with cephalosporins parenterally for pneumococcal pneumonia. *J. Pediatr.* **132**, 174–6.

Buckingham, S.C., King, M.D., and Miller, M.L. (2003). Incidence and etiologies of complicated parapneumonic effusions in children, 1996 to 2001. *Pediatr. Infect. Dis. J.* **22**, 499–504.

Buckingham, S.C., McDougal, L.K., Cathey, L.D., Comeaux, K., Craig, A.S., Fridkin, S.K., and Tenover, F.C. (2004). Emergence of community-associated methicillin-resistant *Staphylococcus aureus* at a Memphis, Tennessee Children's Hospital. *Pediatr. Infect. Dis. J.* **23**, 619–24.

Byington, C.L., Spencer, L.Y., Johnson, T.A., Pavia, A.T., Allen, D., Mason, E.O., Kaplan, S., Carroll, K.C., Daly, J.A., Christenson, J.C., and Samore, M.H. (2002). An epidemiological investigation of a sustained high rate of pediatric parapneumonic empyema: risk factors and microbiological associations. *Clin. Infect. Dis.* **34**, 434–40.

Chartrand, S.A., and McCracken, G.H., Jr. (1982). Staphylococcal pneumonia in infants and children. *Pediatr. Infect. Dis. J.* **1**, 19–23.

Chavez-Bueno, S., Bozdogan, B., Katz, K., Bowlware, K.L., Cushion, N., Cavuoti, D., Ahmad, N., McCracken, G.H., Jr., and Appelbaum, P.C. (2005). Inducible clindamycin resistance and molecular epidemiologic trends of pediatric community-acquired methicillin-resistant *Staphylococcus aureus* in Dallas, Texas. *Antimicrob. Agents Chemother.* **49**, 2283–8.

Chickering, H.T., and Park, J.H. (1919) *Staphylococcus aureus* pneumonia. *JAMA* **72**, 617–26.

Cunha, B.A. (2002). Clinical relevance of penicillin-resistant *Streptococcus* pneumoniae *Semin. Respir. Infect.* **17**, 204–14.

Deeks, S.L., Palacio, R., Ruvinsky, R., Kertesz, D.A., Hortal, M., Rossi, A., Spika, J.S., and Di Fabio, J.L. (1999). Risk factors and course of illness among children with invasive penicillin-resistant *Streptococcus pneumoniae*. The *Streptococcus pneumoniae* Working Group. *Pediatrics.* **103**, 409–13.

Dowell, S.F., Smith, T., Leversedge, K., and Snitzer, J. (1999). Failure of treatment of pneumonia associated with highly resistant pneumococci in a child. *Clin. Infect. Dis.* **29**, 462–3.

Feikin, D.R., Schuchat, A., Kolczak, M., Barrett, N.L., Harrison, L.H., Lefkowitz, L., McGeer, A., Farley, M.M., Vugia, D.J., Lexau, C., Stefonek, K.R., Patterson, J.E., and Jorgensen, J.H. (2000). Mortality from invasive pneumococcal pneumonia in the era of antibiotic resistance, 1995–1997. *Am. J. Public Health.* **90**, 223–9.

File, T.M., Jr. (2002). Appropriate use of antimicrobials for drug-resistant pneumonia: focus on the significance of beta-lactam resistant *Streptococcus pneumoniae*. *Clin. Infect. Dis.* **34**, S17–26.

File, T.M., Jr. (2004). *Streptococcus pneumoniae* and community-acquired pneumonia: a cause for concern. *Am. J. Med.* **117**, Suppl 3A: 39S–50S.

Francis, J.S., Doherty, M.C., Lopatin, U., Johnston, C.P., Sinha, G., Ross, T., Cai, M., Hansel, N.N., Perl, T., Ticehurst, J.R., Carroll, K., Thomas, D.L., Nuermberger, E., and Bartlett, J.G. (2005). Severe community-onset pneumonia in healthy adults caused by methicillin-resistant *Staphylococcus aureus* carrying the Panton-Valentine leukocidin genes. *Clin. Infect. Dis.* **40**, 100–7.

Frank, A.L., Marcinak, J.F., Mangat, P.D., Tjhio, J.T., Kelkar, S., Schreckenberger, P.C., and Quinn, J.P. (2002). Clindamycin treatment of methicillin-resistant *Staphylococcus aureus* infections in children. *Pediatr. Infect. Dis. J.* **21**, 530–4.

Gonzalez, B.E., Martinez-Aguilar, G., Hulten, K.G., Hammerman, W.A., Coss-Bu, J., Avalos-Mishaan, A., Mason, E.O., Jr., and Kaplan, S.L. (2005). Severe Staphylococcal sepsis in adolescents in the era of community-acquired methicillin-resistant *Staphylococcus aureus*. *Pediatrics.* **115**, 642–8.

Gonzalez, B.E., Hulten, K.G., Dishop, M.K., Lamberth, L.B., Hammerman, W.A., Mason, E.O., Jr., and Kaplan, S.L. (2005). Pulmonary manifestations in children with invasive community-acquired *Staphylococcus aureus* infection. *Clin. Infect. Dis.* **41**, 583–90.

Gorak, E.J., Yamada, S.M., and Brown, J.D. (1999). Community-acquired methicillin-resistant *Staphylococcus aureus* in hospitalized adults and children without known risk factors. *Clin. Infect. Dis.* **29**, 797–800.

Guillemot, D., Varon, E., Bernede, C., Weber, P., Henriet, L., Simon, S., Laurent, C., Lecoeur, H., and Carbon, C. (2005). Reduction of antibiotic use in the community reduces the rate of colonization with penicillin G-nonsusceptible *Streptococcus pneumoniae*. *Clin. Infect. Dis.* **41**, 930–8.

Hardie, W., Bokulic, R., Garcia, V.F., Reising, S.F., and Christie, C.D. (1996). Pneumococcal pleural empyemas in children. *Clin. Infect. Dis.* **22**, 1057–63.

Herold, B.C., Immergluck, L.C., Maranan, M.C., Lauderdale, D.S., Gaskin, R.E., Boyle-Vavra, S., Leitch, C.D., and Daum, R.S. (1998). Community-acquired methicillin-resistant *Staphylococcus aureus* in children with no identified predisposing risk. *JAMA* **279**, 593–8.

Hunt, C., Dionne, M., Delorme, M., et al. (1999). Four pediatric deaths from community-acquired methicillin-resistant *Staphylococcus aureus*: Minnesota and North Dakota, 1997–1999. *MMWR* **48**, 707–10.

Kaplan, S.L., Afghani, B., Lopez, P., Wu, E., Fleishaker, D., Edge-Padbury, B., Naberhuis-Stehouwer, S., and Bruss, J.B. (2003). Linezolid for the treatment of methicillin-resistant *Staphylococcus aureus* infections in children. *Pediatr. Infect. Dis. J.* **22 (Suppl)**, S178–85.

Kaplan, S.L., Hulten, K.G., Gonzalez, B.E., Hammerman, W.A., Lamberth, L., Versalovic, J., and Mason, E.O., Jr. (2005). Three-year surveillance of community-acquired *Staphylococcus* aureus infections in children. *Clin. Infect. Dis.* **40**, 1785–91.

Karchmer, A.W. (1991). *Staphylococcus aureus* and vancomycin: the sequel. *Ann. Intern. Med.* **115**, 739–41.

Klugman, K.P. (2002). The successful clone: the vector of dissemination of resistance in *Streptococcus pneumoniae*. *J. Antimicrob. Chemother.* **50 (Suppl) S2**, 1–5.

Kollef, M.H., Rello, J., Cammarata, S.K., Croos-Dabrera, R.V., and Wunderink, R.G. (2004). Clinical cure and survival in Gram-positive ventilator-associated pneumonia: retrospective analysis of two double-blind studies comparing linezolid with vancomycin. *Intensive Care Med.* **30**, 388–94.

Levine, D.P., Fromm, B.S., and Reddy, B.R. (1991). Slow response to vancomycin or vancomycin plus rifampin in methicillin-resistant *Staphylococcus aureus* endocarditis. *Ann. Intern. Med.* **115**, 674–80.

Lowy, F.D. (1998). Medical Progress: *Staphylococcus aureus* Infections. *N. Engl. J. Med.* **339**, 520–32.

Martinez-Aguilar, G., Hammerman, W.A., Mason, E.O., Jr., and Kaplan, S.L. (2003). Clindamycin treatment of invasive infections caused by community-acquired, methicillin-resistant and methicillin-susceptible *Staphylococcus aureus* in children. *Pediatr. Infect. Dis. J.* **22**, 593–8.

Martinez-Aguilar, G., Avalos-Mishaan. A., Hulten, K., Hammerman, W., Mason, E.O. Jr., and Kaplan, S.L. (2004). Community-acquired, methicillin-resistant and methicillin-susceptible *Staphylococcus aureus* musculoskeletal infections in children. *Pediatr. Infect. Dis. J.* **23**, 701–6.

Martin, M.A. (1994). Methicillin-resistant *Staphylococcus aureus*: the persistent resistant nosocomial pathogen. *Curr Clin Top Infect Dis.* **14**, 170–91.

McCracken, G.H., Jr. (1995). Emergence of resistant *Streptococcus pneumoniae*: a problem in pediatrics. *Pediatr. Infect. Dis. J.* **14**, 424–8.

McIntosh, K. (2002). Community-acquired pneumonia in children. *N. Engl. J. Med.* **346**, 429–37.

Michelow, I.C., Olsen, K., Lozano, J., Rollins, N.K., Duffy, L.B., Ziegler, T., Kauppila, J., Leinonen, M., and McCracken, G.H., Jr. (2004). Epidemiology and clinical characteristics of community-acquired pneumonia in hospitalized children. *Pediatrics* **113**, 701–7.

Miller, L.G., Perdreau-Remington, F., Rieg, G., Mehdi, S., Perlroth, J., Bayer. A.S., Tang, A.W., Phung, T.O., and Spellberg, B. (2005). Necrotizing fasciitis caused by community-associated methicillin-resistant *Staphylococcus aureus* in Los Angeles. *N. Engl. J. Med.* **352**, 1445–53.

Pallares, R., Linares, J., Vadillo, M., Cabellos, C., Manresa, F., Viladrich, P.F., Martin, R., and Gudiol, F. (1995). Resistance to penicillin and cephalosporin and mortality from severe pneumococcal pneumonia in Barcelona, Spain. *N. Engl. J. Med.* **333**, 474–80.

Rebhan, A.W., and Edwards, H.E. (1960). Staphylococcal pneumonia: a review of 329 cases. *Can. Med. Assoc. J.* **82**, 513–7.

Sakoulas, G., Moise-Broder, P.A., Schentag, J., Forrest, A., Moellering, R.C., Jr., and Eliopoulos, G.M. (2004). Relationship of MIC and bactericidal activity to efficacy of vancomycin for treatment of methicillin-resistant *Staphylococcus aureus* bacteremia. *J. Clin. Microbiol* **42**, 2398–402.

Schrag, S.J., McGee, L., Whitney, C.G., Beall, B., Craig, A.S., Choate, M.E., Jorgensen, J.H., Facklam, R.R., and Klugman, K.P.; Active Bacterial Core Surveillance Team (2004). Emergence of *Streptococcus pneumoniae* with very-high-level resistance to penicillin. *Antimicrob. Agents Chemother.* **48**, 3016–2.

Sloas, M.M., Barrett, F.F., Chesney, P.J., English, B.K., Hill, B.C., Tenover, F.C., and Leggiadro, R.J. (1992). Cephalosporin treatment failure in penicillin- and cephalosporin-resistant *Streptococcus pneumoniae* meningitis. *Pediatr. Infect. Dis. J.* **11**, 662–6.

Stevens, D.L., Herr, D., Lampiris, H., Hunt. J.L., Batts, D.H., and Hafkin, B. (2002). Linezolid versus vancomycin for the treatment of methicillin-resistant *Staphylococcus aureus* infections. *Clin. Infect. Dis.* **34**, 1481–90.

Tan, T.Q., Mason, E.O., Jr., Wald, E.R., Barson, W.J., Schutze, G.E., Bradley, J.S., Givner, L.B., Yogev, R., Kim, K.S., and Kaplan, S.L. (2002). Clinical characteristics of children with complicated pneumonia caused by *Streptococcus pneumoniae*. *Pediatrics*. **110**, 1–6.

Tsolia, M.N., Psarras, S., Bossios, A., Audi, H., Paldanius, M., Gourgiotis, D., Kallergi. K., Kafetzis, D.A., Constantopoulos. A., and Papadopoulos, N.G. (2004). Etiology of community-acquired pneumonia in hospitalized school-age children: evidence for high prevalence of viral infections. *Clin. Infect. Dis.* **39**, 681–6

Weigelt, J., Itani, K., Stevens, D., Lau, W., Dryden, M., and Knirsch, C. Linezolid CSSTI Study Group. (2005). Linezolid versus vancomycin in treatment of complicated skin and soft tissue infections. *Antimicrob. Agents Chemother.* **49**, 2260–6.

Weisblum, B. (1995). Insights into erythromycin action from studies of its activity as inducer of resistance. *Antimicrob. Agents Chemother.* **39**, 797–80.

Whitney, C.G., Farley, M.M., Hadler, J., Harrison, L.H., Bennett, N.M., Lynfield, R., Reingold. A., Cieslak, P.R., Pilishvili, T., Jackson, D., Facklam, R.R., Jorgensen, J.H., and Schuchat, A (2003). Active Bacterial Core Surveillance of the Emerging Infections Program Network. Decline in invasive pneumococcal disease after the introduction of protein-polysaccharide conjugate vaccine. *N. Engl. J. Med.* **348**, 1737–46.

Whitney, C.G., and Klugman, K.P. (2004), Vaccines as tools against resistance: the example of pneumococcal conjugate vaccine. *Semin. Pediatr. Infect. Dis.* **15**, 86–93.

Wunderink, R.G., Rello, J., Cammarata, S.K., Croos-Dabrera, R.V., and Kollef, M.H. (2003). Linezolid vs vancomycin: analysis of two double-blind studies of patients with methicillin-resistant *Staphylococcus aureus* nosocomial pneumonia. *Chest* **124**, 1789–97.

Yu, V.L., Chiou, C.C., Feldman, C., Ortqvist, A., Rello, J., Morris, A.J., Baddour, L.M., Luna, C.M., Snydman, D.R., Ip, M., Ko, W.C., Chedid, M.B., Andremont, A., and Klugman, K.P. (2003). International Pneumococcal Study Group. An international prospective study of pneumococcal bacteremia: correlation with in vitro resistance, antibiotics administered, and clinical outcome. *Clin. Infect. Dis.* **37**, 230–7.

Yu, V.L., and Baddour, L.M. (2004). Infection by drug-resistant *Streptococcus pneumoniae* is not linked to increased mortality. *Clin. Infect. Dis.* **39**, 1086–7.

12

Diagnosis and Prevention of Pneumococcal Disease

Hanna Nohynek

1. Introduction

Streptococcus pneumoniae, pneumococcus, is one of the major bacterial pathogens of childhood. Infections caused by this organism range from the very severe (e.g., meningitis and septicaemia), to non-bacteremic pneumonia and to less severe upper respiratory tract infections such as acute otitis media and sinusitis. The spread of HIV infection has increased the incidence of pneumococcal disease, especially in many resource poor countries where anti-HIV treatment is not readily available. It is estimated that children infected with HIV/AIDS are 20–40 times more likely to contract pneumococcal disease than those without HIV/AIDS.

Whilst prospects for the prevention of invasive pneumococcal disease have tremendously increased following the licensure of pneumococcal conjugate vaccines (PCV), very little progress in the development of new diagnostics tests, either to confirm etiology in individual cases or to measure the disease burden at the population level, has been made. Because pneumococcal disease burden is difficult to assess, there has been a delay in the introduction of PCV into national programs. In most countries of the world and in most aid agencies, decision-making is strongly linked to available local epidemiological evidence, cost-effectiveness analyses, as well as the public perception of the importance of the disease and by political will.

The proportion of strains of *Streptococcus pneumoniae* that are penicillin resistant is continuously increasing in most areas where surveillance systems have been set up, resource poor countries included (Okeke et al., 2005). What impact PCV introduction might have on antimicrobial resistance trends depends on vaccine effectiveness during large scale use as well as on the phenomenon of serotype replacement (Lipsitch, 2001), discussed in an earlier volume of this book (Dagan, 2004).

Hot Topics in Infection and Immunity in Children, edited by Andrew J. Pollard and Adam Finn.
Springer, New York, 2006

2. From Carriage to Clinical Disease

Human beings are the sole natural hosts of *Streptococcus pneumoniae*. The bacterium is spread from one human to another by droplets. Pneumococci are frequently found in the upper respiratory tract of both healthy children and adults, although carriage rates differ greatly from one age group and population to another. Depending on the geographic location carriage acquisition occurs either earlier or later in life (O'Brien et al., 2003). Risk factors for early carriage are those that increase the chance of transmission such as poor living conditions, crowding, and presence of older siblings in the household. The transmission cycle has lately been the target of intense interest, arising largely in response to vaccine development and also epidemiological observations following introduction of PCV in the United States. In most cases it seems that younger siblings are infected with *Streptococcus pneumoniae* carried by an older sibling or another child living in close vicinity or attending the same day care rather than from his/her parents or other adults (Eskola et al., 2004; O'Brien et al., 2003).

The mechanisms that lead to the development of overt clinical pneumococcal disease following the carriage state are not well understood. Invasion has been associated with high acquisition rates, and high frequency and prolonged duration of pneumococcal carriage rather than the serotype (Smith et al., 1993). Recent studies, however, demonstrate that a recent acquisition is essential for the development of clinical disease, at least for acute otitis media caused by *Streptococcus pneumoniae* (Syrjänen et al., 2005). Predisposing viral infection may also increase the risk of invasion, although no specific virus has been identified (Kleemola et al., 2005). The lack of serotype specific immunity could be an additional invasion-driving factor.

3. Methods to Detect Involvement of *Streptococcus pneumoniae* in Disease

The World Health Organization estimates that more than 1.6 million people die every year from pneumococcal infections – primarily pneumonia and meningitis – including more than 800,000 children under 5 years old. According to these estimates, 40% of total acute lower respiratory tract infection, and 35% of total meningitis is caused by *Streptococcus pneumoniae*. For each invasive, potentially deadly pneumococcal infection case, there are from 10 to over 100-fold more milder clinical infectious cases caused by *Streptococcus pneumoniae*. These entities, although milder, constitute the major pneumococcal disease burden to both individuals, their families and the society both in rich and in resource poor countries. But how are these estimates derived? How reliable are they?

3.1. Direct, Specific Means to Detect Pneumococci

For a clinical case to be classified as definitely caused by *Streptococcus pneumoniae* it needs to be confirmed by culture of blood, CSF, pleural or joint fluid or by culture of middle ear fluid after tympanocentesis depending on the clinical suspicion of the focus of infection. Blood is the most commonly collected sample in seriously ill children. The culture method, however, is affected by the previous use of antimicrobials and the volume of blood obtained. Blood culture is very insensitive in the diagnosis of pneumococcal pneumonia, most of which is only transiently bacteremic or completely nonbacteremic. This was well demonstrated in a recent study in Finland, in which blood culture identified only one tenth of the cases where the causative agent was proven to be *Streptococcus pneumoniae* by PCR from a percutaneous lung tap (Vuori-Holopainen et al., 2002). The same study team also undertook a meta-analysis of 59 published lung tap studies (Vuori-Holopainen et al., 2001) and found considerable variability in the results due to differences in the source population, patient selection criteria and operator skill. The authors urged clinicians to reconsider lung taps as an alternative, more sensitive means to detect the pneumococcus as the agent of pneumonia but many consider lung taps too invasive a technique to justify its use in routine patient care.

Other means of diagnosis which have been explored are culture and/or Gram stain of transtracheal specimens, polymerase chain reaction from blood or pleural fluid, and antigen detection. Of these, detection of pneumococcal DNA from serum has been successful in children, but its interpretation is biased by the high positivity rate detected in the pneumococcal carriage state (Dagan et al., 1998). DNA detection in pleural fluid or lung tissue is a more accepted proof of pneumococcal involvement in infection (Vuori-Holopainen et al., 2001).

Pneumococcal polysaccharide can be detected in body fluids or on epithelial surfaces where pneumococci reside (i.e., nasopharynx, oral cavity, trachea or lung) or in urine where components of bacterial cell wall and other bacterial debris are excreted. Latex agglutination is the most frequently used non-culture method, although other tests, such as counterimmunoelectrophoresis, are available. The antigen detection methods lack specificity when used to identify invasive disease from detection in respiratory secretions since the respiratory tract is not only the source of the causative agent but also of the normal upper respiratory tract flora, which includes the pneumococcus. Urine antigen testing is plagued by the same dilemma: urine is the source of pneumococcal degradation particles in carrier and disease states alike.

3.2. Indirect, Specific Means to Detect Pneumococcal Involvement

Some investigators have used indirect means to demonstrate the involvement of the pneumococcus in infection, either by measuring free or complexed antibodies directed against pneumococcal structures such as capsular or C-polysaccharides, or proteins such pneumolysin (Lankinen et al., 1999) or PsaA (Scott A et al., 2005). These methods increase the diagnostic yield, however, one serum sample is not enough; paired samples with a minimum interval of 7 to 10 days are required to detect a rising level, thus making the method very inconvenient for immediate clinical diagnostics, and better suited for epidemiological studies. Furthermore, the interpretation of serological tests is far from clear-cut. It is not well understood how

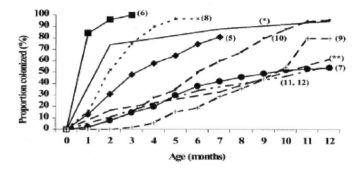

Figure 12.1   Carriage acquisition in different countries. Cumulative prevalence of pneumococcal NP colonization by age in various populations. Numbers in parentheses, references. *, Israeli Bedouins; **, Israeli Jews (R Dagan, unpublished data). Reference 5 is from India, 6 from Papua New Guinea, 7 from the United States, 8 from Australia (Aboriginal site), 9 from Costa Rica, 10 from United States, 11 from Sweden and 12 from Finland (O'Brien et al., 2003).

often bacterial carriage acquisition per se significantly elicits antibodies and whether the change in concentration can be differentiated from that following invasive disease.

3.3. Nonspecific Means to Detect Pneumococcal Involvement

3.3.1. Radiology

Pneumococcal pneumonia causes a lobar alveolar infiltrate on chest X-ray which is distinguishable from the more widely spread interstitial infiltrates that are considered to typically appear in pneumonia caused by viral pathogens. This view has been challenged by those who maintain that the etiology of pneumonia cannot be interpreted from chest radiograms.

The probe method to measure the disease burden of *Haemophilus influenzae* type b (Hib, see below) suggested that chest radiographs could be used as an important adjunct tool to detect pathogen specific causes of pneumonia. This observation led the World Health Organization to standardize the technique of taking and interpreting chest radiographs (WHO, 2001; Cherian et al., 2005) in order to compare measurements across study and trial sites. The emphasis was on the sensitivity, rather than specificity of the measurement. The radiological appearance of possible bacterial pneumonia was defined as "end point consolidation", i.e. "a dense opacity that may be: (a) a fluffy consolidation of a portion or whole of a lobe or of the entire lung, often containing air-bronchograms and sometimes associated with pleural effusion." Also, atelectasis of an entire lobe that produces a dense opacity and a positive silhouette sign with the mediastinal border was considered to be end point consolidation (WHO, 2001). It was hoped that such a definition would capture as many cases of pneumococcal pneumonia as possible. This definition has been used in the Navajo and the Gambian PCV studies (O'Brien et al., 2003; Cutts et al., 2005).

3.3.2. Acute Phase Reactants

There are several infection markers used in clinical pediatric diagnostics to indicate the severity of infection which serve as adjunct tools in decision making for care and monitoring of the course of infection in individual patients. The most commonly referred markers are erythrocyte sedimentation rate, white blood cell count, as well as concentrations of C-reactive protein, procalcitonin, interleukin-6, and S-amyloid A (Prat et al., 2003; Huttunen T et al., 2003). These tests are non-specific. These inflammatory markers typically increase in inflammation, infection, and trauma. Some, such as C-reactive protein are more sensitive to bacterial than viral infection (Peltola et al., 1988), but none are specific for *Streptococcus pneumoniae*. Recently, as a consequence of lessons learnt from Hib studies (see below, Levine et al., 1999), acute phase reactants have been used to determine the overall proportion of pneumococcal infections averted by PCV (Madhi et al., 2005). Using an algorithm that included, in addition to radiology, values of acute phase reactants CRP and PCT (with specified cut off levels; CRP above 120 mg/l and procalcitonin above 5 ng/ml), a South African study group was able to demonstrate a much higher proportion of pneumonia cases prevented (with a maximum vaccine efficacy of 64%), than by standard bacteremia or radiology definitions alone.

4. Measuring the Disease Burden Caused by Pneumococcus

Because of problems related to proving the aetiological diagnosis in pneumonia clinicians may grossly underestimate the disease burden caused by bacterial pathogens including *Streptococcus pneumoniae*. This was first demonstrated in an efficacy study with *Haemophilus influenzae* type b (Hib) conjugate vaccine in The Gambia. In this study the "probe" method revealed that Hib was almost three times as frequent a cause of pneumonia, when cases were defined by radiology, and not by culture alone (Table 12.1) (Mulholland et al., 1997). In Chile, the observation was even more pronounced, i.e. the Hib vaccine prevented 5 times more nonbacteremic than bacteremic Hib cases (Levine et al., 1999) (Table 12.1).

To understand the figures behind the estimates of pneumococcal disease burden it is important to understand care-seeking practices, population demography, criteria for outpatient vs. inpatient attendance and care, indications for and diagnostic methods used, and case definitions. Depending on the sensitivity vs. specificity trade off in case definitions, the disease burden may appear quite different (Greenwood, 2005; Madhi et al., 2005).

Presently the WHO together with the Pneumococcal Accelerated Development and Implementation Program (GAVI ADIP Pneumo) is collecting data from many different nations to derive the best estimates of global and local pneumococcal disease burden.

Table 12.1. Example of *Haemophilus influenzae* type b (Hib) conjugate vaccine studies utilizing the probe method approach

Study	% radiological cases prevented by Hib vaccine	% pneumonia with Hib isolated in culture
The Gambia	21,1	8,0[1]
Chile	22,0	0,7[2]

[1]Blood, pleural, and lung aspirate culture.
[2]Blood culture alone.

5. Immunity to and Prevention of Pneumococcal Diseases

5.1. The Different Pneumococcal Structures and Vaccines Directed Against Them

Pneumococci consist of a family of Gram positive bacteria with over 90 different types of capsular polysaccharide. The capsule is a virulence factor for the bacterium, which alters opsonisation by phagocytes. The capsular polysaccharides are T-cell independent, and only weekly immunogenic; immunity to them develops gradually during childhood. Antibodies against the capsular polysaccharides are protective and prevent pneumococcal disease. The classification of *Streptococcus pneumoniae* into serogroups, types and subtypes is done according to the capsular swelling reaction, which is mostly detected microscopically by naked eye as the quellung reaction or alternatively by counter immunoelectrophoresis or latex agglutination.

The first documented vaccines against *Streptococcus pneumoniae* used either the whole bacterial cell or the capsular polysaccharides (Wright et al., 1914) as the starting point. During the 1970's and 1980's more purified forms of the multivalent polysaccharide vaccines (PPV) were tested against a wide array of pneumococcal infections, but with disappointingly little success in young children (Fedson et al., 2004). In second generation of pneumococcal vaccines, the capsular polysaccharide is linked via a chemical conjugation process to a carrier protein such as an outer membrane protein of *Neisseria meningitidis*, tetanus or diphtheria toxoid, or a 37 kD protein of *Haemophilus influenzae*, thus making the polysaccharide antigen T-cell dependent. These vaccines induce immunological memory at an early stage, i.e. before 2 years of age, unlike the PPVs which are poorly immunogenic for most pneumococcal serotypes at this stage (Eskola et al., 2004).

Pneumococci have several surface proteins such as PspA and PsaA as well as endotoxins such as pneumolysin excreted by the bacterium at the different phases of the host-pathogen life cycle. The third generation of Pnc vaccines, which are in their early phase of clinical development utilize these common Pnc proteins alone without pneumococcal polysaccharides (see Chapter 10).

5.2. The Overall Protective Effect of Pneumococcal Vaccines

Pneumococcal disease can be prevented by a) the direct protective effect of the vaccine on vaccinated individuals (both PPV and PCV) and/or b) an indirect protective effect via reduced transmission of the pathogen to susceptible, nonvaccinated individuals (PCV only, since the mucosal protection provided by PPV is insignificant). The public health benefit arising from both is enforced by the reduction of the incidence of vaccine preventable strains resistant to antimicrobials. Also, the unexpected finding from the Gambian study that the 9-valent PCV prevented 16% of the overall, all-cause mortality in the study children (Cutts et al., 2005), makes the vaccine a very powerful tool both against childhood morbidity and mortality globally.

5.3. The Effect of Conjugated Pneumococcal Vaccines on Different Forms of Pneumococcal Disease

Up to date, 4 different types of PCVs have been developed for large scale clinical trials. They consist of different selection of pneumococcal serotypes ranging from 7 to 11, and different carrier proteins as described above. All of these are immunogenic and safe (Eskola et al., 2004). So far only the 7-valent PCV with mutant diphtheria toxoid as carrier protein has been licensed and introduced for large scale use.

5.3.1. Invasive Disease

Altogether 6 randomized controlled trials have been published to date on the efficacy of PCV for individual healthy infants (Table 12.2): four of these were primarily designed to provide information on PCV efficacy/ effectiveness against invasive pneumococcal disease (IPD) (Lucero et al., 2004; Black et al., 2000; Klugman et al., 2003; O'Brien et al., 2003; Cutts et al., 2005) and two for acute otitis media (Eskola et al., 2001; Kilpi et al., 2003; Prymula et al., 2005). As expected from the experience with conjugated Hib vaccines, the PCV efficacy against IPD is high, even in HIV-positive children, when a 4 or 3-dose schedule is used.

5.3.2. Pneumonia

Of the randomized controlled trials studying the effect of PCV on pneumonia, only the Gambian (Cutts et al., 2005) and the Philippine trials (Lucero et al., 2005) have pneumonia as their primary endpoint. Access to care, case definitions of both clinical and radiological pneumonia and care practices have varied across trials, which may partly explain the differences observed in vaccine efficacy varying from 20 to 37% (Table 12.2). Other reasons for differences in vaccine efficacy relate to differences in disease epidemiology demographics. Pneumonia caused by those serotypes of *Streptococcus pneumonia* chosen for the currently available PCV7 may not be as common as previously thought (Hausdorff et al., 2005; Hausdorff, 2002). In general, however, the observed efficacy of PCV against pneumonia is an outcome

Table 12.2. Randomized controlled trials of pneumococcal conjugate vaccines (PCV)

Study site	# Pnc serotypes in PCV and carrier protein	Number of doses/ and schedule	Number of/ children in study	Study endpoint	Vaccine efficacy	
					%	95 % CI
California, the U.S.	7 CRM	42–4–6 months	37 868	IPD	97	83–100
				Pneumonia	21	4–34
				AOM	7	4–10
Arizona New, Mexico Utah, U.S.*	7 CRM	42–4–6 months	8 292	IPD	77	–9–95
Finland	7 CRM	42–4–6 months	1 662	AOM		
				Vaccine type AOM	57	44–67
				AOM overall	6	–4–16
Finland	7 OMP	42–4–6 months	1 666	AOM		
				Vaccine type AOM	56	44–66
				AOM overall	–1	–12–10
South Africa	9 CRM			AOM overall	6	–4–16
				AOM overall	6	–4–16
				HIV-positive Pneumonia	65	24–86
				HIV-negatives	25	4–40
				HIV-positives	8	–15–27
The Gambia	9 CRM	36–10–14 weeks	17 437	Death (any cause)	16	3–28
				IPD	77	51–90
				Pneumonia	37	25–48
Czech Republic	11 PD	43–4–5–12 months	4 968	AOM		
				Vaccine type AOM	53	35–66
				AOM overall	34	21–44
The Philippines	11 TD	36–10–14 weeks	12 190	Pneumonia	Results available in year 2006	

Vaccine efficacy according to per protocol analysis.

Carrier proteins CRM = mutant diphtheria toxoid, OMP = outer membrane protein of *Neisseria meningitides*, PD = protein D of *Haemopilus influenzae*, TD = tetanus and diphtheria toxoids.

IPD = invasive pneumococcal disease with bacteriological confirmation from body site, where microbial growth usually not present (i.e. blood, CSF, pleural fluid, joint fluid).

Pneumonia = radiologically confirmed.

AOM = acute otitis media.

*the only study with cluster randomization, all other with individual randomization.

of several different factors, i.e. what proportion of pneumonia is bacterial, what proportion of this is caused by pneumococcus, what proportion of pneumococcal pneumonia is caused by the vaccine serotypes and ultimately, what proportion of lung infection caused by the vaccine serotypes is actually preventable by PCV. This chain of thought implies that childhood pneumonia is a heterogeneous continuum of different diseases, of different severity and with different etiologic agents, and pneumococcus plays a role in it to a varying degree. This view is further emphasized by the finding in the South African trial, where PCV was observed to contribute to the prevention of pneumonia with viral involvement (Madhi et al., 2004).

5.3.3. Acute Otitis Media

Three randomized, controlled trials have looked into the effect of PCV on acute otitis media (AOM): the Northern Californian Kaiser Permanente trial in the United States (Black et al., 2000) and the FinOM trial in Finland using two different formulations of a seven-valent PCV (Eskola et al., 2000; Kilpi et al., 2003), and the Czech trial using the eleven-valent PCV with the *Haemophilus influenzae* derived protein D as the carrier protein (Prymula et al., 2005). The case detection and definition of AOM in each of these trials was different; however, the overall vaccine effect against any first episode of AOM in the Californian and Finnish trials was 7%, in the Czech trial as high as 34%, and the Pnc serotype-specific effect 57% in the Finnish and Czech trials, which both used tympanocenthesis and microbiological methods to verify the etiologic cause of AOM.

5.3.4. Nasopharyngeal Carriage

All PCV have proven to reduce nasopharyngeal carriage of vaccine serotype Pnc thus reducing transmission of Pnc. This was first demonstrated by Dagan et al. in 1996 with a heptavalent PCV, and thereafter repeatedly demonstrated by all the different PCV preparations in several different geographic and ethnic settings, although mostly in rich or middle income industrialized countries (Eskola et al., 2004).

5.3.5. Herd Immunity Effect

Vaccines which operate via reducing or blocking transmission of the pathogen have a potential to induce herd immunity on large scale in a population where targeted vaccination takes place. This is the case with PCV via its effect on reduction of nasopharyngeal carriage of vaccine type *Streptococcus pneumoniae* (O'Brien et al., 2003). Herd immunity has not traditionally been the focus for measuring vaccine effects in the pre-licensure phase of clinical development of any product. This is partly because the regulatory authorities in charge of granting a licence to a new vaccine have mostly been interested in individual rather than in population level effects. Thus the classical, individually randomized controlled trials have usually not been methodologically designed to measure the herd effect. In most cases, the herd effect appears after some years of large scale use. For the PCV, this is demonstrated by the experience in the U.S. where 7PCV was introduced in 2000, and the first signs of the 7PCV herd effect (reduction of total invasive Pnc disease in older, unimmunized age groups) was observable in 2003 (Lexau et al., 2005). By the end of 2004, the modeling based on nationwide invasive disease surveillance estimated that two thirds of all the invasive Pnc diseases were prevented by the herd effect (Center for Disease Control, 2005). This effect was especially marked in adults older than 50 years without co-morbidities (Lexau et al., 2005).

Whether carriage reduction, and consequently the herd effect of PCV on older age groups, is as pronounced in the resource poor, more densely populated areas as well as in areas with high prevalence of HIV, will remain to be seen. There already is an indication from the U.S., though, that especially in HIV-positives, the incidence of non-vaccine serotype *Streptococcus pneumoniae* increases (Lexau et al., 2005).

5.4. Introduction of Pneumococcal Vaccines into National Vaccination Programs

The Pnc polysaccharide vaccine is recommended for use in risk groups, including children above 2 years of age in most rich countries (Fedson et al., 2004; Pebody et al., 2005). How well the target groups are being reached is unclear as most countries do not have adequately functioning vaccine registries which could be linked to disease specific registries to understand the association between PPV vaccine coverage and age-specific effectiveness of PPV.

Presently very few countries have introduced PCV universally into their national vaccination program. First was the United States in the year 2000. This decision was backed up by favorable cost-effectivess predictions (Lieu et al., 2000). The first European country to introduce PCV on universal basis was Luxembourg in year 2004 (Pebody et al., 2005). Most other rich countries have recommendations for the use of PCV in risk groups only, the cost of which is covered either by the state or insurance. The reason for this selective approach is both the high cost of the vaccine and the uncertainty of the disease burden which could be prevented as well as concern about the impact of serotype replacement. Accordingly, more dynamic cost-effectiveness models have been constructed to allow for sensitivity analyses which take these different predictions into consideration (Melegaro et al., 2004).

None of the resource poor countries have introduced PCV into their national programs to date. The Global Alliance of Vaccines and Immunizations is working together with WHO to establish new funding mechanisms to cover the high cost of PCV introduction.

6. Future Directions

In summary, *Streptococcus pneumoniae* is a more common causative agent of childhood morbidity and mortality both in rich and resource poor countries than can be demonstrated by the best available data. PCV provides a powerful tool in the prevention of this morbidity and mortality globally. The fact that only one manufacturer presently produces the rather expensive 7-valent PCV is a cause for concern, especially since it is not ideally designed to match the pneumococcal serotypes causing majority of the disease burden globally. Randomized controlled trials take a lot of time and are expensive. Novel ways of thinking are needed to aid authorities in charge of registration and licensure of new vaccines. It has been proposed that serological correlates and other proxys like pneumococcal carriage reduction should be used for licensure purposes given the inherent problems of diagnostic methods to detect pneumococcal disease.

There is evidence that equivalent individual protection against IPD can be obtained by using fewer doses than are currently recommended (Lucero et al., 2004; Barzilay et al., 2005). However it is not known how many doses are needed to provide protection against pneumococcal carriage and thus induce herd immunity in different geographic regions and especially in areas with high rates of HIV infection.

References

Barzilay, E.J., O'Brien, K.L., Kwok, Y.S., Hoekstra, R.M., Zell, E.R., Reid, R., Santosham, M., Whitney, C.G., Feikin, D.R. (2005). Could a single dose of pneumococcal conjugate vaccine in children be effective? Modeling the optimal age of vaccination. *Vaccine* Sep 30; (Epub).

Black, S., Shinefield, H., Fireman, B., Lewis, E., Ray, P., Hansen, J.R., Elvin, L., Ensor, K.M., Hackell, J., Siber, G., Malinoski, F., Madore, D., Chang, I., Kohberger, R., Watson, W., Austrian, R., Edwards, K. (2000). Efficacy, safety and immunogenicity of heptavalent pneumococcal conjugate vaccine in children. Northern California Kaiser Permanente Vaccine Study Center Group. *Pediatr. Infect. Dis. J.* 19:187–95.

Black, S.B., Shinefield, H.R., Ling, S., Hansen, J., Fireman, B., Spring, D., Noyes, J., Lewis, E., Ray, P., Lee, J., Hackell, J. (2002). Effectiveness of heptavalent pneumococcal conjugate vaccine in children younger than five years of age for prevention of pneumonia. *Pediatr. Infect. Dis. J.* 21:810–15.

Centers for Disease Control and Prevention (CDC) (2005). Direct and indirect effects of routine vaccination of children with 7-valent pneumococcal conjugate vaccine on incidence of invasive pneumococcal disease–United States, 1998–2003. *MMWR Morb. Mortal. Wkly Rep.* 54:893–7.

Cherian, T., Mulholland, E.K., Carlin, J.B., Ostensen, H., Amin, R., de Campo, M., Greenberg, D., Lagos, R., Lucero, M., Madhi, S.A., O'Brien, K.L., Obaro, S., Steinhoff, M.C. (2005). Standardized interpretation of paediatric chest radiographs for the diagnosis of pneumonia in epidemiological studies. *Bull WHO.* 83:353–9.

Cutts, F.T., Zaman, S.M., Enwere, G., ym. (2005). Efficacy of nine-valent pneumococcal conjugate vaccine against pneumonia and invasive pneumococcal disease in The Gambia: randomised, double-blind, placebo-controlled trial. *Lancet.* 365:1139–46.

Dagan, R. (2004). The potential of pneumococcal conjugate vaccines to reduce antibiotic resistance. In: Pollard, A.J., McCracken, G.H., and Finn, A. (eds.) Hot topics in infection and immunity in children. *Advances in Experimental Medicine and Biology,* Volume 549. Kluwer Academic/ Plenum Publishers, New York, pp. 211–19.

Dagan, R., Shriker, O., Hazan, I., Leibovitz, E., Greenberg, D., Schlaeffer, F., Levy, R. (1998). Prospective study to determine clinical relevance of detection of pneumococcal DNA in sera of children by PCR. *J. Clin. Microbiol.* 36:669–73.

Dagan, R., Melamed, R., Muallem, M., Piglansky, L., Greenberg, D., Abramson, O., Mendelman, P.M., Bohidar, N., Yagupsky, P. (1996). Reduction of nasopharyngeal carriage of pneumococci during the second year of life by a heptavalent conjugate pneumococcal vaccine. *J. Infect. Dis.* 174:1271–8.

Eskola, J., Kilpi, T., Palmu, A., Jokinen, J., Haapakoski, J., Herva, E., Takala, A., Käyhty, H., Karma, P., Kohberger, R., Siber, G., Mäkelä, P.H.; Finnish Otitis Media Study Group. (2001). Efficacy of a pneumococcal conjugate vaccine against acute otitis media. *N. Engl. J. Med.* 344:403–9.

Eskola, J., Black, S.B., and Shinefield, H. (2004). Pneumococcal conjugate vaccine. In: Plotkin, S., Orenstein, W. (eds.), *Vaccines*, 4th edition, Saunders, Philadelphia, pp. 589–624.

Fedson, D.S., and Musher, D.M. (2004). Pneumococcal polysaccharide vaccine. In: Plotkin, S., Orenstein, W. (eds.), *Vaccines*, 4th edition, Saunders, Philadelphia, pp. 529–88.

Greenwood, B. (2005). Interpreting vaccine efficacy. *Clin. Infect. Dis.* 40:1519–20.

Hausdorff, W.P., Feikin, D.R., Klugman, K.P. (2005). Epidemiological differences among pneumococcal serotypes. *Lancet Infect. Dis.* 5:83–93.

Hausdorff, W.P. (2002). Invasive pneumococcal disease in children: geographic and temporal variations in incidence and serotype distribution. *Eur. J. Pediatr.* 161 Suppl 2:S135–9.

Huttunen, T., Teppo, A.M., Lupisan, S., Ruutu, P., Nohynek, H. (2003). Correlation between the severity of infectious diseases in children and the ratio of serum amyloid A protein and C-reactive protein. *Scand. J. Infect. Dis.* 35:488–90.

Kilpi, T., Åhman, H., Jokinen, J., Lankinen, K.S., Palmu, A., Savolainen, H., Grönholm, M., Leinonen, M., Hovi, T., Eskola, J., Käyhty, H., Bohidar, N., Sadoff, J.C., Mäkelä, P.H.; Finnish Otitis Media Study Group. (2003). Protective efficacy of a second pneumococcal conjugate vaccine against pneumococcal acute otitis media in infants and children: randomized, controlled trial of a 7-valent pneumococcal polysaccharide-meningococcal outer membrane protein complex conjugate vaccine in 1666 children. *Clin. Infect. Dis.* 37:1155–64.

Kleemola, M., Nokso-Koivisto, J., Herva, E., Syrjänen, R., Lahdenkari, M., Kilpi, T., Hovi, T. (2005). Is there any specific association between respiratory viruses and bacteria in acute otitis media of young children? *J. Infect.* Jun 28; ••.

Klugman, K.P., Madhi, S.A., Huebner, R.E., Kohberger, R., Mbelle, N., Pierce, N., and Vaccine Trialists Group. (2003). A trial of a 9-valent pneumococcal conjugate vaccine in children with and those without HIV infection. *N. Engl. J. Med.* 349:1341–8.

Lankinen, K.S., Ruutu, P., Nohynek, H., Lucero, M., Paton, J.C., Leinonen, M. (1999). Pneumococcal conjugate vaccine. In: Plotkin, S., Orenstein, W. (eds.), *Vaccines*, 4th edition, Saunders, Philadelphia, pp. 589–624 pneumonia diagnosis by demonstration of pneumolysin antibodies in precipitated immune complexes: a study in 350 Philippine children with acute lower respiratory infection. *Scand. J. Infect. Dis.* 31:155–61.

Levine, O.S., Cherian, T., Shah, R., and Batson, A. (2004). PneumoADIP: an example of translational research to accelerate pneumococcal vaccination in developing countries. *J. Health Popul. Nutr.* 22:268–74.

Levine, O.S., Lagos, R., Munoz, A., Villaroel, J., Alvarez, A.M., Abrego, P., and Levine, M.M. (1999). Defining the burden of pneumonia in children preventable by vaccination against Haemophilus influenzae type b. *Pediatr. Infect. Dis. J.* 18:1060–4.

Lexau, C.A., Lynfield, R., Danila, R., Pilishvili, T., Facklam, R., Farley, M.M., Harrison, L.H., Schaffner, W., Reingold, A., Bennett, N.M., Hadler, J., Cieslak, P.R., Whitney, C.G.; and Active Bacterial Core Surveillance Team. (2005). Changing epidemiology of invasive pneumococcal disease among older adults in the era of pediatric pneumococcal conjugate vaccine. *J. Amer. Med. Ass.* 294:2043–51.

Lipsitch, M. (2001). Interpreting results from trials of pneumococcal conjugate vaccines: a statistical test for detecting vaccine-induced increases in carriage of nonvaccine serotypes. *Am. J. Epidemiol.* 154:85–92.

Lieu, T.A., Ray, G.T., Black, S.B., Butler, J.C., Klein, J.O., Breiman, R.F., Miller, M.A., Shinefield, H.R. (2000). Projected cost-effectiveness of pneumococcal conjugate vaccination of healthy infants and young children. *J. Amer. Med. Ass.* 283:1460–8.

Lucero, M.G., Dulalia, V.E., Parreno, R.N., Lim-Quianzon, D.M., Nohynek, H., Mäkelä, H., and Williams, G. (2004). Pneumococcal conjugate vaccines for preventing vaccine-type invasive pneumococcal disease and pneumonia with consolidation on x-ray in children under two years of age. *Cochrane Database Syst. Rev.* 4:CD004977.

Lucero M.G., Puumalainen T., Ugpo J.M., Williams G., Käyhty H., Nohynek, H. (2004) Similar antibody concentrations in Filipino infants at age 9 months, after 1 or 3 doses of an adjuvanted, 11-valent pneumococcal diphtheria/tetanus-conjugated vaccine: a randomized controlled trial. *J Infect Dis.* 189:2077–84.

Lucero, M.G. Update on the Philippines ARIVAC effectiveness study of an 11-valent pneumococcal conjugate vaccine (11-PCV). (2005). Global Vaccine Research Forum Conference Abstract Book, Salvador de Bahia, Brazil, June 13, 2005.

Madhi, S.A., Kuwanda, L., Cutland, C., Klugman, K.P. (2005). The impact of a 9-valent pneumococcal conjugate vaccine on the public health burden of pneumonia in HIV-infected and -uninfected children. *Clin. Infect. Dis.* 40:1511–18.

Madhi, S.A., Heera, J.R., Kuwanda, L., and Klugman, K.P. (2005). Use of procalcitonin and C-reactive protein to evaluate vaccine efficacy against pneumonia. *PLoS Med.* 2:e38.

Madhi S.A., Klugman K.P.; Vaccine Trialist Group. (2004). A role for Streptococcus pneumoniae in virus-associated pneumonia. *Nat. Med.* 10:811–13.

Melegaro, A., and Edmunds, W.J. (2004). Cost-effectiveness analysis of pneumococcal conjugate vaccination in England and Wales. *Vaccine.* 22:4203–14.

Mulholland, E.K., Hilton, S., Adegbola, R.A., Usen, S., Oparaugo, A., Omosigho, C., Weber, M., Palmer, A., Schneider, G., Jobe, K., Lahai, G., Jaffar, S., Secka, O., Lin, K., Ethevenaux, C., and Greenwood, B. (1997). Randomized trial of Haemophilus influenzae type b-tetanus protein conjugate vaccine for prevention of pneumonia and meningitis in Gambian infants. *Lancet.* 349:1191–7.

O'Brien, K.L., Dagan, R. (2003). The potential indirect effect of conjugate pneumococcal vaccines. *Vaccine.* 21:1815–25.

O'Brien, K.L., Moulton, L.H., Reid, R., Weatherholtz, R., Oski, J., Brown, L., Kumar, G., Parkinson, A., Hu, D., Hackell, J., Chang, I., Kohberger, R., Siber, G., Santosham, M. (2003). Efficacy and safety of seven-valent conjugate pneumococcal vaccine in American Indian children: group randomised trial. *Lancet.* 362:355–61.

O'Brien, K.L., Nohynek, H., the WHO Pneumococcal Vaccine Trials Carriage Working Group (2003). Report from a WHO Working Group: Standard method for detecting upper respiratory carriage of *Streptococcus pneumoniae. Pediatr. Infect. Dis. J.* 22:133–40.

Okeke, I.N., Laxminarayan, R., Bhutta, Z.A., Duse, A.G., Jenkins, P., O'Brien, T.F., Pablos-Mendez, A., and Klugman, K.P. (2005). Antimicrobial resistance in developing countries. Part I: recent trends and current status. *Lancet. Infect. Dis.* 5:481–93.

Pebody, R.G., Leino, T., Nohynek, H., Hellenbrand, W., Salmaso, S., and Ruutu, P. (2005). Pneumococcal vaccination policy in Europe. *EuroSurveillance Monthly.* 10:10 (Epub).

Peltola, H., and Jaakkola, M. (1988). C-reactive protein in early detection of bacteremic versus viral infections in immunocompetent and compromised children. *J. Pediatr.* 113:641–6.

Pilishvili, T., Farley, M., Vazquez, M., Reingold, A., Nyquist, A., and Stefanek, K. Effectiveness of heptavalent pneumococcal conjugate vaccine in children. In: Proceedings of the 43rd Interscience Conference on Antimicrobial Agents and Chemotherapy Abstracts. Chicago, IL:American Society for Microbiology; 2003. p. 285.

Prat, C., Dominguez, J., Rodrigo, C., Gimenez, M., Azuara, M., Jimenez, O., Gali, N., and Ausina, V. (2003). Procalcitonin, C-reactive protein and leukocyte count in children with lower respiratory tract infection. *Pediatr. Infect. Dis. J.* 22:963–7.

Prymula, R., Peeters, P., Chrobok, V., Kriz, P., Novakova, E., Kohl, I., Lommel, P., Poolman, J., Prieels J-P., Schuerman, L. (2005). An eleven-valent pneumococcal-protein D conjugate (11Pn-PD) vaccine protects against otitis caused by both pneumococci or *Haemophilus influenzae. European Society for Paediatric Infectious Disease abstract book*, Valencia, Spain, May 2005.

Scott, J.A., Mlacha, Z., Nyiro, J., Njenga, S., Lewa, P., Obiero, J., Otieno, H., Sampson, J.S., and Carlone, G.M. (2005). Diagnosis of invasive pneumococcal disease among children in Kenya with enzyme-linked immunosorbent assay for immunoglobulin G antibodies to pneumococcal surface adhesin A. *Clin. Diagn. Lab. Immunol.* 12:1195–201.

Smith, T., Lehmann, D., Montgomery, J., Gratten, M., Riley, I.D., and Alpers, M.P. (1993). Acquisition and invasiveness of different serotypes of *Streptococcus pneumoniae* in young children. *Epidemiol. Infect.* 1:27–39.

Syrjänen, R., Auranen, K., Leino, T., Kilpi, T., and Mäkelä, P.H. (2005). Pneumococcal acute otitis media in relation to pneumococcal nasopharyngeal carriage. *Pediatr. Infect. Dis. J.* 24:801–6.

Vuori-Holopainen, E., Salo, E., Saxen, H., Hedman, K., Hyypiä, T., Lahdenperä, R., Leinonen, M., Tarkka, E., Vaara, M., and Peltola, H. (2002). Etiological diagnosis of childhood pneumonia by use of transthoracic needle aspiration and modern microbiologicalmethods. *Clin. Infect. Dis.* 34:583–90.

Vuori-Holopainen, E., and Peltola, H. (2001). Reappraisal of lung tap: review of an old method for better etiologic diagnosis of childhood pneumonia. *Clin. Infect. Dis.* 32:715–26.

World Health Organization Pneumonia Vaccine Trial Investigators' Group (2001). Standardization of interpretation of chest radiographs for the diagnosis of pneumonia in children. WHO/V&B/01.35, Geneva.

Wright, A.E., and Morgan, W.P. (1914). Observations on prophylactic inoculation against pneumococcal infections, and on the results which have been achieved by it. *Lancet.* i, 1–10.

13

New Antibiotics for Gram-Positive Infections

John S. Bradley

1. Introduction

Gram-positive pathogens cause a considerable number of community-associated and nosocomial infections in children. The increase of antibiotic resistance in Gram-positive cocci to commonly used antibiotic agents has created a challenge to physicians in developing recommendations for both empiric therapy of infections as well as definitive therapy once culture results are available (Appelbaum and Jacobs, 2005; Hancock, 2005; Lundstrom, 2004; Stein, 2005). For hospitalized children who are immune-suppressed, nosocomial infections with antibiotic-resistant organisms can be life-threatening (Brady, 2005; Clark et al., 2004; Jarvis, 2004; Pong and Bradley, 2004; Saiman, 2002).

A number of new antimicrobial agents have been approved by regulatory agencies or have entered into prospective, comparative trials during the past five years. Unfortunately, little information regarding pediatric pharmacokinetics, safety and efficacy are available on most of these newer agents. This chapter will review the need for these agents based on current antibiotic resistance in Gram-positive pathogens, and summarize information on new antibiotics that are in phase 2 trials (prospective, noncomparative treatment trials) or beyond, in adults. The potential clinical use of these agents is discussed.

2. The Pathogens

Staphylococcus aureus has long been associated with infections in children, both community-associated and hospital acquired. Penicillin-resistance which is beta-lactamase mediated, has been widespread for many years. Nosocomial methicillin-resistance (MRSA) caused by altered transpeptidase binding sites was

Hot Topics in Infection and Immunity in Children, edited by Andrew J. Pollard and Adam Finn. Springer, New York, 2006

first reported 30 years ago. However, these nosocomial strains of MRSA did not appear to survive well in the community, competing poorly with methicillin-susceptible strains. However, the recent emergence around the world of community-associated methicillin-resistant strains of *Staphylococcus aureus* (CA-MRSA), apparently associated with fewer antibiotic-resistance genes and often associated with the production of Panton-Valentine leukocidin and perhaps other virulence genes, has occurred at an alarming rate (Holmes et al., 2005; Kaplan, 2005; Kuehnert et al., 2005; Moran et al., 2005; Robert et al., 2005). These organisms appear well-adapted to a community setting, and have spread quickly in both pediatric and adult populations. While vancomycin (and teichoplanin in several countries) has been the mainstay of parenteral antibiotic therapy of these organisms, vancomycin heteroresistance (decreased susceptibility noted within subpopulations of *S. aureus*) and complete resistance (due to vanA genes) have been reported (Cosgrove et al., 2004; Ruef, 2004). In addition, resistance to macrolides has been shown to occur by at least two mechanisms. Methylation of the ribosomal binding site confers resistance not only to macrolides, but also to clindamycin and the quinupristin (a streptogramin B) component of the combination antibiotic containing both quinupristin and dalfopristin (Synercid®) (Woodford, 2005). Efflux pumps are also important in some organisms, leading to the expulsion of macrolides from within the cell (Klaassen and Mouton, 2005). Both mechanisms may be present in the pathogen simultaneously. The need for safe and effective therapy for children that is available in both parenteral and oral formulations is very clear.

Coagulase-negative staphylococci are primarily nosocomial pathogens and represent the most commonly isolated blood-borne pathogens in infants hospitalized in neonatal intensive care units and in immune-suppressed children with indwelling central venous catheters (Weisman, 2004). These organisms are most often resistant to all beta-lactams, often resistant to aminoglycosides, and many are susceptible only to glycopeptide antibiotics such as vancomycin.

Streptococcus pneumoniae has developed increasing resistance to multiple classes of antibiotics over the past decade (Jacobs, 2004). Despite the major success of conjugate vaccines for bacteremic pneumococcal infection in the developed world, the vaccine has had much less impact on otitis media, and only a modest effect on the incidence of pneumonia in children (Bogaert et al., 2004; Lucero et al., 2004; Posfay-Barbe and Wald, 2004). Resistance to beta-lactams is caused by alteration of bacterial transpeptidases (also known as penicillin-binding proteins, or PBP's); resistance to macrolides occurs by the mechanisms very similar to those outlined above for *S. aureus*; resistance to a bactericidal effect of vancomycin has been reported by defective autolysis pathway genes (Mitchell and Tuomanen, 2001); resistance to fluoroquinolones has occurred by alterations in bacterial DNA gyrase and topoisomerase genes (Jacobs, 2004).

Enterococci, particularly *Enterococcus faecium*, are increasingly resistant to multiple antibiotics, including ampicillin, aminoglycosides, and more recently, vancomycin. Although most reported outbreaks primarily involve adult populations, nosocomial infections involving infants have also been reported (Graham, 2002). The vanA gene mechanism of resistance noted in *E. faecalis* was recently described in clinical infections caused by *S. aureus*, as noted above.

Streptococcus pyogenes has remained remarkably susceptible to penicillin; resistance has not occurred in the many decades during which isolates have been collected. While intermittent increases in erythromycin resistance have occurred in many areas throughout the world (Green et al., 2005; Syrogianopoulos et al., 2001), they may be linked to increased, widespread use of macrolide-class antibiotics.

3. Older Antibiotics

Antibiotic therapy until the most recent decade has focused on beta-lactams (penicillins, cephalosporins, cephamycins, monobactams, and carbapenems), aminoglycosides (gentamicin, tobramycin, amikacin), tetracyclines, glycopeptides (vancomycin, teichoplanin), and lincosamides (clindamycin). Erythromycin antibiotics (the macrolide class) have been available for many decades; clarithromycin and azithromycin first became available in the United States 15 years ago, and for purposes of this review are considered older therapies. Quinupristin/dalfopristin (Synercid®) was approved for adults pre-2000 to treat vancomycin-resistant *E. faecium* infections, and will not be discussed in this chapter.

Although fluoroquinolones have been used for adults during the past 20 years, they did not enter into pediatric investigation until 1997 as the need for new antibiotics to treat antibiotic-resistant pneumococcus causing meningitis and otitis media became clear. At present, the only fluoroquinolone approved for use in children, ciprofloxacin, has relatively poor Gram-positive activity, and should not be used for either staphylococcal or streptococcal infections. Information on the value of the newer fluoroquinolones with enhanced pneumococcal activity is reviewed below.

4. Newer Antibiotics

New agents which are active against Gram-positive pathogens are in various stages of development and approval. While many targets have been identified, few candidate compounds have actually progressed through animal studies to phase 1 pharmacokinetic and toxicologic studies in humans. Even fewer antibiotics have reached phase 3 trials where extensive comparative data are collected on safety and efficacy, leading to government approvals.

4.1. Linezolid

Several antibiotics in the oxazolidinone class are under investigation, with linezolid being the first to receive approval by regulatory agencies (see Figure 13.1). The oxazolidinones are protein synthesis inhibitors which bind to the to 23S rRNA of the 50S ribosomal subunit at a site distinct from chloramphenicol and the lincosamides, and prevent formation of the 70S initiation complex (Moellering, 2003). Linezolid is active against a wide range of Gram-positive cocci, including streptococci which are penicillin- and macrolide-resistant. Furthermore, it is active against both coagulase-positive and -negative staphylococci, including those which are

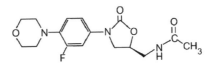

Figure 13.1.   Linezolid.

methicillin- and/or vancomycin-resistant. Linezolid has bacteriostatic activity against susceptible strains of *Staphylococcus aureus* compared to the beta-lactams, glycopeptides, and lipopeptides. Only limited activity exists against Gram-negative organisms.

Pharmacokinetic data are available in children from all age groups, including premature neonates (Jungbluth et al., 2003). Linezolid is available in both parenteral and oral formulations. The oral bioavailability is virtually 100%, allowing clinicians to use an oral dose identical to that which is administered intravenously. Data suggest that absorption of linezolid following ingestion is somewhat delayed in those who are fed concurrently, compared with those who receive antibiotic without food, although the child's total drug exposure is not significantly effected. Protein binding in serum is 31%; the drug is cleared by the kidneys, both unchanged and after oxidation of the parent compound, with an elimination half-life of about 3 hours. Oxidation is the primary mechanism of inactivation of linezolid, therefore no dose reduction is recommended for children with renal insufficiency, and cytochrome P450 system drug-drug interactions are not of concern.

Linezolid has been well-studied in children with documented Gram-positive infections (Kaplan et al., 2003a; Kaplan et al., 2003b; Meissner et al., 2003; Tan, 2004; Yogev et al., 2003). The overall clinical response rates for nosocomial and community-acquired pneumonia and for complicated and uncomplicated skin and skin structure infections were equivalent to comparator agents. The response rates for infections caused by *Staphylococcus aureus* (both methicillin-sensitive and methicillin-resistant strains) were statistically equivalent to active comparator agents. In the studies reported, no vancomycin-intermediate or – resistant strains of *Staphylococcus aureus* were isolated; therefore, clinical efficacy of linezolid against vancomycin non-susceptible strains was not evaluated. In these studies, the rates of clinical and laboratory adverse events between linezolid treated children and vancomycin treated control patients were equivalent. Specifically, the hematologic toxicity profile for both neutropenia and thrombocytopenia were statistically similar to vancomycin in these pediatric studies. From adult reports, however, rare, reversible myelosuppression has been noted. Earlier concerns during the phase 1–2 trials regarding the possibility of hypertension in children receiving linezolid who were also on diets high in tyramine do not appear to have been justified.

Linezolid appears to be effective therapy for pneumococcal and staphylococcal infections in situations in which the resistance profile of the pathogen precludes other agents, or the child cannot tolerate other active agents. The availability of oral therapy for multiply resistant organisms allows for outpatient therapy of serious infections which otherwise would require ongoing parenteral therapy.

4.2. Fluoroquinolones

Fluoroquinolone antibiotics with enhanced Gram-positive activity primarily targeted at *Streptococcus pneumoniae* have been available for use in adults since 1996, but pediatric clinical trials did not start with any enhanced-activity fluoroquinolones for either respiratory tract or pneumococcal meningitis until a few years later. Early animal toxicology studies demonstrated a dose-dependent cartilage toxicity, primarily in weight-bearing joints in certain juvenile animals. Until the need for fluoroquinolones for resistant pathogens was demonstrated, no significant clinical investigation had been undertaken in infants and children outside of those treated for cystic fibrosis. Currently, published or presented data exist for the efficacy of trovafloxacin for meningitis (Saez-Llorens et al., 2002) (although this antibiotic may not be widely available), gatifloxacin for acute otitis media, and levofloxacin for otitis media and community-acquired pneumonia (Saez-Llorens et al., 2005; Sher et al., 2005; Soley et al., 2005; Abstract 1297 of the 45th Interscience Conference, 2005). None of these antibiotics are currently approved for pediatric indications. However, the manufacturer of levofloxacin is currently moving forward with plans for FDA approval, and a suspension formulation of levofloxacin is currently available in the United States. From presented data, the efficacy of levofloxacin and gatifloxacin for respiratory tract infections is equivalent to comparator agents. With respect to safety, acknowledging the limitations of relatively small numbers of children who have been treated with fluoroquinolones, no credible tendon/joint toxicity signal has yet been detected.

The fluoroquinolones offer the potential for bactericidal oral therapy of invasive pneumococcal infections in situations in which beta-lactams, macrolides, and clindamycin cannot be used. The small and as yet undefined risk of transient, reversible cartilage toxicity is likely to be offset by the benefits for this very small population of children with extremely difficult to treat pneumococcal infections. As the safety of the fluoroquinolones becomes better defined, the number of infected children who may benefit from this class of antibiotic may increase.

4.3. Daptomycin

Structurally, daptomycin is a lipopeptide which contains 13 amino acids linked together in a circular configuration, with the addition of a lipophilic tail which inserts into the cell membrane (see Figure 13.2). The mechanism of action of daptomycin is poorly understood, but it appears that depolarization of the membrane occurs as the antibiotic polymerizes, leading to the formation of channels which result in leakage of cell contents, inhibition of protein, DNA and RNA synthesis, and cell death (Carpenter, 2004; Hancock, 2005; Tedesco and Rybak, 2004). The antibiotic is rapidly bactericidal, with concentration-dependent activity. Daptomycin is highly protein bound (92%), with a prolonged serum half-life of 8.1 hours which provides justification for once daily administration. The metabolism of daptomycin has not been well defined; 52% of the antibiotic is found in the urine as unchanged drug during the first 6 days following administration. The dosage should be reduced in patients with decreased renal function, or in those receiving hemo- or

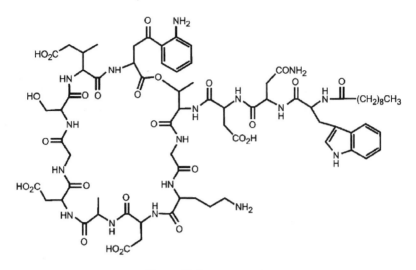

Figure 13.2. Daptomycin.

peritoneal dialysis. Given the unique mechanism of action, daptomycin is active against many vancomycin-resistant pathogens, including *Staphylococcus aureus* and coagulase-negative staphylococci, as well as enterococci and pneumococci.

Although the antibiotic is reasonably well tolerated in current clinical studies, early experience with the antibiotic when administered every 8 hours suggested skeletal muscle toxicity. However, using a dosing schedule from every 12 to every 24 hours, elevated serum concentrations of creatine phosphokinase occur in only 2.8% of prospectively studied adults, compared to 1.8% of vancomycin-treated control patients. The mechanism of muscle toxicity is unknown, but current guidelines suggest following CPK concentrations on a weekly basis. Adverse events which occurred in greater than 5% of adults include constipation, nausea, headache, diarrhea, vomiting, and injection site reactions, although these side effects occurred in control patients with a statistically similar frequency (Stein, 2005; Tedesco and Rybak, 2004).

The antibiotic is approved for skin and skin structure infections in adults. No data are available in children. Studies in adults with staphylococcal pneumonia suggested clinical response rates which were not equivalent to vancomycin, raising the possibility that the distribution of unbound antibiotic in bronchial-alveolar lining fluid and lung parenchyma was insufficient to achieve the required antibiotic exposure in those tissues (Silverman et al., 2005).

4.4. New Glycopeptides

The glycopeptide antibiotics vancomycin and teichoplanin (not approved for use in the US) have been available for many years. With the increasing need for antibiotics with activity against vancomycin-nonsusceptible staphylococci and enterococci, chemical modifications to the basic glycopeptide structure have produced a series of new compounds with altered pharmacokinetics and/or enhanced activity against Gram-positive pathogens.

Oritavancin (see Figure 13.3), like vancomycin, inhibits both transglycosylation and transpeptidation in bacteria by binding to the terminal D-ala-D-ala portion of the pentapeptide building blocks preventing elongation and cross-linking of peptidoglycan polymers during the formation of the cell wall. However, unlike vancomycin, oritavancin becomes a dimer and is anchored to the cell membrane by a lipophilic side chain, enhancing binding to the pentapeptide precursor. Oritavancin demonstrates concentration-dependent bactericidal activity. It is highly protein-bound (86%), which may play a role in the extended serum beta-elimination half life of 200–300 hours, allowing the antibiotic to be administered daily for 3–7 days to achieve antibiotic exposure for an entire treatment course, or to be administered once weekly. It is active against both MRSA and VRE as well as VRSA. Clinical efficacy in therapy of skin and skin structure infections has been shown to be equivalent to vancomycin. The side effect profile is also statistically equivalent to vancomycin from data presented to date (Bhavnani et al., 2004; Guay, 2004).

Dalbavancin (see Figure 13.4) displays many characteristics of the other glycopeptides in the inhibition of transglycosylation (and indirectly, transpeptidation), but appears to exert additional antibiotic effects similar to that of the lipopeptides in destabilizing the bacterial cell membrane, providing some degree of dual mechanism activity. The dalbavancin is active against staphylococci, including methicillin-resistant strains, as well as streptococci. With respect to enterococci, however, *E. faecalis* stains which carry the vanA resistance genes are not susceptible, in contrast to *E. faecium* strains which carry the vanB resistance determinant. The antibiotic demonstrates concentration-dependent bactericidal activity, is highly protein-bound (98%) with a prolonged serum beta half life of 257 hours using a

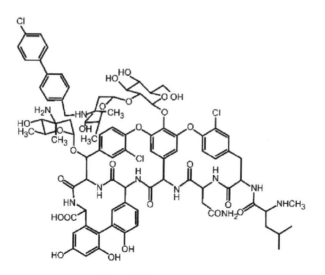

Figure 13.3. Oritavancin.

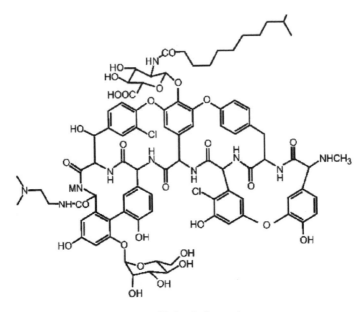

Figure 13.4. Dalbavancin.

renal excretion pathway of drug elimination. Dosing for skin and skin structure infections in adults has occurred once weekly in current clinical trials. Safety assessments in phase 2 trials suggest an equivalent profile to vancomycin, but patient numbers are too small for statistically valid conclusions. Dalbavancin may be useful in the therapy of MRSA and some strains of VRE, and weekly parenteral therapy will be of great value in facilitating outpatient parenteral therapy of serious infections (Leighton et al., 2004; Raad et al., 2005; Seltzer et al., 2003; Streit et al., 2005).

Telavancin (see Figure 13.5) also displays multiple mechanisms of action, including both inhibition of transglycosylation in cell wall synthesis, as well as destabilization of the bacterial cell membrane. The activity of telavancin is also similar to other extended spectrum glycopeptides in that it includes virtually all strains of MRSA, VRSA, VRE, as well as strains of pneumococcus. Telavancin exhibits concentration-dependent bactericidal activity, with a serum half-life of 7–9 hours which justified once daily dosing in clinical trials. In limited adult studies of skin and skin structure infection, telavancin was as effective as comparators, with fewer side effects compared to vancomycin. The histamine-release side effects seen with more rapid infusions of vancomycin were also noted with telavancin. As with the other new glycopeptides, telavancin may offer an effective and safe alternative to vancomycin in the treatment of multiply-resistant Gram positive pathogens (Hegde et al., 2004; Higgins et al., 2005; King et al., 2004; Shaw et al., 2005; Stryjewski et al., 2005).

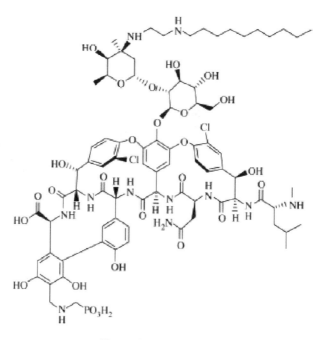

Figure 13.5. Telavancin.

4.5. Ceftobiprole

Beta-lactam antibiotics have compiled an excellent safety profile in pediatrics during the past four decades. Both enzymatic degradation and altered transpeptidase binding sites have resulted in resistance to penicillins, cephalosporins, and carbapenems. However, with structural modifications of the beta-lactam molecule, both of these mechanisms of resistance in *Staphylococcus aureus* which produce MRSA have been overcome (see Figure 13.6). Ceftobiprole is stable to the current staphylococcal beta-lactamases, similar to many cephalosporins. However, unlike any of the currently available beta-lactam antibiotics, ceftobiprole is also capable of binding to the PBP2a transpeptidase to inhibit growth of *Staphylococcus aureus*. It also

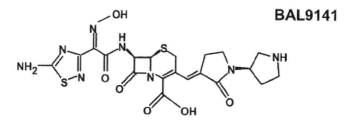

Figure 13.6. Ceftobiprole.

exhibits extended activity against Gram-negative pathogens, closely resembling that of cefepime. It is bactericidal in ways characteristic of beta-lactams, with antibacterial activity based on time-above-MIC pharmacodynamics. It demonstrates a serum elimination half-life of 3.4 hours, protein binding of 38%, and is excreted by the kidney, suggesting a dosing interval of every 12 hours. Thus far, the toxicity profile is similar to other cephalosporins. Should subsequent clinical trials support the development of ceftobiprole for clinical use, the safety profile of cephalosporins for parenteral therapy of serious MRSA infections would be an advantage for children (Issa et al., 2004; Jones et al., 2002; Kosowska et al., 2005; Schmitt-Hoffmann et al., 2004).

4.6. Telithromycin

The increase in macrolide resistance in pneumococcus, group A streptococcus, and *Staphylococcus aureus* has lead to discovery efforts to produce a macrolide antibiotic which is not affected by the alterations in the binding site which leads to macrolide-streptogramin B-lincosamide resistance. Substitutions on the 14-member macrolide structure have produced the ketolides (see Figure 13.7). One compound of this class has recently been approved for use in respiratory tract infections in

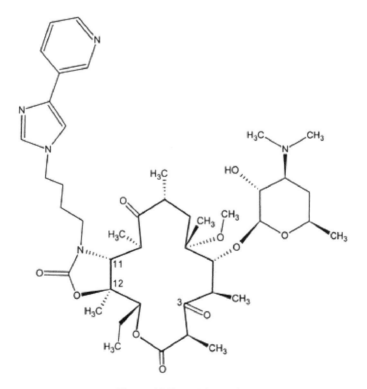

Figure 13.7. Telithromycin.

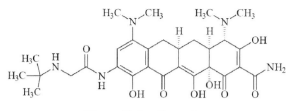

Figure 13.8. Tigecycline.

adults. Telithromycin binds to the ribosome at two sites. Erythromycin binds at a single site in domain V, which can be methylated to prevent macrolide binding. However, a second binding site of telithromycin in domain II allows the antibiotic to persist in binding to the active site, thus preventing protein synthesis. The antibiotic is bactericidal, with pharmacodynamics activity best described by the ratio of the antibiotic AUC (area under the serum antibiotic concentration vs. time curve) to the MIC of the organism. The antibiotic is only available in an oral form, with excellent bioavailability of 57%. Telithromycin is metabolized by the cytochrome P450 system (both CYP 3A4 and non-CYP3A4 isoenzymes), yielding a serum half-life of 9.8 hours. Protein binding is moderate at 65%. The drug is well-tolerated in adults, with diarrhea and nausea occurring in 8–10% of those in prospective clinical trials involving respiratory tract infections including pneumonia and sinusitis. Telithromycin is currently approved for use in adults in certain countries, but data on safety and efficacy in children are lacking (Fogarty et al., 2005; Roos et al., 2005; Shi et al., 2005; Wellington and Noble, 2004).

4.7. Tigecycline

Those who treat children are well aware of the toxicities of tetracycline antibiotics, leading to a general contraindication for their use in children less than 8 years of age. However, as with any antibiotic, the benefits of therapy may outweigh the risks in certain situations. Tigecycline is a tetracycline derivative, administered intravenously, with a broader spectrum of activity than seen with most previously available agents in this class. The mechanism of action and the toxicities, including staining of bones in experimental animals, is present with tigecycline as with other tetracyclines. The antibiotic is not being actively pursued in children at this time, but was recently approved for use in adults for complicated skin and skin structure infections, and complicated intraabdominal infections. Tigecycline has activity against most Gram-positive cocci, including methicillin-resistant *Staphylococcus aureus* and enterococcus, in addition to demonstrating activity against *Pseudomonas aeruginosa*. The structure of the antibiotic is very similar to minocycline, with the addition of a glycylamido moiety (see Figure 13.8). The antibiotic is bacteriostatic, with relatively high protein binding (80%) and relatively long serum elimination half-life of 30 hours, being excreted primarily into bile (60%), but with significant renal excretion (35%). Dosing every 12 hours was investigated in clinical trials. As with other tetracyclines, it inhibits protein synthesis in bacteria by binding to the 30S ribosomal subunit, blocking entry of amino-acyl tRNA molecules into the ribosome. For the child with a life-threatening infection whose organism is resistant to alternative therapy, or in the child who cannot tolerate alternative therapy, tigecycline should provide a reasonable option (Babinchak et al., 2005; Ellis-Grosse et al., 2005; Hoban et al., 2004; Livermore, 2005; Meagher et al., 2005; Noskin, 2005).

5. In the Pipeline

Many candidate antibiotics are being evaluated prior to entering clinical trials. Once a chemical entity is shown to have in vitro activity against Gram-positive pathogens, animal studies on drug handling and toxicology set the stage for human phase 1 clinical trials in adults, in which preliminary dosing parameters and safety are established. Trials then begin to investigate efficacy under well-defined conditions of infection in otherwise healthy adults. As most of the clinical drug development occurs in the pharmaceutical industry, public knowledge of which drugs are currently under investigation, particularly in early-stage development, is limited. However, for some smaller companies, favorable news of drug development often attracts the large investments required at latter stages of clinical drug development. A recent review of small pharmaceutical company drug development programs provides some degree of optimism that, for the 41 companies listed, the drug pipeline is not "dry," and includes many new chemical entities rather than modifications of previously available antibiotics (Boggs and Miller, 2004).

References

Abstract 1297 of the 45th Interscience Conference on Antimicrobial Agents and Chemotherapy, Dec 15–18, 2005, Washington DC, USA.

Appelbaum, P.C., Jacobs, M.R. (2005). Recently approved and investigational antibiotics for treatment of severe infections caused by Gram-positive bacteria. *Curr. Opin. Microbiol.*

Babinchak, T., Ellis-Grosse, E., Dartois, N., Rose, G.M., Loh E. (2005). The efficacy and safety of tigecycline for the treatment of complicated intra-abdominal infections: analysis of pooled clinical trial data. *Clin. Infect. Dis.* 41 Suppl 5:S354–67.

Bhavnani, S.M., Owen, J.S., Loutit, J.S., Porter, S.B., Ambrose, P.G. (2004). Pharmacokinetics, safety, and tolerability of ascending single intravenous doses of oritavancin administered to healthy human subjects. *Diagn. Microbiol. Infect. Dis.* 50(2):95–102.

Bogaert, D., Hermans, P.W., Adrian, P.V., Rumke, H.C., de Groot, R. (2004). Pneumococcal vaccines: an update on current strategies. *Vaccine* 22(17–18):2209–20.

Boggs, A.F., Miller, G.H. (2004). Antibacterial drug discovery: is small pharma the solution? *Clin. Microbiol. Infect.* 10 Suppl 4:32–6.

Brady, M.T. (2005). Health care-associated infections in the neonatal intensive care unit. *Am. J. Infect. Control* 33(5):268–75.

Carpenter, C.F., Chambers, H.F. (2004). Daptomycin: another novel agent for treating infections due to drug-resistant gram-positive pathogens. *Clin. Infect. Dis.* 38(7):994–1000.

Clark, R., Powers, R., White, R., Bloom, B., Sanchez, P., Benjamin, D.K., Jr. (2004). Prevention and treatment of nosocomial sepsis in the NICU. *J. Perinatol.* 24(7):446–53.

Cosgrove, S.E., Carroll, K.C., Perl, T.M. (2004). Staphylococcus aureus with reduced susceptibility to vancomycin. *Clin. Infect. Dis.* 39(4):539–45.

Ellis-Grosse, E.J., Babinchak, T., Dartois, N., Rose, G., Loh, E. (2005). The efficacy and safety of tigecycline in the treatment of skin and skin-structure infections: results of 2 double-blind phase 3 comparison studies with vancomycin-aztreonam. *Clin. Infect. Dis.* 41 Suppl 5:S341–53.

Fogarty, C.M., Patel, T.C., Dunbar, L.M., Leroy, B.P. (2005). Efficacy and safety of telithromycin 800 mg once daily for 7 days in community-acquired pneumonia: an open-label, multicenter study. *BMC Infect. Dis.* 5(1):43.

Graham, P.L., 3rd. (2002). Staphylococcal and enterococcal infections in the neonatal intensive care unit. *Semin. Perinatol.* 26(5):322–31.

Green, M., Allen, C., Bradley, J., Dashefsky, B., Gilsdorf, J.R., Marcon, M.J., Schultze, G.E., Smith, C., Walter, E., Martin, J.M., Edwards, K.A., Barbadora, K.A., Rumbaugh, R.M., Wald, E.R. (2005). In vitro activity of telithromycin against macrolide-susceptible and macrolide-resistant pharyngeal isolates of group A streptococci in the United States. *Antimicrob. Agents Chemother.* 49(6):2487–9.

Guay, D.R. (2004). Oritavancin and tigecycline: investigational antimicrobials for multidrug-resistant bacteria. *Pharmacotherapy* 24(1):58–68.

Hancock, R.E. (2005). Mechanisms of action of newer antibiotics for Gram-positive pathogens. *Lancet Infect. Dis.* 5(4):209–18.

Hegde, S.S., Reyes, N., Wiens, T., Vanasse, N., Skinner, R., McCullough, J., Kaniga, K., Pace, J., Thomas, R., Shaw, J.P., Obedencio, G., Judice, J.K. (2004). Pharmacodynamics of telavancin (TD-6424), a novel bactericidal agent, against gram-positive bacteria. *Antimicrob. Agents Chemother.* 48(8):3043–50.

Higgins, D.L., Chang, R., Debabov, D.V., Leung, J., Wu, T., Krause, K.M., Sandvik, E., Hubbard, J.M., Kaniga, K., Schmidt, D.E.Jr., Gao, Q., Cass, R.T., Karr, D.E., Benton, B.M., Humphrey, P.P. (2005). Telavancin, a multifunctional lipoglycopeptide, disrupts both cell wall synthesis and cell membrane integrity in methicillin-resistant Staphylococcus aureus. *Antimicrob. Agents Chemother.* 49(3):1127–34.

Hoban, D.J., Bouchillon, S.K., Johnson, B.M., Johnson, J.L., Dowzicky, M.J. (2004). In vitro activity of tigecycline against 6792 Gram-negative and Gram-positive clinical isolates from the global Tigecycline Evaluation and Surveillance Trial (TEST Program, 2004). *Diagn. Microbiol. Infect. Dis.* 52(3):215–27.

Holmes, A., Ganner, M., McGuane, S., Pitt, T.L., Cookson, B.D., Kearns, A.M. (2005). Staphylococcus aureus isolates carrying Panton-Valentine leucocidin genes in England and Wales: frequency, characterization, and association with clinical disease. *J. Clin. Microbiol.* 43(5): 2384–90.

Issa, N.C., Rouse, M.S., Piper, K.E., Wilson, W.R., Steckelberg, J.M., Patel, R. (2004). In vitro activity of BAL9141 against clinical isolates of gram-negative bacteria. *Diagn. Microbiol. Infect. Dis.* 48(1):73–5.

Jacobs, M.R. (2004). Streptococcus pneumoniae: epidemiology and patterns of resistance. *Am. J. Med.* 117 Suppl 3A:3S–15S.

Jarvis, W.R. (2004). Controlling healthcare-associated infections: the role of infection control and antimicrobial use practices. *Semin. Pediatr. Infect. Dis.* 15(1):30–40.

Jones, R.N, Deshpande, L.M., Mutnick, A.H., Biedenbach, D.J. (2002). In vitro evaluation of BAL9141, a novel parenteral cephalosporin active against oxacillin-resistant staphylococci. *J. Antimicrob. Chemother.* 50(6):915–32.

Jungbluth, G.L, Welshman, I.R., Hopkins, N.K. (2003). Linezolid pharmacokinetics in pediatric patients: an overview. *Pediatr. Infect. Dis. J.* 22(9 Suppl):S153–7.

Kaplan, S.L. (2005). Treatment of community-associated methicillin-resistant Staphylococcus aureus infections. *Pediatr. Infect. Dis. J.* 24(5):457–8.

Kaplan, S.L., Afghani, B., Lopez, P., Wu, E., Fleishaker, D., Edge-Padbury, B., Naberhuis-Stehouwer, S., Bruss, J.B. (2003a). Linezolid for the treatment of methicillin-resistant Staphylococcus aureus infections in children. *Pediatr. Infect. Dis. J.* 22(9 Suppl):S178–85.

Kaplan, S.L., Deville, J.G., Yogev, R., Morfin, M.R., Wu, E., Adler, S., Edge-Padbury, B., Naberhuis-Stehouwer, S., Bruss, J.B., Linezolid pediatric Study Group (2003b). Linezolid versus vancomycin for treatment of resistant Gram-positive infections in children. *Pediatr. Infect. Dis. J.* 22(8):677–86.

King, A., Phillips, I., Kaniga, K. (2004). Comparative in vitro activity of telavancin (TD-6424), a rapidly bactericidal, concentration-dependent anti-infective with multiple mechanisms of action against Gram-positive bacteria. *J. Antimicrob. Chemother.* 53(5):797–803.

Klaassen, C.H., Mouton, J.W. (2005). Molecular detection of the macrolide efflux gene: to discriminate or not to discriminate between mef(A) and mef(E). *Antimicrob. Agents Chemother.* 49(4):1271–8.

Kosowska, K., Hoellman, D.B., Lin, G., Clark, C., Credito, K., McGhee, P., Dewasse, B., Bozdogan, B., Shapiro, S., Appelbaum, P.C. (2005). Antipneumococcal activity of ceftobiprole, a novel broad-spectrum cephalosporin. *Antimicrob. Agents Chemother.* 49(5):1932–42.

Kuehnert, M.J., Hill, H.A., Kupronis, B.A., Tokars, J.I., Solomon, S.L., Jernigan, D.B. (2005). Methicillin-resistant-Staphylococcus aureus hospitalizations, United States. *Emerg. Infect. Dis.* 11(6):868–72.

Leighton, A., Gottlieb, A.B., Dorr, M.B., Jabes, D., Mosconi, G., VanSaders, C., Mroszczak, E.J., Campbell, K.C., Kelly, E. (2004). Tolerability, pharmacokinetics, and serum bactericidal activity of intravenous dalbavancin in healthy volunteers. *Antimicrob. Agents Chemother.* 48(3):940–5.

Livermore, D.M. (2005). Tigecycline: what is it, and where should it be used? *J. Antimicrob. Chemother.*

Lucero, M.G., Dulalia, V.E., Parreno, R.N., Lim-Quianzon, D.M., Nohynek, H., Makela, H., Williams, G. (2004). Pneumococcal conjugate vaccines for preventing vaccine-type invasive pneumococcal disease and pneumonia with consolidation on x-ray in children under two years of age. *Cochrane Database Syst. Rev.* (4):CD004977.

Lundstrom, T.S., Sobel, J.D. (2005). Antibiotics for gram-positive bacterial infections: vancomycin, quinupristin-dalfopristin, linezolid, and daptomycin. *Infect. Dis. Clin. North Am.* 18(3):651–68.

Meagher, A.K., Ambrose, P.G., Grasela, T.H., Ellis-Grosse, E.J. (2005). The pharmacokinetic and pharmacodynamic profile of tigecycline. *Clin. Infect. Dis.* 41 Suppl 5:S333–40.

Meissner, H.C., Townsend, T., Wenman, W., Kaplan, S.L., Morfin, M.R., Edge-Padbury, B., Naberhuis-Stehouwer, S., Bruss, J.B. (2003). Hematologic effects of linezolid in young children. *Pediatr. Infect. Dis. J.* 22(9 Suppl):S186–92.

Mitchell, L., Tuomanen, E. (2001). Vancomycin-tolerant Streptococcus pneumoniae and its clinical significance. *Pediatr. Infect. Dis. J.* 20(5):531–3.

Moellering, R.C. (2003). Linezolid: the first oxazolidinone antimicrobial. *Ann. Intern. Med.* 138(2):135–42.

Moran, G.J., Amii, R.N., Abrahamian, F.M., Talan, D.A. (2005). Methicillin-resistant Staphylococcus aureus in community-acquired skin infections. *Emerg. Infect. Dis.* 11(6):928–30.

Noskin, G.A. (2005). Tigecycline: a new glycylcycline for treatment of serious infections. *Clin. Infect. Dis.* 41 Suppl 5:S303–14.

Pong, A., Bradley, J.S. (2004). Clinical challenges of nosocomial infections caused by antibiotic-resistant pathogens in pediatrics. *Semin. Pediatr. Infect. Dis.* 15(1):21–9.

Posfay-Barbe, K.M., Wald, E.R. (2004). Pneumococcal vaccines: do they prevent infection and how? *Curr. Opin. Infect. Dis.* 17(3):177–84.

Raad, I., Darouiche, R., Vazquez, J., Lentnek, A., Hachem, R., Hanna, H., Goldstein, B., Henkel, T., Seltzer, E. (2005). Efficacy and safety of weekly dalbavancin therapy for catheter-related blood-stream infection caused by gram-positive pathogens. *Clin. Infect. Dis.* 40(3):374–80.

Robert, J., Etienne, J., Bertrand, X. (2005). Methicillin-resistant Staphylococcus aureus producing Panton-Valentine leukocidin in a retrospective case series from 12 French hospital laboratories, 2000–2003. *Clin. Microbiol. Infect.* 11(7):585–7.

Roos, K., Tellier, G., Baz, M., Leroy, B., Rangaraju, M. (2005). Clinical and bacteriological efficacy of 5-day telithromycin in acute maxillary sinusitis: a pooled analysis. *J. Infect.* 50(3): 210–20.

Ruef, C. (2004). Epidemiology and clinical impact of glycopeptide resistance in Staphylococcus aureus. *Infection* 32(6):315–27.

Saez-Llorens, X., McCoig, C., Feris, J.M., Vargas, S.L., Klugman, K.P., Hussey G.D., Frenck, R.W., Falleiros-Carvalho, L.H., Arguideas, A.G., Bradley, J., Arrieta, A.C., Wald, E.R., Pancorbo, S., McCracken, G.H.Jr., Marques, S.R., Trovan meningitis Study Group. (2002). Quinolone treatment for pediatric bacterial meningitis: a comparative study of trovafloxacin and ceftriaxone with or without vancomycin. *Pediatr. Infect. Dis. J.* 21(1):14–22.

Saez-Llorens, X., Rodriguez, A., Arguedas, A., Hamed, K.A., Yang, J., Pierce, P., Echols, R. (2005). Randomized, investigator-blinded, multicenter study of gatifloxacin versus amoxicillin/clavulanate treatment of recurrent and nonresponsive otitis media in children. *Pediatr. Infect. Dis. J.* 24(4):293–300.

Saiman, L. (2002). Risk factors for hospital-acquired infections in the neonatal intensive care unit. *Semin. Perinatol.* 26(5):315–21.

Schmitt-Hoffmann, A., Nyman, L., Roos, B., Schleimer, M., Sauer, J., Nashed, N., Brown, T., Man, A., Weidekamm, E. (2004). Multiple-dose pharmacokinetics and safety of a novel broad-spectrum cephalosporin (BAL5788) in healthy volunteers. *Antimicrob. Agents Chemother.* 48(7):2576–80.

Seltzer, E., Dorr, M.B., Goldstein, B.P., Perry, M., Dowell, J.A., Henkel, T. (2003). Once-weekly dalbavancin versus standard-of-care antimicrobial regimens for treatment of skin and soft-tissue infections. *Clin. Infect. Dis.* 37(10):1298–303.

Shaw, J.P., Seroogy, J., Kaniga, K., Higgins, D.L., Kitt, M., Barriere, S. (2005). Pharmacokinetics, serum inhibitory and bactericidal activity, and safety of telavancin in healthy subjects. *Antimicrob. Agents Chemother.* 49(1):195–201.

Sher, L., Arguedas, A., Husseman, M., Pichichero, M., Hamed, K.A., Biswas, D., Pierce, P., Echols, R. (2005). Randomized, investigator-blinded, multicenter, comparative study of gatifloxacin versus amoxicillin/clavulanate in recurrent otitis media and acute otitis media treatment failure in children. *Pediatr. Infect. Dis. J.* 24(4):301–8.

Shi, J., Montay, G., Bhargava, V.O. (2005). Clinical pharmacokinetics of telithromycin, the first ketolide antibacterial. *Clin. Pharmacokinet.* 44(9):915–34.

Silverman, J.A., Mortin, L.I., Vanpraagh, A.D., Li, T., Alder, J. (2005). Inhibition of daptomycin by pulmonary surfactant: in vitro modeling and clinical impact. *J. Infect. Dis.* 191(12):2149–52.

Soley, C., Arguedas, A., Porras, W., Guevara, S., Loaiza, C., Perez, A., Rincon, G., Schultz, M., Arguedas, J., Brilla, R. (2005). In vitro activities of levofloxacin and comparable agents against middle ear fluid, nasopharyngeal, and oropharyngeal pathogens obtained from Costa Rican children with recurrent otitis media or failing other antibiotic therapy. *Antimicrob. Agents Chemother.* 49(7):3056–8.

Stein, G.E. (2005). Safety of newer parenteral antibiotics. *Clin. Infect. Dis.* 41 Suppl 5:S293–302.

Streit, J.M., Sader, H.S., Fritsche, T.R., Jones, R.N. (2005). Dalbavancin activity against selected populations of antimicrobial-resistant Gram-positive pathogens. *Diagn. Microbiol. Infect. Dis.*

Stryjewski, M.E., O'Riordan, W.D., Lau, W.K., Pien, F.D., Dunbar, L.M., Vallee, M., Fowler, V.G.Jr., Chu, V.H., Spencer, E., Barriere, S.L., Kitt, M.M., Cabell, C.H., Corey, G.R., FAST Investigator Group. (2005). Telavancin versus standard therapy for treatment of complicated skin and soft-tissue infections due to gram-positive bacteria. *Clin. Infect. Dis.* 40(11):1601–7.

Syrogiannopoulos, G.A., Grivea, I.N., Fitoussi, F., Doit, C., Katopodis, G.D., Bingen, E., Beratis, N.G. (2001). High prevalence of erythromycin resistance of Streptococcus pyogenes in Greek children. *Pediatr. Infect. Dis. J.* 20(9):863–8.

Tan, T.Q. (2004). Update on the use of linezolid: a pediatric perspective. *Pediatr. Infect. Dis. J.* 23(10):955–6.

Tedesco, K.L., Rybak, M.J. (2004). Daptomycin. *Pharmacotherapy* 24(1):41–57.

Weisman, L.E. (2004). Coagulase-negative staphylococcal disease: emerging therapies for the neonatal and pediatric patient. *Curr. Opin. Infect. Dis.* 17(3):237–41.

Wellington, K., Noble, S. (2004). Telithromycin. *Drugs* 64(15):1683–94; discussion 1695–6.

Woodford, N. (2005). Biological counterstrike: antibiotic resistance mechanisms of Gram-positive cocci. *Clin. Microbiol. Infect.* 11 Suppl 3:2–21.

Yogev, R., Patterson, L.E., Kaplan, S.L., Adler, S., Morfin, M.R., Martin, A., Edge-Padbury, B., Naberhuis-Stehouwer, S., Bruss, J.B. (2003). Linezolid for the treatment of complicated skin and skin structure infections in children. *Pediatr. Infect. Dis. J.* 22(9 Suppl):S172–7.

14

Clinical Manifestations of Nontuberculous Mycobacteria

Robert S. Heyderman and Julia Clark

1. Introduction

Tuberculosis and leprosy have been widely recognized for over 2000 years and their causative organisms were identified in the nineteenth century when microscopy first became available. In contrast, nontuberculous mycobacteria (NTM), also known as atypical mycobacteria or mycobacteria other than *Mycobacterium tuberculosis* (MOTT), were not widely appreciated as human pathogens until the 1950's when Timpe and Runyon first reported evidence of human disease (Timpe and Runyon, 1954). Subsequently, an increasing range of NTM have been described (Table 14.1) and their impact on immunocompromised children has become widely recognized. *Mycobacterium ulcerans*, the cause of a disfiguring necrotic lesion known as Buruli ulcer, is responsible for a huge but neglected burden of childhood disease in Africa, Asia, the Pacific and South America (Thangaraj et al., 2003; Thangaraj et al., 1999). The relatively recent clinical observation that NTM may cause severe, disseminated infection in otherwise healthy children with no obvious immune disorder has led to the identification of several novel inherited immunodeficiencies. These have provided important insights into the immune defense pathways involved in the containment of intracellular infections (Dupuis et al., 2000; Levin et al., 1995; Newport et al., 1996; Dorman et al., 2004). This chapter will focus on the clinical manifestations of NTM in healthy and immunocompromised children, principally in industrialized countries.

2. Epidemiology

NTM are ubiquitous organisms and can be isolated from soil, house dust, water, food and animals. Transmission is by inhalation, ingestion or direct contact with a contaminated environmental source. Person to person spread is very rare.

Hot Topics in Infection and Immunity in Children, edited by Andrew J. Pollard and Adam Finn.
Springer, New York, 2006

Table 14.1.   Clinical syndromes and their causative NTM

Syndrome	Organisms	Time required for growth
Lymphadenitis	M. avium complex	10 days to 4 weeks
	M. scrofulaceum	
	M. malmoense	
[a]Otomastoiditis	M. abscessus	3 to 7 days
	M. chelonae	
	M. fortuitum	
Cutaneous and soft tissue infections	[b]M. ulcerans	4 to 8 weeks
	[c]M. marinum	7 to 10 days
	M. abscessus	3 to 7 days
	M. chelonae	
	M. fortuitum	
	M. kansasii	10 days to 4 weeks
	M. avium complex	
Catheter-related infections	M. abscessus	3 to 7 days
	M. chelonae	
	M. fortuitum	
Pneumonia	M. avium complex	10 days to 4 weeks
	M. kansasii	
	M. abscessus	3 to 7 days
	M. fortuitum	
	M. chelonae	
Disseminated infection	M. avium complex	10 days to 4 weeks

[a]Chronic otorrhea in children with tympanotomy tubes or chronic perforation.
[b]Causes Buruli ulcer (see text).
[c]Causes fish tank granuloma.

Children with NTM lymphadenitis (see section 4.1) are thought to acquire the infection through gingival or oropharyngeal abrasions resulting from tooth eruptions or foreign bodies in the mouth.

   Prevalence and incidence rates for NTM infection are inexact and there is little reliable data in children. A nationwide survey estimated that among cohorts born in Sweden in the period 1975–85, the cumulative incidence rate for NTM infection before 5 years of age was 26.8 per 100,000 among non-BCG-vaccinated children and 4.6 among those BCG-vaccinated (Romanus et al., 1995). This was almost exclusively lymph node or soft tissue infections, usually due to *Mycobacterium avium-intracellulare*. In the Netherlands, the annual incidence of NTM infection was estimated as 77 cases per 100,000 children, again *M. avium* was the most common NTM isolated (Haverkamp et al., 2004).

3. Laboratory Diagnosis and Antimicrobial Resistance Testing

   Good communication between the clinician and the laboratory is essential to a rapid and accurate diagnosis. More than 50 species of NTM have the potential to cause disease and have been categorized on the basis of colony morphology, growth rate, and pigmentation (Runyon classification) (Saiman, 2004). The growth rate on solid media in particular has proved a practical way to group species within the laboratory. Typically rapid growers require less than 7 days to produce visible colonies, while slow growers may require up to 8 weeks (Table 14.1). Detailed descriptions of the taxonomy and diagnostic methodologies are available elsewhere (Subcommittee of the Joint Tuberculosis Committee of the British Thoracic Society, 2000; American Thoracic Society, 1997; Saiman, 2004; Brown-Elliott and Wallace, 2002). Primary samples from non-sterile sites are decontaminated and processed using the methods applied to the diagnosis of *M. tuberculosis*. Ziehl-Neelsen or auraminephenol fluorochrome staining methods are used to detect mycobacteria. Molecular techniques (PCR-based methods and DNA probes) may be used to detect mycobacteria and distinguish some NTM species and exclude *M. tuberculosis* (Chemlal and Portaels, 2003). However, culture on solid media (Lowenstein-Jensen or Middlebrook 7H10 and 7H11) to enable semi-quantification and into automated liquid culture systems to improve speed of isolation are essential to effective diagnosis. Antimicrobial resistance testing of NTM remains problematic. For several NTM there is a poor correlation between *in vitro* sensitivity testing and the clinical response to treatment. The American Thoracic Society has made "temporary" recommendations regarding how, when and to which agents NTM should be tested (American Thoracic Society, 1997). Routine testing of all NTM is discouraged and recommendations differ for different groups or species of NTM (e.g. slow growing vs. rapid growing NTM). New agents such as the tetracycline derivative tigecycline and the oxazolidinone, linezolid have begun to be used for the treatment of some NTM on an ad-hoc basis (Wallace et al., 2002; Brown-Elliott et al., 2001) but it is unclear at present how useful resistance testing is for these agents.

4. NTM in Otherwise Healthy Children

4.1. NTM Lymphadenitis

Lymph node disease is the most common manifestation of NTM infection in children and has been extensively described (Schaad et al., 1979; Colville, 1993; Hazra et al., 1999; Saggese et al., 2003; Panesar et al., 2003; Mandell et al., 2003; Haverkamp et al., 2004). There are many retrospective case series making comparisons of different surgical treatments in children with significant lymphadenopathy. However, evidence on appropriate drug treatment is confined to small case series. Most NTM isolated from lymph nodes are *M. avium*, with *M. malmoense* increasingly identified in the UK (Colville, 1993).

Typically NTM infection occurs in young children, between 1 and 8 years, as unilateral, chronic, cervical or submandibular lymphadenopathy. Affected individuals are generally well, apyrexial, with no systemic upset, no local node tenderness and do not respond to antimicrobial agents that target streptococci and staphylococci (Schaad et al., 1979; Panesar et al., 2003). The lymph node(s) is initially firm and mobile, may appear suddenly or gradually over a few weeks, and with time there is often progression to overlying skin involvement with a reddish purple color.

Occasionally NTM-associated lymphadenopathy may become very large and disfiguring. If left untreated there will eventually be spontaneous resolution of NTM infection over months or years, though most, before resolving, will progress to abscess formation, discharge and sinus formation before healing with scarring (Mandell et al., 2003; Schaad et al., 1979).

Assessment should include investigation for other possible differential diagnoses. Although an underlying immuncomprimising illness should be considered, routine immune studies are not generally indicated. An acute bacterial infection is usually clinically apparent but a normal full blood count and C-reactive protein may be reassuring. Potential viral etiologies such as Epstein-Barr virus, adenovirus, cytomegalovirus and mumps should be evaluated. *Bartonella henselae* and *Toxoplasma gondii* should be considered particularly in those children exposed to cats. In patients from endemic areas or with a contact history, TB should be excluded. A chest radiograph is usually normal in both TB or NTM lymphadenitis. Almost all healthy children with TB adenitis will have a positive tuberculin reaction, however, up to 30% with NTM lymphadenitis will have 10 mm+ of induration with PPD. Diagnosis maybe aided by comparing tuberculin with NTM skin tests (Daley and Isaacs, 1999; Saggese et al., 2003) but these are not standardized nor widely commercially available. Novel *in vitro* T cell stimulation tests (ELISPOT or whole blood assays) may offer better discrimination in the future (Rolinck-Werninghaus et al., 2003).

CT and MRI scanning of patients with NTM lymphadenitis may show ring-enhancing lesions with minimal inflammatory stranding of the subcutaneous fat (Nadel et al., 1996; Hazra et al., 1999). Although imaging does not reliably distinguish NTM from TB infection, these may be valuable in defining the extent of disease and planning surgical intervention (see below).

Diagnosis should initially be based on clinical suspicion but there is frequently a delay. In many cases a diagnosis may only be arrived at after suggestive histology (caseating granulomas with or without acid-alcohol fast bacilli) or definitive culture of NTM following surgical intervention. Where tissue for histology and culture is acquired by biopsy or incision and drainage, there is often poor wound healing, continuing discharge and complicated by sinus formation (Schaad et al., 1979; Mandell et al., 2003; Panesar et al., 2003). Biopsy or incision and drainage should therefore be avoided whenever possible and are usually only undertaken where there are difficulties in initial diagnosis. There is published experience with fine needle aspiration not producing sinus formation but this remains controversial (Alessi and Dudley, 1988; Mandell et al., 2003). Many authorities therefore advocate complete surgical excision which is associated with cure rates of 81 to 92% (Saggese et al., 2003; Panesar et al., 2003). Surgery, however, is not without risks and may not be feasible in extensive disease or where there is facial nerve or parotid gland involvement. It has been argued that the evidence in favor of surgical treatments is biased by an over representation of the most severe cases (Haverkamp et al., 2004). Medical approaches to treatment using macrolides (azithromycin or clarithromycin) either alone or in combination with rifabutin ± ethambutol have been evaluated in some centers. However, antimicrobial therapy alone, particularly monotherapy, has not been entirely successful (Panesar et al., 2003; Hogan et al., 2005;

Hazra et al., 1999; Haverkamp et al., 2004; Saggese et al., 2003) and may be associated with significant drug side effects. Because the natural history of NTM mycobacterial lymphadenopathy is that of resolution, comparative trials require careful design. A randomized, open-label cohort trial, comparing surgery versus 3 months medical treatment has recently been completed in the Netherlands and will be reported in the near future (Jaap T. van Dissel, personal communication). It is important to recognize that whatever the initial therapeutic approach, scarring, sinus tract formation and mycobacterial reactivation of residual infection may occur.

4.2. Other Soft Tissue Infections

Although not common, sporadic cases and outbreaks of cellulitis, soft-tissue abscesses and rarely extra-cutaneous disease associated with rapidly growing NTM have been documented in healthy individuals. These have occurred in both nosocomial (contaminated surgical or clinical devices contaminated with water) and community settings (contamination of traumatic wounds with soil or water). Recently, outbreaks of cutaneous NTM infection in older children and adults have been linked to nail salon whirlpool footbaths (Winthrop et al., 2002; Gira et al., 2004). Such infections are often insidious and therefore affected individuals may not seek medical attention in the initial stages. Although some resolve without intervention, infection may result in a severe, protracted and potentially scarring furunculosis (Winthrop et al., 2004).

5. NTM in Immunocompromised Children

5.1. Patients with Cystic Fibrosis

Amongst children with cystic fibrosis (CF) mortality rates due to respiratory illness have fallen markedly in the last three decades (Panickar et al., 2005) and the long term outlook has much improved. Perhaps as a result of this improvement, European and North American CF centers have recently recorded NTM infection prevalence rates of 2–28% (mostly *Mycobacterium avium intracellulare*) compared with just 16 CF case reports prior to 1990. When a clinical specimen raises the possibility of NTM disease in the context of progressive CF, it is often difficult to resolve whether this infection represents colonization or clinically significant disease. In a comprehensive US prevalence study in which approximately 10% of the US CF population over 10 years was enrolled, the prevalence of NTM infection was 13% (Olivier et al., 2003b). No genetic association or clustering of NTM infection was observed. However, NTM infection was strongly associated with age (approximately 50% of those affected were older than 40yrs), was associated with preserved lung function and there was no decline in lung function at 15 month follow-up (Olivier et al., 2003b; Olivier et al., 2003a). Diagnostic clinical criteria for the diagnosis of NTM lung disease are available from the American Thoracic Society (American Thoracic Society, 1997) and the British Thoracic Society (2000). However, these criteria were not specifically designed for CF patients and relate best to infections with *M. avium intracellulare*, *M. abscessus*, and *M. kansasii* (Table 14.2).

Specific therapy should be considered in any NTM infected CF patient with deteriorating pulmonary function, unexplained systemic features such as weight loss, night sweats and fevers and particularly infection with rapidly growing aggressive organisms such as *M. abscessus* (Olivier et al., 2003a). Even young patients with CF may develop clinically significant N TM infection (Esther et al., 2005). However, before embarking on long courses of potentially toxic multi-drug regimens, it is important to exclude other causes of deterioration. Multiple isolates from nonsterile sites are required whereas one positive culture from a sterile site, particularly where there is supportive histopathology, is usually sufficient. Findings suggestive of NTM infection on high-resolution chest CT scan include cystic and/or cavitary parenchymal lung disease, subsegmental parenchymal consolidation, pulmonary nodules, and tree-in-bud opacities (Olivier et al., 2003b). Following initiation of therapy, sputum should be monitored monthly and careful surveillance undertaken for side effects. If available, monitoring of drug levels should be considered. It is recommended that treatment for NTM is continued for at least 12 months after sputum clearance but it is important to recognize that eradication may not be possible and it is not infrequent for disease to recur after cessation of therapy (Quittell, 2004).

Table 14.2. Clinical, radiographic and microbiological criteria for the diagnosis of NTM lung disease (adapted from the American Thoracic Society guidelines (American Thoracic Society, 1997))

Clinical criteria	Radiographic criteria	Microbiological criteria
Cough Fatigue Fever Weight loss Haemoptysis Dyspnea Exclusion of other disease Adequate treatment of underlying chronic disease	Infiltrates +/− nodules Cavitation Multiple nodules *High resolution CT:* Cystic and/or cavitary parenchymal lung disease, subsegmental parenchymal consolidation, pulmonary nodules, and tree-in-bud opacities	A) If three sputum/BAL results available from the previous 12 months: 3 positive cultures with negative AFB smear results or 2 positive cultures with 1 positive smear *or* B) If only one bronchial washing available: 1 BAL (no sputum available) smear ++ with positive culture or positive culture ++ on solid media *or* C) Tissue biopsy with typical histological features and one or more sputum or BAL cultures positive even in low numbers

5.2. Patients with Leukemia or Cancer

Central venous catheter (CVC)-associated and soft tissue NTM infections have been reported in association with leukemias, lymphomas and solid tumors (Reilly and McGowan, 2004; Levendoglu-Tugal et al., 1998). Among hematopoietic stem cell transplant recipients, adult series suggest that the incidence of NTM infection is 50–600 times greater than in the general population (Doucette and Fishman, 2004). Usually these infections are due to rapid growing NTM but *M. avium intracellulare* has also been reported (Reilly and McGowan, 2004; Levendoglu-Tugal et al., 1998). Occasionally NTM in blood cultures may be mistaken for *Corynebacterium* or *Nocardia* species which leads to a delay in appropriate management or the initiation of inappropriate therapy (Levendoglu-Tugal et al., 1998). Affected patients are not necessarily neutropenic but are frequently lymphopenic (total lymphocyte count <1000/mm^3) and may also have lung involvement. Some children have been effectively treated with CVC removal alone. However, standard treatment includes line removal, treatment with at least 2 anti-mycobacterial drugs for 2–12 weeks for localized disease and 6 months or longer for widespread disease (Reilly and McGowan, 2004; Levendoglu-Tugal et al., 1998). The intensity of immunosuppressive therapy may need to be modified and for tunnel or exit site infections, surgical excision of surrounding tissues may also be necessary (Burns et al., 1997). NTM infection should be considered in all persistently febrile children with cancer or leukemia, particularly those who don't respond to conventional antibiotics or antifungals. Individuals with lymphopenia and profound immunocompromise are at high risk. Although conventional blood cultures will detect rapidly growing NTM, they are unlikely to detect *M. avium intracellulare* or other slow-growing mycobacteria. Specific blood cultures are required for this purpose (Reilly and McGowan, 2004; Levendoglu-Tugal et al., 1998).

5.3. Patients with Familial Disorders

As summarized in "Hot topics in infection and immunity in children II" (Picard and Casanova, 2005), clinical observation of children with unusual deep-seated or disseminated NTM infections lead to the identification of a defect in up-regulation of TNFα by macrophages in response to interferonγ (IFNγ) (Levin et al., 1995). This defect was subsequently located to point mutation at nucleotide 395 IFNγ receptor 1 resulting in a truncated protein (no transmembrane or cytoplasmic domain) (Newport et al., 1996; Jouanguy et al., 1996). Following this initial discovery, 22 patients with recessive complete IFNR1 deficiency and 38 with dominant partial deficiency have been identified (Dorman et al., 2004) and numerous other defects in the IFN-γ, STAT1 signaling and IL-12 pathways have been uncovered (Figure 14.1). In most of these single gene defects a close relationship between genotype, cellular immune/inflammatory phenotype and the clinical disease manifestations has been observed (Dorman et al., 2004). NTM infections appear to predominate but infections with non-typhoidal salmonellae and several herpes viruses have also been problematic. Prior to these discoveries, familial clustering, racial differences in incidence, and twin studies had suggested that genetic factors play a

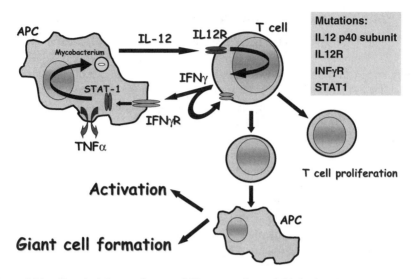

Figure 14.1.   **Genetic defects and susceptibility to mycobacterial infection.** Contact with intracellular pathogens (e.g., NTM) cause antigen presenting cells (APC) such as macrophages to secrete IL12. This leads to the release of interferonγ (IFNγ) which activates the APC to kill the pathogen. Persistent macrophage activation results in differentiation into epithelioid cells and the formation of giant cells. Mutations in the IL12 p40 subunit, the IL12 receptor (IL12R), the INFγ receptor (INFγR) and the STAT1 signalling pathway have been identified. These lead to inadequate containment of intracellular pathogens such as NTM.

role in susceptibility to tuberculosis. There is now an intensified research effort to identify more subtle defects in the IFNγ pathway which predispose to *M. tuberculosis* infection in larger populations.

6. Principles of Drug Therapy

An accurate diagnosis and good microbiology are essential to effective therapy. As will be appreciated from the previous discussions, there is very little evidence to guide the treatment of NTM infections. Outside studies of HIV infected individuals, there are no large published clinical trials of the different regimens recommended and expert opinion is based on the results of small series (Subcommittee of the Joint Tuberculosis Committee of the British Thoracic Society, 2000; American Thoracic Society, 1997). Confusion has arisen from the inappropriate extension of experience of the treatment of *M. tuberculosis* to the management of NTM. As discussed, resistance testing may be unhelpful. Ideally, a combination of at least two anti-mycobacterial agents is needed to prevent resistance from emerging. In the approach to an individual patient it is important to identify immunocompromise where present and remove the focus of infection where feasible. Surveillance for side effects of vision including visual acuity or color discrimination (ethambutol), the presence of eye pain or uveitis (rifabutin); hepatitis (isoniazid, rifampin, ethionamide, clarithromycin, rifabutin); renal impairment or auditory dysfunction (streptomycin, amikacin); central nervous system dysfunction (cycloserine, ethionamide); and hematologic abnormalities (sulfonamides, cefoxitin, rifabutin) is vital.

Acknowledgements

We would like to thank Adam Finn, Michel Erlewyn-Lajeunesse, Jaap van Dissel and Mike Levin for useful discussions.

References

Alessi, D.P. and Dudley, J.P. (1988) Atypical mycobacteria-induced cervical adenitis. Treatment by needle aspiration. *Arch Otolaryngol Head Neck Surg*, **114,** 664–666.

American Thoracic Society (1997) Diagnosis and treatment of disease caused by nontuberculous myco-bacteria. *Am J Respir Crit Care Med*, **156,** S1–25.

Brown-Elliott, B.A. and Wallace, R.J., Jr. (2002) Clinical and taxonomic status of pathogenic non-pigmented or late-pigmenting rapidly growing mycobacteria. *Clin Microbiol Rev*, **15,** 716–746.

Brown-Elliott, B.A., Wallace, R.J., Jr., Blinkhorn, R., Crist, C.J. and Mann, L.B. (2001) Successful treatment of disseminated Mycobacterium chelonae infection with linezolid. *Clin Infect Dis*, **33,** 1433–1434.

Burns, J.L., Malhotra, U., Lingappa, J. and Smith, S. (1997) Unusual presentations of nontuberculous mycobacterial infections in children. *Pediatr Infect Dis J*, **16,** 802–806.

Chemlal, K. and Portaels, F. (2003) Molecular diagnosis of nontuberculous mycobacteria. *Curr Opin Infect Dis*, **16,** 77–83.

Colville, A. (1993) Retrospective review of culture-positive mycobacterial lymphadenitis cases in children in Nottingham, 1979–1990. *Eur J Clin Microbiol Infect Dis*, **12,** 192–195.

Daley, A.J. and Isaacs, D. (1999) Differential avian and human tuberculin skin testing in non-tuberculous mycobacterial infection. *Arch Dis Child*, **80,** 377–379.

Dorman, S.E., Picard, C., Lammas, D., Heyne, K., van Dissel, J.T., Baretto, R., Rosenzweig, S.D., Newport, M., Levin, M., Roesler, J., Kumararatne, D., Casanova, J.L. and Holland, S.M. (2004) Clinical features of dominant and recessive interferon gamma receptor 1 deficiencies. *Lancet*, **364,** 2113–2121.

Doucette, K. and Fishman, J.A. (2004) Nontuberculous mycobacterial infection in hematopoietic stem cell and solid organ transplant recipients. *Clin Infect Dis*, **38,** 1428–1439.

Dupuis, S., Doffinger, R., Picard, C., Fieschi, C., Altare, F., Jouanguy, E., Abel, L. and Casanova, J.L. (2000) Human interferon-gamma-mediated immunity is a genetically controlled continuous trait that determines the outcome of mycobacterial invasion. *Immunol Rev*, **178,** 129–137.

Esther, C.R., Jr., Henry, M.M., Molina, P.L. and Leigh, M.W. (2005) Nontuberculous mycobacterial infection in young children with cystic fibrosis. *Pediatr Pulmonol*, **40,** 39–44.

Gira, A.K., Reisenauer, A.H., Hammock, L., Nadiminti, U., Macy, J.T., Reeves, A., Burnett, C., Yakrus, M.A., Toney, S., Jensen, B.J., Blumberg, H.M., Caughman, S.W. and Nolte, F.S. (2004) Furuncu-losis due to Mycobacterium mageritense associated with footbaths at a nail salon. *J Clin Microbiol*, **42,** 1813–1817.

Haverkamp, M.H., Arend, S.M., Lindeboom, J.A., Hartwig, N.G. and van Dissel, J.T. (2004) Nontuber-culous mycobacterial infection in children: a 2-year prospective surveillance study in the Netherlands. *Clin Infect Dis*, **39,** 450–456.

Hazra, R., Robson, C.D., Perez-Atayde, A.R. and Husson, R.N. (1999) Lymphadenitis due to nontuber-culous mycobacteria in children: presentation and response to therapy. *Clin Infect Dis*, **28,** 123–129.

Hogan, M., Price, D., Burrage, K. and Pushpanathan, C. (2005) Atypical mycobacterial cervical lymph-adenitis with extensive local spread: a surgical disease. *Pediatr Surg Int*, 1–3.

Jouanguy, E., Altare, F., Lamhamedi, S., Revy, P., Emile, J.F., Newport, M., Levin, M., Blanche, S., Seboun, E., Fischer, A. and Casanova, J.L. (1996) Interferon-gamma-receptor deficiency in an infant with fatal bacille Calmette-Guerin infection. *N Engl J Med*, **335**, 1956–1961.

Levendoglu-Tugal, O., Munoz, J., Brudnicki, A., Fevzi Ozkaynak, M., Sandoval, C. and Jayabose, S. (1998) Infections due to nontuberculous mycobacteria in children with leukemia. *Clin Infect Dis*, **27**, 1227–1230.

Levin, M., Newport, M.J., D'Souza, S., Kalabalikis, P., Brown, I.N., Lenicker, H.M., Agius, P.V., Davies, E.G., Thrasher, A., Klein, N. and et al. (1995) Familial disseminated atypical mycobacterial infection in childhood: a human mycobacterial susceptibility gene? *Lancet*, **345**, 79–83.

Mandell, D.L., Wald, E.R., Michaels, M.G. and Dohar, J.E. (2003) Management of nontuberculous mycobacterial cervical lymphadenitis. *Arch Otolaryngol Head Neck Surg*, **129**, 341–344.

Nadel, D.M., Bilaniuk, L. and Handler, S.D. (1996) Imaging of granulomatous neck masses in children. *Int J Pediatr Otorhinolaryngol*, **37**, 151–162.

Newport, M.J., Huxley, C.M., Huston, S., Hawrylowicz, C.M., Oostra, B.A., Williamson, R. and Levin, M. (1996) A mutation in the interferon-gamma-receptor gene and susceptibility to mycobacterial infection. *N Engl J Med*, **335**, 1941–1949.

Olivier, K.N., Weber, D.J., Lee, J.H., Handler, A., Tudor, G., Molina, P.L., Tomashefski, J. and Knowles, M.R. (2003a) Nontuberculous mycobacteria. II: nested-cohort study of impact on cystic fibrosis lung disease. *Am J Respir Crit Care Med*, **167**, 835–840.

Olivier, K.N., Weber, D.J., Wallace, R.J., Jr., Faiz, A.R., Lee, J.H., Zhang, Y., Brown-Elliot, B.A., Handler, A., Wilson, R.W., Schechter, M.S., Edwards, L.J., Chakraborti, S. and Knowles, M.R. (2003b) Nontuberculous mycobacteria. I: multicenter prevalence study in cystic fibrosis. *Am J Respir Crit Care Med*, **167**, 828–834.

Panesar, J., Higgins, K., Daya, H., Forte, V. and Allen, U. (2003) Nontuberculous mycobacterial cervical adenitis: a ten-year retrospective review. *Laryngoscope*, **113**, 149–154.

Panickar, J.R., Dodd, S.R., Smyth, R.L. and Couriel, J.M. (2005) Trends in death from respiratory illness in children in England and Wales from 1968 to 2000. *Thorax*, **[Epub ahead of print]**.

Picard, C. and Casanova, J.L. (2005) In *Hot topics in infection and immunity in children II* (Eds, Pollard, A.J. and Finn, A.) Springer, New York, pp. 89–99.

Quittell, L.M. (2004) Management of non-tuberculous mycobacteria in patients with cystic fibrosis. *Paediatr Respir Rev*, **5 Suppl A**, S217–219.

Reilly, A.F. and McGowan, K.L. (2004) Atypical mycobacterial infections in children with cancer. *Pediatr Blood Cancer*, **43**, 698–702.

Rolinck-Werninghaus, C., Magdorf, K., Stark, K., Lyashchenko, K., Gennaro, M.L., Colangeli, R., Doherty, T.M., Andersen, P., Plum, G., Herz, U., Renz, H. and Wahn, U. (2003) The potential of recombinant antigens ESAT-6, MPT63 and mig for specific discrimination of Mycobacterium tuberculosis and M. avium infection. *Eur J Pediatr*, **162**, 534–536.

Romanus, V., Hallander, H.O., Wahlen, P., Olinder-Nielsen, A.M., Magnusson, P.H. and Juhlin, I. (1995) Atypical mycobacteria in extrapulmonary disease among children. Incidence in Sweden from 1969 to 1990, related to changing BCG-vaccination coverage. *Tuber Lung Dis*, **76**, 300–310.

Saggese, D., Compadretti, G.C. and Burnelli, R. (2003) Nontuberculous mycobacterial adenitis in children: diagnostic and therapeutic management. *Am J Otolaryngol*, **24**, 79–84.

Saiman, L. (2004) The mycobacteriology of non-tuberculous mycobacteria. *Paediatr Respir Rev*, **5 Suppl A**, S221–223.

Schaad, U.B., Votteler, T.P., McCracken, G.H., Jr. and Nelson, J.D. (1979) Management of atypical mycobacterial lymphadenitis in childhood: a review based on 380 cases. *J Pediatr*, **95**, 356–360.

Subcommittee of the Joint Tuberculosis Committee of the British Thoracic Society (2000) Management of opportunist mycobacterial infections: Joint Tuberculosis Committee Guidelines 1999. *Thorax*, **55**, 210–218.

Thangaraj, H.S., Evans, M.R. and Wansbrough-Jones, M.H. (1999) Mycobacterium ulcerans disease; Buruli ulcer. *Trans R Soc Trop Med Hyg*, **93**, 337–340.

Thangaraj, H.S., Phillips, R.O., Evans, M.R. and Wansbrough-Jones, M.H. (2003) Emerging aspects of Buruli ulcer. *Expert Rev Anti Infect Ther*, **1**, 217–222.

Timpe, A. and Runyon, E.H. (1954) The relationship of atypical acid-fast bacteria to human disease; a preliminary report. *J Lab Clin Med*, **44**, 202–209.

Wallace, R.J., Jr., Brown-Elliott, B.A., Crist, C.J., Mann, L. and Wilson, R.W. (2002) Comparison of the in vitro activity of the glycylcycline tigecycline (formerly GAR-936) with those of tetracycline, minocycline, and doxycycline against isolates of nontuberculous mycobacteria. *Antimicrob Agents Chemother*, **46,** 3164–3167.

Winthrop, K.L., Abrams, M., Yakrus, M., Schwartz, I., Ely, J., Gillies, D. and Vugia, D.J. (2002) An outbreak of mycobacterial furunculosis associated with footbaths at a nail salon. *N Engl J Med*, **346,** 1366–1371.

Winthrop, K.L., Albridge, K., South, D., Albrecht, P., Abrams, M., Samuel, M.C., Leonard, W., Wagner, J. and Vugia, D.J. (2004) The clinical management and outcome of nail salon-acquired Mycobacterium fortuitum skin infection. *Clin Infect Dis*, **38,** 38–44.

15

Kingella kingae: An Emerging Pediatric Pathogen

Pablo Yagupsky

1. Introduction

If not promptly diagnosed and adequately treated, septic arthritis in children may be a devastating disease, causing prolonged morbidity and crippling long-term sequelae. Isolation and identification of the causative organism and performance of antibiotic susceptibility testing are crucial factors in making the right therapeutic choice (Trujillo and Nelson, 1997).

In the preantibiotic era, Gram-positive organisms and especially *Staphylococcus aureus* and *Streptococcus pyogenes* were the most common bacteria isolated from patients with joint infections (Almquist, 1970). With the advent of antimicrobial drugs the incidence of ß-hemolytic streptococci declined, leaving *S. aureus* as the predominant skeletal system pathogen. In the 1960's, introduction of chocolate-agar plates for the routine culture of synovial fluid specimens resulted in the recognition of *Haemophilus influenzae* type b as the most common etiology of septic arthritis in young children (Almquist, 1970). Unfortunately, even when chocolate-agar media were routinely employed, the etiology of septic arthritis remained undetermined in a substantial proportion of cases. In 5 large series of pediatric patients with septic arthritis published between 1984 and 1999, no pathogen was isolated in 346 of 1,042 patients (33.2%) (Peltola and Vahvanen, 1984; Speiser et al., 1985; Welkon et al., 1986; Barton et al., 1987; Trujillo and Nelson, 1997).

Failure to isolate the causative organism may be the result of an incorrect diagnosis or previous administration of antibiotic therapy. However, the possibility that some cases of "culture-negative" septic arthritis are caused by fastidious organisms that are not detected by conventional bacteriological methods should also be entertained.

Hot Topics in Infection and Immunity in Children, edited by Andrew J. Pollard and Adam Finn.
Springer, New York, 2006.

2. Detection of *K. kingae*

2.1. Detection by Culture

In the late 1980's, pediatricians working at the Soroka University Medical Center (SUMC) in southern Israel decided to inoculate synovial fluid aspirates from patients with presumptive septic arthritis into blood culture vials, in an attempt to improve the bacteriological diagnosis (Yagupsky et al., 1992). Shortly after the adoption of this unorthodox approach, a Gram-negative coccobacillus growing in pairs or short chains was detected in several blood culture vials seeded with joint exudates of young children (Figure 15.1). Based on the characteristic Gram stain, ß-hemolysis production, positive oxidase test, and glucose and maltose utilization, the organism was identified as *Kingella kingae*, an obscure bacterium first described by Elizabeth King in the 1960's, and rarely isolated from patients with skeletal system infections or endocarditis (Henriksen and Bøvre, 1968).

Attempts to isolate the organism from synovial fluid exudates seeded onto routine solid media succeeded in only 2 of 12 patients, whereas inoculation of these specimens into aerobic blood culture vials yielded, after a median incubation of 4 days, *K. kingae* in all cases. When the positive blood culture vials were then sub-cultured onto blood-agar plates, *K. kingae* grew without difficulties, demonstrating that routine solid media are able to support its nutritional requirements. It is postulated that pus exerts an inhibitory effect upon *K. kingae*, and that decreasing the concentration of detrimental factors by diluting the specimen in a large broth volume

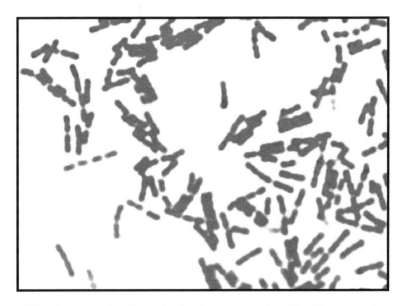

Figure 15.1.   Gram stain of *K. kingae* showing short gram-negative rods with tapered ends arranged in pairs or short chains. Reprinted from *Lancet Infectious Diseases*, Vol 4, Yagupsky et al., "Kingella Kingae: from medical rarity to an emerging paediatric pathogen," 332–41, 2004, with permission from Elsevier.

is a key factor in improving the recovery of the organism (Yagupsky et al., 1992; Yagupsky, 2004).

When synovial fluid specimens were systematically inoculated into blood culture vials, *K. kingae* appeared as the most common etiology of septic arthritis in Israeli children younger than 2 years, causing half of all culture-proven cases (Yagupsky et al., 1995a). This observation has been confirmed in France (Moumile et al., 2003; Moumile et al., 2005) and the USA (Lundy and Kehl, 1998; Luhmann and Luhmann, 1999; Moylett, 2000) indicating that *K. kingae* is a much more common pathogen that it has been traditionally appreciated.

2.2. Detection by Nucleic Acid Amplification Techniques

Because of the importance of rapid bacteriologic diagnosis of infections in normally-sterile body sites and the difficulties in recovering *K. kingae* in cultures, attempts have been made to use novel nucleic acid amplification techniques to improve detection of the organism. The usual strategy consisted of performing broad-range bacterial PCR amplification using universal primers complementary to constant regions from a part of the gene coding for 16S rRNA, followed by direct sequencing of the amplicon (Stahelin et al., 1998; Moumile et al., 2003). Alternatively, the amplicon was subjected to a new DNA amplification using species-specific primers designed for the simultaneous detection of those organisms most commonly associated with invasive infections (Reekmans et al., 2000). In a recent study by Moumile et al. (2003), use of the PCR method demonstrated *K. kingae*-specific DNA sequences in joint fluid aspirates from 5 of 10 children in whom bacterial cultures, including those performed using blood culture vials, were sterile. These results suggest that the recovery of *K. kingae* by culture methods remains sub-optimal and the organism is probably responsible for a substantial fraction of cases of culture-negative septic arthritis.

3. Clinical Presentation of *K. kingae* Infections

Most of the old medical literature on *K. kingae* infections consisted of reports of single cases or small series of patients in which unusual clinical manifestations were probably overrepresented. Based on the large experience with the organism accumulated at the SUMC over the years, a more accurate picture can be drawn. Between 1988 and 2005, a total of 96 patients with invasive *K. kingae* disease were diagnosed. Infections of the skeletal system infections were the most common presentation and were diagnosed in 55 of 96 (57.3%) patients, followed by bacteremia without focus (occult bacteremia) in 36 (37.5%), bacteremic lower respiratory tract infection in 3 (3.1%), and endocarditis in 2 (2.1%) (Figure 15.2).

Children with skeletal infections caused by *K. kingae* typically present in good general condition, have low-grade fever, show symptoms of URI, have varicella-induced buccal erosions or primary herpetic stomatitis, and have minimal or no laboratory abnormalities, requiring, thus, a high index of clinical suspicion (Yagupsky, 2004).

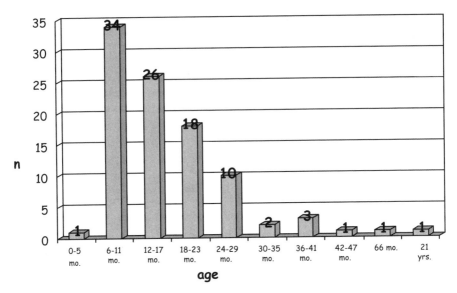

Figure 15.2. Age distribution of 96 patients with invasive *K. kingae* infections diagnosed in southern Israel in the 17-year period 1988–2004. Reprinted from *Lancet Infectious Diseases*, Vol 4, Yagupsky et al., "Kingella Kingae: from medical rarity to an emerging paediatric pathogen," 332–41, 2004, with permission from Elsevier.

Kingella kingae osteoarthritis usually affects the large weight-bearing bones such as the femur or tibia, and joints such as the knee, ankle, or hip. Involvement of short bones such as the calcaneus or talus and small joints such as the sternocla-vicular or metacarpo-phalangeal, which are rarely invaded by other organisms, is also seen (Yagupsky et al., 2003; Yagupsky, 2004).

Four afebrile patients, referred to the Emergency Room due to limping, grew *K. kingae* in the blood culture. Skeletal symptoms resolved in all four without anti-biotic therapy, suggesting an abortive skeletal infection (Yagupsky and Press, 2004). This observation indicates the need to obtain routine blood cultures from young children presenting with skeletal complaints even in the absence of fever, constitu-tional symptoms or leukocytosis.

When appropriately treated with antibiotics, patients with *K. kingae* osteoar-thritis recover without functional sequelae, and no evolution to chronicity has been ever reported in children with osteomyelitis (Yagupsky, 2004).

Hematogenous invasion of the intervertebral disc by *K. kingae* has been reported in 27 children, all but 3 aged <5 years and in one adult (Yagupsky, 2004). In a large French study, *K. kingae* was recovered from 6 of 22 (27.3%) children with culture-proven spondylodiskitis and was the second most common bacterium iso-lated after *Staphylococcus aureus* (Garron et al., 2002). Children with *K. kingae* diskitis usually present with minimal constitutional changes, limping, or refusal to sit-up or stand. The disease generally affects the lumbar intervertebral space and the diagnosis can be readily made by an MRI study or a [99]technetium scan. *Kingella kingae* diskitis runs a favorable clinical course, leaving no orthopedic or neurologic sequelae (Yagupsky, 2004).

Kingella kingae bacteremia without evidence of endocarditis has been observed mostly in children (Yagupsky and Dagan, 1994) and in a few adults (Yagupsky, 2004). Concomitant infections of the skeletal, respiratory or central nervous system were present in some patients but no focus (occult bacteremia) was noted in most others. A maculopapular rash, resembling disseminated meningococcal or gonococcal infection has been described in a few bacteremic patients (Yagupsky, 2004). Because of the serious implications of overlooking the diagnosis of endocarditis, it has been strongly advocated to evaluate all adult patients with *K. kingae* bacteremia to exclude endocardial invasion (Dayan et al., 1989).

Kingella kingae stands for the K of the HACEK acronym. As opposed to occult bacteremia and osteoarthritis, which are especially common in young children, *K. kingae* endocarditis affects older children and adults (Yagupsky, 2004). Predisposing valvular damage or systemic lupus erythematosus is common. Bacterial endocarditis caused by *K. kingae* frequently runs an unusually severe course with frequent complications such as septic shock, mycotic aneurisms and cerebrovascular accidents, and has a high mortality rate (Yagupsky, 2004).

4. Epidemiology of *K. kingae* Infections

4.1. Age and Gender Distribution of Patients with Invasive *K. kingae* Disease

Sixty-one of the 96 (63.5%) patients diagnosed at the SUMC during a 17-year period were males. The age-distribution of these patients shows that only one was younger than 6 months, 78 (81.3%) patients were aged 6–24 months, and overall 94 of 96 (97.9%) patients were younger than 4 years (Figure 15.3). It should be pointed-out that the remaining two cases that included the oldest child in the series (aged

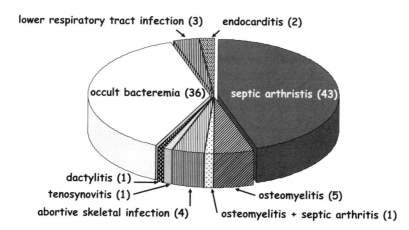

Figure 15.3. Clinical presentation of 96 patients with invasive *K. kingae* infections diagnosed in southern Israel. Reprinted from *Lancet Infectious Diseases*, Vol 4, Yagupsky et al., "Kingella Kingae: from medical rarity to an emerging paediatric pathogen," 332–41, 2004, with permission from Elsevier.

66 months) and the only adult were the only patients with endocarditis detected during the entire 17-year period. The calculated annual incidence of invasive *K. kingae* disease in southern Israel represented between 1/3 and 1/5 of that of *H. influenzae* type b in the same population prior to the introduction of the conjugate vaccine (Slonim et al., 2003).

4.2. Carriage of *K. kingae*

Although the niche of the organism was initially unknown, it was suspected that the bacterium was a respiratory tract commensal. This assumption was supported by several lines of evidence: 1: other members of the *Neisseriaceae* family are part of the normal respiratory flora; 2: *K. kingae* has been found in 1% of nasal and throat cultures; 3: children with invasive *K. kingae* infections frequently present with concomitant signs of upper respiratory tract infections; 4: *K. kingae* has been isolated from blood or respiratory secretions of patients with pneumonia, pleural empyema or laryngo-tracheo-bronchitis (Yagupsky, 2004).

To investigate the niche of *K. kingae* in the respiratory tract, separate tonsillar and nasopharyngeal cultures were obtained twice-a-week from 48 children aged 14 to 24 months attending a daycare center (DCC) in southern Israel (Yagupsky et al., 1995b). During an 11-month period, *K. kingae* was isolated from 109 of 624 (27.5%) fortnightly tonsillar cultures. The organism was not recovered from any of the nasopharyngeal cultures, indicating that *K. kingae* colonizes a rather narrow anatomical niche.

Overall, 35 of 48 (72.9%) attendees had at least one positive culture for the organism during the study period, a prevalence rate within the range of carriage of *Streptococcus pneumoniae* in DCCs. Among the 27 children who had two or more positive *K. kingae* cultures, continuous and intermittent patterns of carriage were observed. None of the colonized attendees experienced a documented invasive *K. kingae* infection during the 11-month period. Despite continuous exposure to the organism, *K. kingae* was not recovered from the DCC caretakers, suggesting that adults are more resistant to colonization by the organism (Yagupsky et al., 1995b).

When *K. kingae* isolates from the DCC attendees were studied by a combination of typing methods, it was shown that 8 different strains, defined by a unique combination of immunoblotting, ribotyping, and pulsed-field gel electrophoresis (PFGE) patterns, circulated in the DCC during the study period (Slonim et al., 1998). Two distinct strains represented 46% and 28% of all DCC organisms typed. When 60 *K. kingae* isolates derived from epidemiologically unrelated individuals, including pharyngeal carriers and patients with invasive infections, were studied using the same typing methods, 43 distinct strains were identified and none of them represented more than 5% of all organisms. These results strongly support the person-to-person transmission of *K. kingae* among DCC attendees, and suggest that the organism is disseminated, like many other respiratory pathogens, by large droplets or shared toys coated with saliva and respiratory secretions.

Children frequently harbored the same strain continuously or intermittently for weeks or months. Organisms were periodically replaced by new strains, indicating that carriage of *K. kingae* is a dynamic process with frequent turnover of organism, similar to that observed in other pathogens of respiratory origin such as pneumococci, *Haemophilus influenzae* or *Moraxella catarrhalis* (Slonim et al., 1998). In addition, a few strains recovered from DCC attendees were genotypically indistinguishable from organisms isolated from patients with invasive infections, suggesting that some children may be asymptomatic carriers of potentially virulent *K. kingae* strains (Slonim et al., 1998).

Because of the peculiar age distribution of children with invasive *K. kingae* infections, the age-related carriage of the organism among different child populations, including healthy infants aged 2–4 months, as well as older individuals was further investigated (Yagupsky et al., 1995b). The results of the study showed a good correlation between the prevalence of *K. kingae* in the pharynx and the age-related morbidity caused by the organism. *Kingella kingae* was not found in infants aged 2–4 months, among whom the disease appears to be exceptionally rare, while the prevalence of respiratory carriage was 10.0% in children aged 6 months to 4 years, coinciding with the age at which most cases of disease are diagnosed. The carriage rate in children older than 4 years, among whom the incidence of invasive *K. kingae* disease is low, was only 6.0% (Yagupsky et al., 1995b).

4.3. Seasonal Distribution of Invasive *K. kingae* Infections

Examination of the monthly distribution of the 96 cases of invasive disease diagnosed in southern Israel since 1988 shows that *K. kingae* infections are significantly more common in the second half of the year (69.8%) of all cases. To elucidate the relationship between seasonal carriage of *K. kingae* and occurrence of invasive infection, the prevalence of the organism in the pharynx was determined in children and adults between the months of February and May (when invasive diseases are rare), and between October and December (at the time of the year when most cases occur) (Yagupsky et al., 2002). The carriage rate was similar in the two periods, indicating that the increased attack rate of invasive disease observed in the second half of the year is not the result of a higher prevalence of *K. kingae* in the respiratory tract and suggesting the existence of possible co-factors such as seasonal viral infections.

4.4. Outbreaks of Invasive *K. kingae* Infections in Daycare Centers

Since the first description of *K. kingae*, cases of invasive disease caused by the organism have been sporadic. In 2003, the first recognized outbreak of skeletal *K. kingae* infections was reported in a Minnesota DCC (Centers for Disease Control and Prevention (CDC), 2004). Within a short period, 3 attendees developed osteoarthritis, and *K. kingae* was isolated from 2 of them. Pharyngeal cultures obtained from children attending the same classroom revealed a 45.0% colonization rate, whereas only 6.3% of youngsters attending other classes carried *K. kingae*. Typing of the DCC isolates showed that the same strain caused skeletal infections as well as pharyngeal colonization. These findings indicate that this particular strain was not only a highly virulent and transmissible organism, but also a very successful respiratory colonizer. In March 2005, another outbreak of skeletal *K. kingae* infections was detected among young children attending a DCC in Israel, indicating that the bacterium is an emerging cause of invasive disease in children in daycare.

5. Immune Response to Respiratory Carriage and Invasive Infection

The high rates of *K. kingae* carriage and infection observed in the first two years of life followed by a sharp drop in older children, the scarcity of cases among adults, and the occurrence of disease in immunocompromised patients suggest that an acquired immune response is necessary to protect from colonization and infection. To study the immune response to acute infection, paired acute-phase and convalescent-phase sera were obtained from 19 children with culture-proven *K. kingae* septic arthritis, osteomyelitis or occult bacteremia and studied by an ELISA immunoassay using *K. kingae* outer-membrane proteins (OMP's) as the coating antigen (Slonim et al., 2003). Although the specific function of *K. kingae* OMP's has not been investigated yet, studies conducted with other respiratory organisms have shown that surface-exposed OMP's express adhesins and other virulence factors that interact with mucosal surfaces and the immune system and, as such, are frequent targets for antibodies that protect against mucosal and invasive infections. In most patients, a significant increase in the optic density values was measured in the convalescent sample implying that acute invasive infections elicit an immune response.

Using the same serological essay, the age-related incidence of invasive infections in the population was correlated with the serum antibody levels in a cohort of children followed prospectively for the first 2 years of life (Figure 15.4). Below the age of 6 months, the incidence of disease was extremely low coinciding with high levels of IgG antibodies. Low levels of IgA antibodies, which do not cross the placenta, found in this period suggested that the IgG-dependent immunity had a maternal origin. Decreasing antibody levels in children aged 6–18 months overlapped with

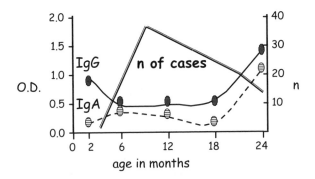

Figure 15.4. Average levels of IgG and IgA antibodies to *K. kingae* OMPs of 10 infants followed during the first 2 years of life (antibody levels were measured by an ELISA test and expressed in optic density (O.D.) units, and number of cases of invasive disease by age intervals.

the increasing incidence of invasive *K. kingae* disease in this age-group and indicated fading vertically-acquired immunity. Increasing antibody levels in older children and declining incidence of disease suggested cumulative experience with *K. kingae* antigens induced by carriage and infection (Slonim et al., 2003).

6. Pathogenesis of *K. kingae* Infections

Our current knowledge of the course of invasive *K. kingae* infections suggests that virulent organisms carried asymptomatically in the pharynx, penetrate the respiratory mucosa and that this penetration is facilitated by non-specific viral infections and stomatitis (Amir and Yagupsky, 1998) (Figure 15.5). *Kingella kingae* organisms may then propagate to the lower respiratory tract causing pneumonia, or invade the bloodstream. Occult bacteremia may follow, or the bacterium may be seeded to remote sites, such as the skeletal system or the endocardium, for which *K. kingae* shows a striking and still unexplained affinity, or more rarely to the eyes and meninges. In a minority of cases, the organism is cleared from the skeletal system by an effective immune response resulting in an abortive infection, but the vast majority of patients develop hematogenous arthritis, osteomyelitis, diskitis and other suppurative focal complications (Yagupsky, 2004).

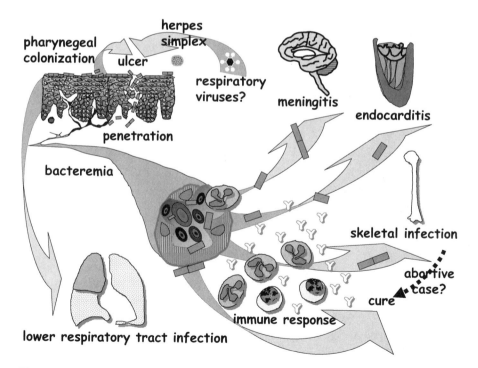

Figure 15.5. Proposed pathogenesis of *K. kingae* infections. Reprinted from the *Expert Review of Anti-Infective Therapy*, Vol 2, Yagupsky et al., "Kingella kingae infections of the skeletal system in children: diagnosis and therapy," 787–94, 2004, with permission from Future Drugs.

7. Management of *K. kingae* Infections

7.1. Antibiotic Susceptibility of *K. kingae*

Despite the fact that *K. kingae* organisms carried in the respiratory tract of young children are repeatedly exposed to antibiotics, the bacterium remains susceptible to a large variety of antimicrobial drugs, and only a few β-lactamase-positive isolates have been reported so far (Yagupsky, 2004). *Kingella kingae* shows, however, decreased susceptibility to cloxacillin and clindamycin, and all isolates are resistant to glycopeptides (Yagupsky et al., 2001). These favorable *in-vitro* data explain the excellent clinical response achieved with penicillins and cephalosporin antibiotics that are empirically administered to children with skeletal system infections or suspected bacteremia.

7.2. Antibiotic Treatment of *K. kingae* Infections

Because of the lack of specific guidelines for the treatment of *K. kingae* infections, patients with invasive disease have been treated with a wide array of different antibiotics according to protocols developed for specific conditions such as septic arthritis, osteomyelitis, endocarditis or meningitis caused by more familiar pathogens.

Initial drug therapy for skeletal infections in children usually consists of intravenous administration of second or third-generation cephalosporins (Yagupsky, 2004). This therapy is frequently changed to penicillin, ampicillin, or cefuroxime once the culture results and antibiotic susceptibility of the isolate are known. Clinical improvement, decreasing erythrocyte sedimentation rate and C-reactive protein values, as well as serum bactericidal levels, are used to guide switching to an oral ß-lactam drug and determine duration of antibiotic therapy (Yagupsky, 2004). Treatment has varied from 17 days to 3 months for arthritis, from 3 weeks to 6 months for osteomyelitis, and from 3 to 12 weeks for spondylodiscitis. Surgical drainage of the affected joint is usually reserved for children with septic arthritis involving the hip or shoulder. In patients with osteomyelitis, abscess formation, persistent bacteremia or development of a bone sequestrum are considered indications for surgery. In patients with spondylodiscitis, detection of a para-spinal abscess or signs of neurological deficit make surgical drainage and decompression mandatory (Garron et al., 2002).

Patients with occult bacteremia have been generally given intravenous β-lactam antibiotics followed by oral therapy once the clinical condition improved. Total duration of therapy ranged from 7 days to 2 weeks in most cases (Yagupsky, 2004).

Patients with *K. kingae* endocarditis were usually administered intravenous β-lactam antibiotics alone or in combination with an aminoglycoside for 4 to 7 weeks. Early surgical interventions have been necessary in individuals with life-threatening complications unresponsive to medical therapy (Yagupsky, 2004).

8. Conclusions

As the result of routine drawing of blood cultures from young febrile children, use of the blood culture vial method for culturing synovial fluid exudates, and development of nucleic acid amplification essays, *K. kingae* is being increasingly recognized as an important cause of bacteremia and skeletal system infections below the age of 2 years. *Kingella kingae* is carried in the pharynx by asymptomatic children and is transmitted from child-to-child in young daycare center attendees among whom the organism may cause outbreaks of invasive disease. Concomitant non-specific viral infections and stomatitis appear to facilitate penetration of *K. kingae* into the bloodstream, resulting in occult bacteremia and seeding of the organism to joints, bones, intervertebral disks, endocardium and other remotes sites. Maternal antibodies appear to protect from respiratory colonization and invasive infections during the first 6 months of life. Fading of vertically-transmitted immunity results in a high prevalence of the organism in the pharynx and increased incidence of bacteremia and skeletal system infections between the ages of 6 months and 2 years. Induction of immunity by prolonged carriage results in low incidence of invasive infections in older children.

The clinical presentation of the disease is often subtle and acute phase reactants are frequently within normal limits, emphasizing the need for a high index of clinical suspicion, drawing blood cultures in young children with joint or bone complaints even in the absence of fever, and inoculation of exudates into blood culture bottles. With the exception of endocarditis, *K. kingae* infections usually run a benign clinical course and promptly respond to antibiotic therapy and especially to ß-lactam drugs, leaving no long-term sequelae.

References

Almquist, E.E. (1970). The changing epidemiology of septic arthritis in children. *Clin. Orthop. Relat. Res.* 68:96–99.

Amir, J., and Yagupsky, P. (1998). Invasive *Kingella kingae* infection associated with stomatitis in children. *Pediatr. Infect. Dis. J.* 17:757–758.

Barton, L.L., Dunkle, L.M., and Habib, F.H. (1987). Septic arthritis in childhood. A 13-year review. *Am. J. Dis. Child.* 141:898–900.

Centers for Disease Control and Prevention (CDC). (2004). Osteomyelitis/septic arthritis caused by *Kingella kingae* among day care attendees-Minnesota, 2003. *MMWR Morb. Mortal. Wkly. Rep.* 53:241–243.

Dayan, A., Delclaux, B., Quentin, R., Rabut, H., Lavandier, M., and Goudeau, A. (1989). The isolation of *Kingella kingae* by hemoculture must always suggest the diagnosis of endocarditis. *Presse Med.* 18:1340–1341.

Garron, E., Viehweger, E., Launay, F., Guillaume, J.M., Jouve, J.L., and Bollini, G. (2002). Nontuberculous spondylodiscitis in children. *J. Pediatr. Orthop.* 22:321–328.

Henriksen, S.D., and Bøvre, K. (1968). *Moraxella kingii* sp. nov., a hemolytic saccharolytic species of the genus *Moraxella*. *J. Gen. Microbiol.* 51:377–385.

Luhmann, J.D., and Luhmann, S.J. (1999). Etiology of septic arthritis in children: an update for the 1990s. *Pediatr. Emerg. Care* 15:40–42.

Lundy, D.W., and Kehl, D.K. (1998). Increasing prevalence of *Kingella kingae* in osteo-articular infections in young children. *J. Pediatr. Orthop.* 18:262–267.

Moylett, E.H., Rossmann, S.N., Epps, H.R., and Demmler, G.J. (2000). Importance of *Kingella kingae* as a pediatric pathogen in the United States. *Pediatr. Infect. Dis. J.* 19:263–265.

Moumile, K., Merckx, J., Glorion, C., Berche, P., and Ferroni, A. (2003). Osteoarticular infections caused by *Kingella kingae* in children; contribution of polymerase chain reaction to the microbiologic diagnosis. *Pediatr. Infect. Dis.* 22:837–839.

Moumile, K., Merckx, J., Glorion, C., Pouliquen, J.C., Berche, P., and Ferroni, A. (2005). Bacterial aetiology of acute osteoarticular infections in children. *Acta Paediatr.* 94:419–422.

Peltola, H., and Vahvanen, V. (1984). A comparative study of osteomyelitis and purulent arthritis with special reference to aetiology and recovery. *Infection* 12:75–79.

Reekmans, A., Noppen, M., Naessens, A., and Vincken, W. (2000). A rare manifestation of *Kingella kingae* infection. *Eur. J. Intern. Med.* 11:343–344.

Slonim, A., Walker, E.S., Mishori, E., Porat, N., Dagan, R., and Yagupsky, P. (1998). Person-to-person transmission of *Kingella kingae* among day care center attendees. *J. Infect. Dis.* 178:1843–1846.

Slonim, A., Steiner. M., and Yagupsky, P. (2003). Immune response to invasive *Kingella kingae* infections, age-related incidence of disease, and levels of antibody to outer-membrane proteins. *Clin. Infect. Dis.* 37:521–527.

Speiser, J.C., Moore, T.L., Osborn, T.G., Weiss, T.D., and Zuckner, J. (1985). Changing trends in pediatric septic arthritis. *Semin. Arthritis Rheum.* 15:132–138.

Stahelin, J., Goldenberger, D., Gnehm, H.E., and Altwegg, M. (1998). Polymerase chain reaction diagnosis of *Kingella kingae* arthritis in a young child. *Clin. Infect. Dis.* 27:1328–1329.

Trujillo, M., and Nelson, J.D. (1997). Suppurative and reactive arthritis in children. *Semin. Pediatr. Infect. Dis.* 8:242–249.

Yagupsky, P. (2004). *Kingella kingae*: from medical rarity to an emerging paediatric pathogen. *Lancet Infect. Dis.* 4:358–367.

Yagupsky, P., and Dagan, R. (1994). *Kingella kingae* bacteremia in children. *Pediatr. Infect. Dis. J.* 13:1148–1149.

Yagupsky, P., and Dagan, R. (2000). Population-based study of invasive *Kingella kingae* infections. *Emerg. Infect. Dis.* 6:85–87.

Yagupsky, P., and Press, J. (2004). Unsuspected *Kingella kingae* infections in afebrile children with mild skeletal symptoms: the importance of blood cultures. *Eur. J. Pediatr.* 163:563–564.

Yagupsky, P., Dagan, R., Howard, C.B., Einhorn, M., Kassis, I., and Simu, A. (1992). High prevalence of *Kingella kingae* in joint fluid from children with septic arthritis revealed by the BACTEC blood culture system. *J. Clin. Microbiol.* 30:1278–1281.

Yagupsky, P., Howard, C.B., Einhorn, M., and Dagan, R. (1993). *Kingella kingae* osteomyelitis of the calcaneus in young children. *Pediat. Infec. Dis. J.* 12:540–541.

Yagupsky, P., Bar-Ziv, Y., Howard, C.B., and Dagan, R. (1995a). Epidemiology, etiology and clinical features of septic arthritis in children younger than 24 months. *Arch. Pediatr. Adolesc. Med.* 149:537–540.

Yagupsky, P., Dagan, R., Prajgrod, F., and Merires, M. (1995b). Respiratory carriage of *Kingella kingae* among healthy children. *Pediatr. Infect. Dis. J.* 14:673–678.

Yagupsky, P., Katz, O., and Peled, N. (2001). Antibiotic susceptibility of *Kingella kingae* isolates from respiratory carriers and patients with invasive infections. *J. Antimicrob. Chemother.* 47:191–193.

Yagupsky, P., Peled, N., and Katz, O. (2002). Epidemiological features of invasive *Kingella kingae* infections and respiratory carriage of the organism. *J. Clin. Microbiol.* 40:4180–4184.

Welkon, C.J., Long, S.S., Fisher, M.C., and Alburger, P.D. (1986). Pyogenic arthritis in infants and children: a review of 95 cases. *Pediatr. Infect. Dis.* 5:669–676.

16

How to Treat Acute Musculoskeletal Infections in Children

Dr. N.G. Hartwig

1. Summary

Timely diagnosis of musculoskeletal infections is essential to prevent severe complications of this disease. Important symptoms that lead to the diagnosis include high fever, malaise, local pain and loss of function of the involved extremity. In the case of arthritis, swelling of the affected joint may develop quickly in the course of disease.

Imaging techniques include ultrasound and MRI. Both are useful in determining the exact site of infection. Infectious parameters, C-reactive protein and/or sedimentation rate of erythrocytes (ESR) are used in following disease activity during treatment.

The gold standard for diagnosing musculoskeletal infection is microbiological isolation of the organism. Bone-biopsy of an involved metaphysis or aspiration of joint-fluid gives the highest chance of a positive culture. Usually *S. aureus* is the causative micro-organism, although specific underlying diseases may predispose to other bacterial infections.

Antibiotic treatment is aimed at the most likely causative micro-organism. In some instances surgical treatment is necessary, especially in cases of focal and chronic osteomyelitis and when the adjacent joint is involved in the inflammatory process.

Collaboration between paediatrician and orthopaedic surgeon is essential to achieve the best treatment for the patient.

2. Introduction

Although seen as a significant cause of morbidity in infants, infections of the musculoskeletal system are relatively unusual during childhood. The estimated incidence is about 1 in 5000 children in the age group of 0 to13 years (Nelson,

Hot Topics in Infection and Immunity in Children, edited by Andrew J. Pollard and Adam Finn.
Springer, New York, 2006

1997). When present, osteomyelitis and septic arthritis are challenging diseases for the treating physician. Causes, clinical presentation and empirical treatment depend on the age of the patient, the pathogenesis of the infection, duration of the infection and the presence of predisposing factors.

3. Aetiology and Pathogenesis

Based upon clinical manifestations, osteomyelitis can be subdivided into the following subtypes:

1. Acute or haematogenous osteomyelitis
2. Subacute or focal osteomyelitis
3. Chronic osteomyelitis
4. Postoperative osteomyelitis
5. Neonatal osteomyelitis

Acute or haematogenous osteomyelitis is the most frequently encountered type of osteomyelitis during childhood (Nelson, 1997; Gutierrez, 2005). Usually it is seen in children under the age of 5 years who have no specific medical history. They present with acute systemic complaints as a result of bacteria or bacterial products reaching the bloodstream. In the majority of cases, *S. aureus* is found as the causative micro-organism. Sometimes group A streptococci or pneumococci are isolated, especially in younger children. Before routine vaccination against Haemophilus influenzae type b (Hib), this micro-organism played an important role. However through high vaccination coverage, Hib is rarely isolated in cases of osteomyelitis nowadays.

When patients have underlying diseases, other micro-organisms may be isolated, e.g. association between sickle cell anaemia and Salmonella (Burnett et al., 1998).

The bone is infected by bacteria that have spread through the bloodstream. In 30% of cases a history of trauma is present shortly before the onset of the infection. Sometimes boils or other staphylococcal skin-infections might have preceded osteomyelitis. Acute osteomyelitis is sometimes seen in association with chickenpox and secondary impetigo. In these cases Group A streptococci are frequently isolated.

The primary focus of infection is usually the metaphysis. Several factors are responsible for this preference. Firstly, the metaphyses of growing children are well supplied with blood and blood vessels. Secondly, the slow blood flow through the strongly curved capillary network gives the micro-organisms the opportunity to bind to the endothelium and penetrate through the fenestrated endothelial cells into the bone tissue. Additionally, the normal phagocytosing function of the endothelial cells is absent in the metaphyseal blood vessels. Thirdly, there are hardly any immune-cells (e.g., macrophages) present in the metaphysis to respond to the incoming micro-organisms (Sonnen and Henry, 1996). Furthermore, *S. aureus* possesses an adherence factor for fibronectin, a protein that is widely distributed within bone and connective tissue. Using this adherence factor *S. aureus* may easily stick to bone tissue in the metaphyseal region.

Characteristic of the course of osteomyelitis is the spread of the inflammatory exudate to the subperiostal areas through the canals of Volkmann and the Haversian system (Sonnen and Henry, 1996). When the subperiostal pressure increases, the blood flow to the metaphysis and the bone cortex becomes endangered. This eventually leads to bone necrosis and the formation of a sequestrum.

Since there is a close relationship between the metaphysis and the adjacent joint, the infection may easily spread to the joint as well. Sometimes arthritis is the clinical presentation of an osteomyelitis in the adjacent bone, especially when *S. aureus* is isolated.

Subacute or focal osteomyelitis usually occurs after human or animal bites or puncturing of the skin by sharp materials (broken glass or nails). Pathogens reach the bone tissue directly from the surface. The spectrum of bacteria involved is broader and includes *Pseudomonas aeruginosa*, *S. aureus* and anaerobic micro-organisms. Clinical symptoms are usually limited to local swelling, pain and redness. General symptoms are frequently lacking (Jacobs et al., 1989).

Chronic osteomyelitis is a specific type of osteomyelitis in which the primary infection has not been diagnosed or has been insufficiently treated. Smouldering inflammation affects the bone and leads to its destruction. Formation of bone sequestrum, pathological fractures and severe growth retardation are the consequence. Surgical treatment is an essential part of the treatment.

Postoperative osteomyelitis is a type of osteomyelitis that occurs in superficial localised bones (e.g., sternum) after a surgical procedure. Micro-organisms that live on or in the skin cause the infection. In addition to pathogens such as *S. aureus* and *P. aeruginosa*, other micro-organisms may be encountered like *S epidermidis*. Patients with diabetes mellitus or corticosteroid treatment do have a higher risk of developing this type of osteomyelitis.

Neonatal osteomyelitis is caused by the same pathogens as found in neonatal sepsis. Frequently, Group B streptococci are isolated, but *E. coli*, *S. aureus* or Candida albicans are also found as causative pathogens. Communication between blood vessels of the metaphysis and the epiphysis lead to progressive spread of the infection to the epiphysis itself and to the adjacent joint. In comparison with osteomyelitis in the older child, this type is more aggressive in its clinical course. In 40% of cases multiple foci of infections are demonstrated on Tc-scans (Asmar, 1992).

Septic arthritis is an infection of the synovial space. As in osteomyelitis, three main routes of infection exist: haematogenous spread through the capillary vascular network of the synovium, by contiguous spread (e.g., osteomyelitis) or after trauma or surgery. It is a condition that needs immediate treatment. Delay in diagnostic procedures and treatment leads to destruction of the joint cartilage and cause irreparable damage.

4. Signs and Symptoms

In acute osteomyelitis general and systemic symptoms are most prominent. These symptoms reflect concurrent bacteraemia: high fever, general malaise and sometimes circulatory problems. More specific localising symptoms follow a little later. In 60–70% of the cases the infection is present in one of the long bones of an extremity, e.g. femur, tibia and humerus. Sporadically other bones are affected. Local pain during palpation of the extremities is a reliable early symptom of osteomyelitis.

Of the local signs, pain and decreased use of the involved extremity are the most prominent complaints at all ages. Later in the course of disease, when the infection has spread to the periosteal region, the classical symptoms of local swelling, redness and heat are observed.

In older children (above 6 months) osteomyelitis generally is a unifocal disease without spread to a joint.

In focal osteomyelitis and chronic osteomyelitis general signs and symptoms are absent and only local signs are observed: swelling of the involved area, redness and fistulae.

Neonatal osteomyelitis is restricted to the infants aged 2 to 8 weeks. As in acute osteomyelitis systemic symptoms like fever and irritability are prominent. Neonates with osteomyelitis develop local signs more quickly, since the bone cortex is thin and the inflammatory exudate will more easily reach the periostal area. The diagnosis of neonatal osteomyelitis should be suspected in the presence of pseudoparesis of one of the extremities. In neonatal osteomyelitis the infection spreads readily to the joint or the epiphysis. In addition it frequently presents as a multifocal infection.

As with acute osteomyelitis, septic arthritis has few specific symptoms during the early stages of disease. However, local signs such as swelling and pain in the joint will develop rapidly. Passive movements of the extremity evoke pain.

5. Diagnosis

The diagnosis of osteomyelitis or septic arthritis should be suspected in any child who presents with acute or chronic pain in one of the bones. In 50% of cases local pain may indicate the site of infection. The differential diagnosis is dependant on age and comprises greenstick fractures, bacterial sepsis, leukaemia and malignancies of the bone. In case of joint swelling possible diagnoses include haemarthrosis, juvenile idiopathic arthritis (JIA), rheumatic fever, coxitis fugax, Perthes disease or reactive arthritis.

Since individuals with subacute and chronic osteomyelitis present with focal pain and swelling without general symptoms, the differential diagnosis includes infection of the soft tissue or malignancies.

Plain X-ray in early stages of acute osteomyelitis seldom shows abnormalities, since bone destruction will take some time. After some weeks more specific abnormalities may be found. Periostal reaction are a late radiological manifestation of osteomyelitis (Gold et al., 1991).

A three-phase ^{99}Technetium scan (Tc-scan) may demonstrate abnormalities early during the course of disease due to the high influx of leukocytes. Tc-scan is also suitable for localising the infectious site when physical examination gives no exact localisation. However in neonates and individuals with chronic symptoms the Tc-scan is sometimes falsely negative.

Magnetic resonance imaging (MRI) appears to be a very sensitive method to identify osteomyelitis during the early stages of disease (Gold et al., 1991) although it is not specific. Oedema, due to the inflammatory reaction, is made visible with MRI. This technique is also very useful in visualising infections of the spine or to discriminate between vertebral osteomyelitis and discitis.

CT-scan may show alterations to the bone as consequence of the osteomyelitis. It appears less sensitive than MRI and is used for preoperative planning or to discriminate osteomyelitis from osteoid osteoma or chondroblastoma.

Ultrasound may be used if osteomyelitis or arthritis is suspected. Characteristic for osteomyelitis is thickening of the periosteum and hypo-echogenic zones within the bone (Kaiser et al., 1998). Demonstration of excess of fluid in the joint may leave a relative broad differential diagnosis, though it pinpoints to the problem during an early stage of disease, especially in the case of hip problems.

Infectious parameters (C-reactive protein, sedimentation rate of erythrocytes (ESR) and leukocyte count) are sensitive though non-specific. These parameters are useful in following the treatment response (Unkila-Kallio et al., 1994).

If 3 of the following 4 parameters are present: fever, reduced use of extremity, leukocytosis and elevated ESR, the probability of osteomyelitis or arthritis is predicted in 93.1%. When all four parameters are found the diagnosis of osteomyelitis or arthritis is correctly predicted in 99.6% of the cases (Kocher et al., 1999).

Microbiological examination includes blood cultures and culture of material obtained by biopsy, fine needle aspiration or surgical drainage. Blood cultures yield positive results in 30–70% of the cases. Culture of biopsy material yields a positive result in 70–85% of cases and is therefore preferred above blood cultures (Nelson, 1997). A positive microbiological result is the gold standard for defining osteomyelitis or septic arthritis.

6. Treatment

Treatment of patients with osteomyelitis or arthritisshould be undertaken by a paediatrician and orthopaedic surgeon in collaboration. When osteomyelitis or arthritis is considered in the differential diagnosis, one should consider these conditions as a working-diagnosis and treat in conformity. The risk that retrospectively the working-diagnosis appeared to be incorrect does not compensate for the disadvantages of delayed treatment, especially in the case of septic arthritis. Before starting antimicrobial therapy microbiological diagnostics should be performed. Positive results are helpful in adjusting antimicrobial therapy during the switch from intravenous to oral therapy. Bone biopsies have a higher probability of positive culture than bloodculture (Karwowska et al., 1998). However, in the case of osteomyelitis, many physicians have the opinion that without a bone biopsy antimicrobial therapy may be started. In their opinion bone biopsies should only be performed when no adequate response to antimicrobial therapy is observed (Steer and Carapetis, 2004).

In the case of septic arthritis, surgical treatment is preferred since the affected joint can be considered as a cavity filled with pus. Opening or draining of the joint is diagnostic as well as therapeutic. Antimicrobial therapy is still a necessary adjuvant to the surgical procedure.

Joints that are easily reached usually are drained by an arthroscopic procedure, while more deeply situated joints, like the hip joint, are approached by arthrotomy. Specimens of the synovial fluid or biopsies of synovial membrane for culture and pathology can be taken when the joint is flushed with saline solution. Aspiration of synovial fluid alone is insufficient treatment (Dabney and Bowen, 1995) and may lead to irreversible damage to the affected joint. Morbidity following arthroscopy is considered low while the time to cure the inflammatory response is reduced by the procedure leading to reduced costs with respect to the hospital stay (Smith, 1986). Drainage procedures using silastic drains that will stay in for some time, are not used anymore. When relapse occurs a second arthroscopic drainage may be performed.

Septic arthritis and osteomyelitis are frequently associated (Perlman et al., 2000). When one has doubts about the presence of concurrent osteomyelitis, a bone biopsy of the metaphyseal area can be performed during the drainage procedure of the joint.

Surgical drainage of osteomyelitis is indicated when abscesses or sequestra are demonstrated by radiological techniques.

Empirical antimicrobial therapy (Table 16.1) is based on the most likely micro-organism, with respect to the age of the patient, the clinical presentation and, when available, the results of the Gram-stain and the culture of the biopsy material (Hartwig et al., 2005).

In older children with the clinical presentation of acute haematogenous osteomyelitis without spread to a joint, empirical treatment can be restricted to flucloxacillin or a second generation cephalosporin, since almost always *S. aureus* is found as the causative micro-organism. When there is β-lactam hypersensitivity, clindamycin is a good alternative. Vancomycin is a large molecule and does not penetrate well into bone tissue. It is therefore less suitable for treating osteomyelitis. However in some instances, when *S epidermidis* is isolated, it is the only treatment option.

For young children or children with haemoglobinopathies (e.g., Sickle cell disease) empirical treatment should include cover for Gram-negative micro-organisms. Therefore 3rd generation cephalosporins are more suitable. When children appear toxic, an aminoglycoside can be added to the treatment. This will lead to faster sterilisation of the bloodstream. Based on cultures and sensitivity testing, it may be possible to switch to antibiotics with a narrower spectrum of activity (Table 16.2) (Hartwig et al., 2005).

The intravenous route is preferred for initial therapy. A switch to oral therapy may be considered if the patient responds well, the CRP has fallen to approximately 1/3 of the value at presentation, and the micro-organism and sensitivity testing is available (see Table 16.3). The optimal duration of treatment for osteomyelitis and arthritis is not well defined. Previously, the total treatment duration wase 6 weeks or more. However, shorter treatment-courses seem to be equally effective. The risk

Table 16.1. Empirical antibiotic therapy in osteomyelitis/arthritis

Host factors	Medication*	Daily doses
Acute haematogenous osteomyelitis/arthritis age <6 months or insufficient Hib vaccination	1. Amoxicillin/clavulanic acid 2. Cefuroxime (+ AMG optional)	150/15 mg/kg in 3× 150 mg/kg in 3× (5 mg/kg in 1×)
Acute haematogenous osteomyelitis/arthritis age >6 months	1. Flucloxacillin 2. Cefuroxime	150 mg/kg in 3× 150 mg/kg in 3×
Acute haematogenous osteomyelitis/arthritis haemoglobinopathy	1. Cefotaxime 2. Ceftriaxone	150 mg/kg in 3× 100 mg/kg in 1×
Subacute/focal osteomyelitis	1. Ceftazidime + AMG (tobramycin or gentamicin)	150 mg/kg in 3× 5 mg/kg in 1×
Chronic osteomyelitis	1. Flucloxacillin 2. Cefuroxime	150 mg/kg in 3× 150 mg/kg in 3×
Post-operative osteomyelitis/ arthritis	1. Flucloxacillin + AMG (tobramycin or gentamicin)	150 mg/kg in 4× 5 mg/kg in 1×
Neonatal osteomyelitis/arthritis	1. Amoxicillin/clavulanic acid 2. Cefotaxime + Flucloxacillin 3. Ceftriaxone + Flucloxacillin	150/15 mg/kg in 4× 150 mg/kg in 3× 150 mg/kg in 4× 100 mg/kg in 1× 150 mg/kg in 4×
Spondylodiscitis	1. Flucloxacillin	150 mg/kg in 3×
Septic arthritis Sexual activity	1. Cefotaxime 2. Ceftriaxone	150 mg/kg in 3× 100 mg/kg in 1×

*In the column "Medication" the given choices are equal alternatives.

of relapse is increased when treatment duration falls below 3 weeks (Dich et al., 1975). When the patient responds quickly to treatment and no complications are present, duration of treatment is generally advised to be 3 to 4 weeks.

When osteomyelitis is due to puncture-wounds, *P. aeruginosa* and *S. aureus* are the most frequently encountered micro-organisms. In the literature several combinations of antibiotics are proposed in these situations. In case of Pseudomonas infections surgical exploration and cleansing of the wound is part of the treatment, followed by 1 to 2 weeks of antibiotic treatment (Jacobs et al., 1989). Quinolones are not often used for this indication in children due to concerns about the safety of their use. Although quinolones penetrate well into bone tissue, *P. aeruginosa* may become resistant during treatment.

Osteomyelitis after bites or in relation to odontogenous infections should include cover for against *S. aureus* and anaerobes. Empirical treatment with amoxicillin/clavulanic acid or clindamycin is preferred.

Empirical antimicrobial treatment for neonatal osteomyelitis should cover Group B streptococci, *E.coli* and *S. aureus* (Jacobs et al., 1989). Treatment with amoxicillin/clavulanic acid (in combination with an aminoglycosid) or a third

Table 16.2. Antibiotic treatment based on isolated micro-organism and antibiotic
sensitivity testing

advice for children >4 weeks of age

Pathogen	Medication*	Daily doses	Duration of treatment**
Group A streptococci	Benzylpenicillin	200.000 U/kg in 6×	3–4 weeks
Group B streptococci	Benzylpenicillin	200.000 U/kg in 6×	4–6 weeks
E. coli	Cefotaxime + AMG (tobramycin or gentamicin)	150 mg/kg in 3× / 5 mg/kg in 1×	4–6 weeks
H. influenzae b (β lactamase neg.)	Amoxicillin	150 mg/kg in 4×	4 weeks
H. influenzae b (β lactamase pos.)	Amoxicillin/clavulanic acid	150/15 mg/kg in 4×	4 weeks
K. kingae	Benzylpenicillin + AMG (tobramycin or gentamycin)	200.000 E/kg in 6× / 5 mg/kg in 1×	3–4 weeks
N gonorrhoeae	Benzylpenicillin	200.000 E/kg in 6×	3 weeks
S aureus	Flucloxacillin	150 mg/kg in 4×	3–4 weeks
S pneumoniae	Benzylpenicillin	200.000 E/kg in 6×	3–4 weeks
Salmonella sp.	Cefotaxime	150 mg/kg in 4×	4 weeks
P aeruginosa	Ceftazidime + AMG (tobramycine of gentamicine)	150 mg/kg in 3× / 5 mg/kg in 1×	4–6 weeks
Candida albicans	Fluconazole	15–20 mg/kg in 1×	6 weeks

*Choice of medications is preference although dependent on the sensitivity testing.
**Duration is dependant on clinical response and fall in infectious parameters.

Table 16.3. Options for switch to oral therapy when pathogen is sensitive

Medication	Daily dose	Maximum daily dose	Remarks
Amoxicillin	50 mg/kg in 3×	4000 mg	Bactericidal, resorption app. 70–90% T1/2: 1–2 hours
Amoxicillin/ clavulanic acid	50/12.5 mg/kg in 3×	4000/1000 mg	Bactericidal, resorption amoxicillin 70–90%, clavulanic acid 60% T1/2: 1–2 hours
Clindamycin	25 mg/kg in 4×	1800 mg	Bacteriostatic, resorption >90% T1/2: 2–3 hours
Co-trimoxazole	36 mg/kg in 2×	1600/320 mg	Bactericidal, resorption >90% T1/2: 8–17 hours
Flucloxacillin	50 mg/kg in 4×	4000 mg	Bactericidal, resorption app. 50% T1/2: 1 hour
Fucidin acid	50 mg/kg in 3×	1500 mg	Resorption app. 90% T1/2: 9 hours
Linezolid	30 mg/kg in 3×	1200 mg	Resorption >90% T1/2: 5–7 hours

generation cephalosporin in combination with fluxcloxacillin is the preferred treatment. No information is present on switch-therapy in neonates. Because of the multifocal presentation and the severity of the disease a treatment duration of 4 to 6 weeks of intravenous antibiotics is usually prescribed.

7. Complications

Relapse, pathological fracture and growth disturbance as consequence of damage to the growth plate are the most common complications. Sometimes when the epiphysis is involved malformations of the joint may occur. Despite adequate therapy, relapse rates are 4%, and result in chronic osteomyelitis (Dich et al., 1975). Unfortunately, these relapses are difficult to treat and respond poorly on antibiotic treatment. Eventually, these patients develop chronic fistulae that drain pus. Extensive surgical treatment with bone resection and sometimes bone transplants may be necessary.

Septic arthritis may lead to cartilage damage and evolve to form an arthrosis. An immobile joint may also be the consequence of adhesions. Continuous passive movement on a continuous passive motion machine early during treatment may prevent this complication. In severe cases arthroscopic adhesiolysis may be necessary, followed by intensive physiotherapeutic treatment.

Orthopaedic follow-up for a prolonged period is desirable in order to anticipate limb-length discrepancies, even in seemingly uncomplicated cases.

8. Prognosis

An early diagnosis, antibiotic treatment and, when necessary, surgical drainage are the most important determinants of a good prognosis. Most complications are encountered in neonatal osteomyelitis or in cases where the osteomyelitis/arthritis has affected hip or shoulder joints.

9. Prevention

Vaccination has contributed to the elimination of Hib as a causative pathogen of osteomyelitis and arthritis. A further reduction in rates of osteomyelitis and arthritis may be expected after the introduction of pneumococcal vaccination. At present no vaccination has been developed against *S. aureus*, the most common pathogen. Except for standard hygiene measures and the prevention of secondary wound infections, no other preventive strategies are available.

The need for tetanus prophylaxis should always be considered.

Antibiotic prophylaxis for surgical procedures will reduce the chance of postsurgical osteomyelitis. In general, one dose of antibiotics at the start of the surgical procedure will suffice (Boxma et al., 1996). Prophylaxis with mupirocin ointment applied in the nasal vestibulum seems to be as effective as an intravenous dose of antibiotics in the prevention of sternal infections after thoracotomy (Kluytmans et al., 1996).

References

Asmar BI. Osteomyelitis in the neonate. Infect Dis Clin North Am 1992;6:117–32.

Boxma H, Broekhuizen T, patka P, et al.. Randomised controlled trial of single-dose antibiotic prophylaxis in surgical treatment of closed fractures: The Dutch Trauma Trial. Lancet 1996; 347:1133–7.

Burnett MW, Bass JW, Cook BA. Aetiology of osteomyelitis complicating sickle cell disease. Pediatrics 1998;12:228–33.

Dabney KW, Bowen JR. Complications of musculoskeletal infections. In: Epps CH, Bowen JR. Complications in pediatric orthopaedic surgery. Philadelphia: Lippincott, 1995;751–87.

Dich VQ, Nelson JD, Haltalin KC: Osteomyelitis in infants and children. A review of 163 cases. Am J Dis Child 1975;129:1273–8.

Gold RH, Hawkins RA, Katz RD. Bacterial osteomyelitis: findings on plain radiography, CT, MR and scintigraphy. Am J Roentgenol 1991;157:365–70.

Gutierrez K. Bone and joint infections in children. Pediatr Clin N Am 2005;52:779–94.

Hartwig NG. Antibacteriele therapie. In: Hartwig NG, de Laat PCJ, Hanff LM. Vademecum Pediatrische Antimicrobiele Therapie (ISBN 90-75340-10-9). 2005:24–58.

Jacobs RF, McCarthy RE, Elser JM. Pseudomonas osteochondritis complicating puncture wounds of the foot in children: a 10-year evaluation. J Infect Dis 1989;160:657–61.

Kaiser S, Jorulf H, Hirsch G. Clinical value of imaging techniques in childhood osteomyelitis. Acta Radiol 1998;39:523–31.

Karwowska A, Davies HD, Jadavji T. Epidemiology and outcome of osteomylitis in the era of sequential intravenous-oral therapy. Pediatr Infect Dis J 1998;17:1021–6.

Kluytmans JA, Mouton JW, VandenBergh MF, et al.. Reduction of surgical-site infections in cardiothoracic surgery by elimination of nasal carriage of *StaphylococcuS. aureus*. Infect Control Hosp Epidemiol 1996;17:780–5.

Kocher MS, Zurakowski D, Kasser JR. Differentiating between septic arthritis and transient synovitis of the hip in children. An evidence-based clinical prediction algorithm. J Bone Joint Surg Am 1999;81:1662–70.

Nelson JD. Toward simple but safe management of osteomyelitis. Paediatrics 1997;99:883–4.

Perlman MH, Patzakis MJ, Kumar PJ, Holthom P. The incidence of joint involvement with adjacent osteomyelitis in pediatric patients. J Pediatr Orthop 2000;20:40–3.

Sectie Pediatrische Infectieziekten. Empirische antibacteriele therapie op basis van vermoedelijke diagnose. In: Hartwig NG, Kornelisse R, en Verduin CM. Blauwdruk Pediatrische Antimicrobiele Therapie. 2001:22–8.

Smith MJ. Arthroscopic treatment of the septic knee. Arthroscopy 1986;2:30–4.

Sonnen GM, Henry NK. Pediatric bone and joint infections. Pediatr Clin North Am 1996;43:933–47.

Steer AC, Carapetis JR. Acute hematogenous osteomyelitis in children: recognition and management. Pediatr Drugs 2004;6:333–46.

Unkila-Kallio L, Kallio MJ, Eskola J, et al.. Serum C-reactive protein, erythrocyte sedimentation rate, and white blood cell count in acute haematogenous osteomyelitis of children. Pediatrics 1994;93:59–62.

17

Prevention of Transmission of HIV-1 from Mothers to Infants in Africa

Hoosen Coovadia and Derseree Archary

1. Introduction

It is not often appreciated that prevention of mother-to-child-transmission (pMTCT) of HIV is one of the great success stories in the rather patchy and dismal record of measures tested to halt and reverse this global disaster. Prevention of mother-to-child-transmission (pMTCT) is a highly cost-effective intervention (<$75 per Disability Adjusted Life Year (DALY) gained), exceeded only by blood safety, and targeted condom distribution (<$1 per DALY gained) (Creese et al., 2002).

The gains made in this field have been almost entirely based on interventions (antiretrovirals; planned caesarean section; safer delivery; avoidance of breastfeeding in richer communities) directed at reducing transmission of HIV from the mother to the baby (Dabis and Ekpini, 2002). This narrow focus on the perinatal period ignores a more comprehensive approach. A wider approach, promoted by the World Health Organization (WHO), includes:

1. Primary prevention of HIV in young women.
2. Reducing unintended pregnancies among HIV-infected women.
3. Preventing vertical transmission.
4. Providing care, treatment and support to HIV infected women and their families.

A recent cost-effectiveness study of pMTCT in eight African countries (Sweat et al., 2004) confirms the validity of this idea. Lowering HIV prevalence among women by 1.25% or reducing unintended pregnancies among HIV-infected women by 16%, yielded an equivalent reduction in HIV-infected infant transmissions to that achieved by pMTCT using single-dose Nevirapine (sdNVP). Therefore small reductions in maternal HIV prevalence or in unintended pregnancy by HIV infected women have equivalent impacts as pMTCT; the latter reinforces the need to integrate family planning into pMTCT programs in developing countries (Duerr et al., 2005).

Hot Topics in Infection and Immunity in Children, edited by Andrew J. Pollard and Adam Finn. Springer, New York, 2006

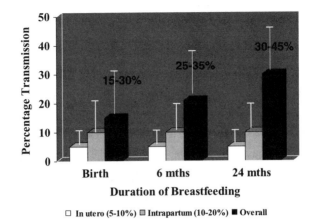

Figure 17.1. Risks and timing of mother-to-child transmission of HIV-1.

2. Success, Failures & Barriers

The number of children being newly infected by Human Immunodeficiency Virus (HIV) every year is staggering (UNAIDS, 2004a). Everyday roughly 1800 children are infected with HIV throughout the world. In 2004, 640000 (range: 570000–750000) children under the age of 15 years had been newly infected worldwide; most were the result of mother-to-child transmission (MTCT) of the virus, and 510000 (460000–600000) of these children died in that year (UNAIDS, 2004a). At the end of 2004, it was estimated that 1.9 million children (range 1.7–2.3 million) were living with HIV in Sub-Saharan Africa (UNAIDS, 2004a). The tragedy is that today we have the knowledge, we have the means, and we have willing recipients of our preventive measures, to cut down drastically this terrible burden of infections.

HIV infected newborns are becoming uncommon in the USA and Europe. In 2003, less than 1000 children were estimated to have become infected with HIV in North America and Western Europe and less than 100 in Australia and New Zealand (UNAIDS, 2003). In a recent New York Times report, in the USA there were 2000 HIV infected infants in 1990, by 2003 the number was 200; in New York City the figures were 321 in 1990 and only 5 in 2003 (Santora, 2005). These achievements can be repeated in developing countries provided some critical challenges can be overcome. The challenges in implementing ARV programs in Africa, including pMTCT programs, are: coping with the huge unmet need and growing demand for prevention and treatment; the financial resource gap; ensuring equitable access; efficient and effective prevention and treatment programs; prevention efforts combined with treatment and care programs; coordination, collaboration, and harmonization of all efforts by funders, and other stakeholders; strengthening health systems, and longer term strategies (UNAIDS, 2004b).

Prevention of MTCT programs has also succeeded in some developing countries such as well known examples like Thailand and others including Brazil (Kreitchmann et al., 2004), Western Cape Province in South Africa (Colvin et al., 2005), and Cameroon (Ayouba et al., 2003). One of the reasons for the lack of public and professional appreciation of these successes is the negative publicity which has attended the use of sdNVP to reduce MTCT. Single-dose NVP (a tablet to the mother during labor, a dose to the baby soon after birth) (Guay et al., 1999; Jackson et al., 2003) is a wonderfully simple and safe regimen which is both affordable and effective in developing countries. The criticisms have often been politically inspired and related to presumed toxicities of the drug; reliability of the scientific evidence of efficacy; and conduct of the key trial in Uganda (HIVNET012) (Institute of Medicine Report, April 2005). None of these are valid and have obscured the large scale benefits which have resulted from wide-scale implementation of the sdNVP regimen in Africa. National sdNVP programs have provided the critical spark to initiation and extension of HIV services (especially antiretrovirals) to mothers, their partners and their families. Despite the appropriateness and practicality of the sdNVP regimen, bottlenecks in delivery have often hindered progress and reduced effectiveness. These factors account for the failure of pMTCT programs in parts of South Africa (Colvin et al., 2005) and Kenya (Temmerman et al., 2003; Quaghebeur et al., 2004). In one area in Kenya, the 14 week transmission rate before the prevention program was 21.7% and it was 18.1% after the program (Quaghebeur et al., 2004).

3. Risks of Mother-to-Child-Transmission of HIV-1.

The timing and transmission of HIV-1 from Mother-to Infant is shown in the Table 17.1 (Cock et al., 2000). Transmission takes place in utero, intrapartum, and postnatally through breastfeeding. The in utero rate is 5–10%, intrapartum rate is 10–20%, and the postnatal (breastfeeding) rates vary according to the duration of breastfeeding. Avoidance of breastfeeding which works well in richer populations, is not a realistic option for developing countries because the dangers of formula feeding often outweigh the hazards of breastfeeding transmission of HIV

Table 17.1.   Timing and risks of transmission

Timing and risks of transmission	Transmission Rate
During pregnancy	5–10%
During labor and delivery	10–20%
During breastfeeding	5–20%
Overall without breastfeeding	15–30%
Overall with breastfeeding till 6 months	25–35%
Overall with breastfeeding till 18 to 24 months	30–45%

(De Cock et al., 2000.).

Table 17.2. Transmission rates (%)/month for prevailing (presumed Exclusive# and Mixed*) breastfeeding practice

Study	Rate of transmission (%)/month
BHITS Meta-analysis (BHITS Study Group, 2004)	0.71
Nairobi, Kenya Study (Nduati et al., 2000)	0.7
Durban, South African Study (Coutsoudis et al., 2001)	1.1

#Exclusive breastfeeding: an infant receives only breastmilk, and no other liquids or solids, not even water, with the exception of drops or syrups consisting of vitamins, mineral supplements or medicines.
*Mixed Feeding: feeding both breast milk and other foods or liquids (WHO, 2004a).

(Coutsoudis et al., 2002). We have had least success in providing convincing evidence of interventions to reduce breastfeeding transmission; one major contributing factor to this, is uncertainty about the precise rate of breastfeeding transmission. The data has improved recently (see Table 17.2). The best evidence obtained from a pooled analysis of most of the breastfeeding studies available suggests that the rate of breastfeeding transmission is about 0.71% per month (BHITS Study Group, 2004); this figure is the consequence of mostly mixed feeding (where breastmilk is supplemented by other fluids, milks or foods). Studies from Kenya (Nduati et al., 2000) and South Africa (Coutsoudis et al., 2001) show close similarities in transmission rates, while the latter also reveals the effect of exclusive breastfeeding (the recommended type of breastfeeding) in reducing the risk of transmission (see also Table 17.3). A recent study from Zimbabwe (ZVITAMBO: Iliff et al., 2005) strongly supports the finding of lower rates with exclusive breastfeeding for 6 months; the rates of transmission are 0.29% per month for exclusive breastfeeding, and 0.98% for mixed feeding. Compared with exclusive breastfeeding, early mixed breastfeeding was associated with a 4 fold greater risk of postnatal transmission at 6 months (Iliff et al., 2005).

Table 17.3. Cumulative transmission and transmission rates (%)/month for Exclusive and Mixed breastfeeding

Study	Cumulative Transmission (%)		Transmission Rates (%)/month	
	Exclusive	Mixed	Exclusive	Mixed
Durban Study				
6 weeks–15 months	9.7	14	–	–
ZVITAMBO				
6 weeks–6 months	1.3	4.4	0.29	0.98
6 weeks–18 months	6.94	13.92	0.42	0.84

4. Prevention Trials

4.1. Industrialized Countries

The experience in industrialized countries had previously shown that there is a gradient in transmission rates determined by the number of ARVs provided: in those given no drugs the MTCT rate was 20%; with monotherapy it was 8%–10.4%; dual therapy gave a rate of 3.8%; and with combination ARVs (>2 ARVs) and Highly Active Antiretroviral Therapy (HAART) the figure was down to 1.2%–3% (Cooper et al., 2002; Peters et al., 2003).

The landmark PACTG 076 trial with an efficacy of 68% had established the principle and value of ARVs (AZT) in markedly reducing MTCT (Conner et al., 1994). However, there were a number of reasons why the results were inapplicable to poorer populations in Africa and Asia. These, inter alia, included affordability (the price of ARVs were very high at the time), health services required for implementation (especially antenatal and delivery services), and avoidance of breastfeeding (breastfeeding is the norm in developing countries). With the subsequent precipitous fall in the price of ARVs after the year 2000 the cost of these drugs fell away as an inhibitory element.

4.2. Developing Countries

Shorter regimens, using cheaper drugs, and combination ARVS suitable for developing countries, were tested in Africa and Thailand (See Tables 17.4; 17.5 & 17.6).

Among non-breastfeeding populations the efficacy of AZT ranged from 70% (Lallemant et al., 2000) to 50% (Shaffer et al., 1999). In breastfeeding populations, the important indicator of success is the transmission at the time when breastfeeding is discontinued; this is often between 18 months to 24 months among poor women. The higher efficacies obtained at 6 weeks of age, 63% in the PETRA trial (Arm A) (Petra Study Team 2002) and 47% in HIVNET012 (Guay et al., 1999), and between 3 months and 6 months in the West African Studies of 36–48% (Wiktor et al., 1999; Dabis et al., 1999; LeRoy et al., 2002) had decreased between 18–24 months to 41% in HIVNET012 (Jackson et al., 2003), 33% in PETRA (Arm A) (Petra Study Team, 2002), and 23–28% in the West African studies (Leroy et al., 2002).

For women who come too late to antenatal clinic to receive one of the standard antenatal and/or intrapartum regimens, the use of post-exposure ARVs to the neonate only, can reduce MTCT. A dose of NVP to the newborn results in a transmission rate at 6–8 weeks of 20.9%, whilst the addition of AZT to the NVP reduces transmission further to 15.3% (Taha et al., 2003). If the mother had also received a dose of NVP, the addition of AZT to the infant NVP produces no benefit. These results have to be considered in terms of the excellent results obtained with the studies described above, and therefore are not as impressive. They serve as a back-up for the minority of women who miss the antenatal and/or intrapartum opportunity for an optimum course of ARVs for prevention of MTCT.

Table 17.4. Antiretroviral interventions to reduce mother-to-child transmission of HIV: Non-breastfeeding

Study	Mother			Infant	Transmission rate: Active vs placebo	Relative efficacy
	Antepartum	Intrapartum	Postpartum			
PACTG 076 (Connor et al., 1994) *Zidovudine*	100 mg 5×/day (po) (from week 14–34)	At onset: 2 mg/kg (iv) for 1 hr, then 1 mg/kg/hr	No	2 mg/kg qid (po) (for 6 weeks)	At 18 months: 8.3% vs 25.5%	68%
Harvard University – Thai-PHPT-1 (Lallemant et al., 2000) Arm LL *Zidovudine*	300 mg 2×/day (po) (from week 28)	300 mg 3 hourly	No	2 mg/kg qid (for 6 weeks)	At 6 months Kaplan Meier 6.5%	
Arm LS *Zidovudine*	300 mg 2×/day (po) (from week 28)	300 mg 3 hourly	No	2 mg/kg qid (for 3 days) 2 mg/kg qid (for 6 weeks) 2 mg/kg qid (for 3 days) No	4.7%	
Arm SL *Zidovudine*	300 mg 2×/day (po) (from week 35)	300 mg 3 hourly	No		8.6%	

Arm SS *Zidovudine*	300 mg 2×/day (po) (from week 35)	300 mg 3 hourly	No		10.50%	50.1%
CDC Thai (Shaffer et al., 1999) *Zidovudine*	300 mg 2×/day (po) from week 36	At onset 300 mg, then 3 hourly (po)	No		At 6 months 9.4% for ZDV vs 18.9% Placebo	
Placebo	Placebo	Placebo				
Harvard University Thai-PHPT-2 (Lallemant et al., 2004) *Zidovudine*	300 mg 2×/day (po) from week 28	300 mg 3 hourly	No	2 mg/kg qid (1–6 weeks)	At 6 months	
NVP		200 mg		6 mg	1.1%	
Placebo		Placebo		Placebo	6.3%	
Thailand (Chalermchokcharoenkit et al., 2004) *Zidovudine*	300 mg 2×/day (po) (week 34–36)	300 mg 3 hourly	No	2 mg/kg (for 4 weeks)	At 4 months Kaplan Meier: 4.6%	
NVP		200 mg		2 mg/kg at 2–3 days		

Table 17.5. Antiretroviral interventions to reduce mother-to-child transmission of HIV: breastfeeding – initial studies

	Treatment				Transmission rate: Active vs placebo	Relative efficacy
	Mother			Infant		
	Antepartum	Intrapartum	Postpartum			
Ivory Coast (Wiktor et al., 1999) *Zidovudine*	300 mg 2×/day (po) (from week 36)	At onset: 300 mg, then 300 mg 3 hourly (po)	No	No	At 3 months 15.7% vs 24.9%	37%
Ivory Coast (Dabis et al., 1999) / **Burkina Faso** (LeRoy et al., 2002) *Zidovudine*	300 mg 2×/day (po) (from week 36–38)	At onset 600 mg single dose (po)	300 mg 2×/day (po) (for 1 week)	No	At 6 months 18.0% vs 27.5% At 24 months 21.5% vs 30.6%	36–48% 23–28%
Petra Trial (Petra Study Team, 2002) Arm A *Zidovudine*	300 mg 2×/day (po) (from week 36)	At onset 300 mg 3 hourly (po)	150 mg 2×/day (po) (for 1 week)	4 mg/kg 2×/day (po) (for 1 week)	At 6 weeks 5.7% vs 15.3%	63%
3TC *Lamivudine*	150 mg 2×/day (po) (from week 36)	150 mg 12 hourly (po)	300 mg 2×/day (po) (for 1 week)	2 mg/kg 2×/day (po) (for 1 week)	At 18 months 14.9% vs 22.2%	33%

Arm B *Zidovudine*	No	300 mg 3 hourly (po)	300 mg 2×/day (po) (for 1 week)	4 mg/kg 2×/day (po) (for 1 week)	At 6 weeks 8.9% vs 15.3%	42%
3TC *Lamivudine*	No	150 mg 12 hourly (po)	150 mg 2×/day (po) (for 1 week)	2 mg/kg 2×/day (po) (for 1 week)	At 18 months 18.1% vs 22.2%	7%
HIVNET 012 (Guay et al., 1999; Jackson et al., 2003)					At 14 to 16 weeks (NVP vs ZDV)	
Nevirapine	No	At onset 200 mg (po)	No	2 mg/kg within 72 hours of birth	13.1% vs 25.1%	47%
Zidovudine	No	600 mg (po), then 3 hourly until delivery	No	4 mg/kg (po) bid for 7 days after delivery	At 18 months (NVP vs ZDV) 15.7% vs 25.8%	41%
SAINT (Moodley et al., 2003)		At onset:			ZDV/3 TC vs NVP At 8 weeks	
Arm1 *Zidovudine*	No	600 mg (po), then 300 mg 3 hourly until delivery	300 mg (po) bid for 1 week	>2 kg: 12 mg(po) bid for 1 week		
3TC *Lamivudine*	No	150 mg (po), then 150 mg every 12 hours until delivery	150 mg (po) bid for 1 week	<2 kg: 4 mg/kg bid for 1 week	9.3% (ZDV/3TC)	
					vs	
Arm2 *Nevirapine*	No	200 mg (po) (additional dose 48 hours later if still in labor)	200 mg (24–48 hours)	>2 kg: 6 mg (po) bid for 1 week <2 kg: 4 mg/kg bid for 1 week 6 mg – (24–48 hrs)	12.3% (NVP)	

Table 17.6. Antiretroviral interventions to reduce mother-to-child transmission of HIV: breastfeeding- recent studies

Study	Mother			Infant	Transmission rate: Active vs placebo
	Antepartum	Intrapartum	Postpartum		
Ditrame Plus 1.0 (Dabis et al., 2005) (2001–2002)		At onset			At 6 weeks
					14.7%#
Zidovudine	300 mg × 2 day (po) from week 36	600 mg	300 mg × 2 day for 1 week	2 mg/kg qid (7 days)	6.5%
NVP		200 mg		2 mg/kg on day 2	
Ditrame Plus 1.1 (Dabis et al., 2005) (2002–2003)					At 6 weeks
Zidovudine	300 mg × 2 day (po) from week 32	600 mg	300 mg × 2 day for 3 days	2 mg/kg qid (7 days)	4.7%
Lamivudine *3TC*	150 mg bid from week 32	300 mg	150 mg bid for 3 days		
NVP		200 mg		2 mg/kg on day 2	
MASHI (Shapiro et al., 2005)					At 7 months
Zidovudine	300 mg × 2 day (po) from week 34	600 mg		Breastfed: 2 mg qid for 6 months Formula-fed: 2 mg qid for 1 month	Breastfed: 9.1% Formula fed: 5.6%
±NVP *±Placebo*		200 mg Placebo		2 mg/kg Placebo	1–7 months Breastfed: 4.5%*

#Transmission based on historical data.
*cf BHITS: 4.2% and ZVITAMBO: 4.4%; ZDV alone.

It became clear at this stage that while shorter courses significantly reduced transmission and therefore were suitable for developing countries, ARV prophylaxis which extended from the antepartum period through delivery and postpartum, thereby reducing viral load, was the most effective. Longer antepartum courses (from about 28 weeks of pregnancy) were better than shorter antenatal courses, and combination ARVs improved outcome over single ARVs. sdNVP however remained extremely useful because of its efficacy (similar to PETRA ARM B), affordability, and ease of implementation.

In a review of prevention of MTCT trials in developing countries, the Ghent IAS Working Group in 2004 (38) drew the following conclusions;

1. The most frequently used ARVs for MTCT prevention were AZT, AZT and 3TC, and NVP.
2. The MTCT rates over 24 months without ARVs were between 30% to 45%.
3. The MTCT rates over 24 months with ARVs were between 16% to 23%.

A recent pooled analysis compared the efficacy of different ARV regimens in reducing the risk of vertical transmission at 6 weeks in breastfeeding African populations (Leroy et al., 2005). Compared to Placebo the PETRA Arm A was the most effective, next was short-course antenatal AZT, and then sdNVP. Compared to sdNVP only the longest regimen of PETRA (Arm A) was significantly more effective.

4.3. Recent Trials of Combination ARVs in Non-Breastfeeding Populations

More recently the addition of sdNVP to short course ARV regimens, in both formula-fed and breastfed groups, significantly improved efficacy. A study in Thailand (Lallemant et al., 2004) has shown that adding sdNVP to short-course AZT (beginning at 28 weeks pregnancy) can achieve efficacies comparable to the best reported from anywhere in the world. The MTCT rate was 6.3% among 348 women and their infants who received the standard-of-care short course regimen of AZT, and 1.1% among 353 women who received the standard-of-care AZT plus sdNVP (Table 17.4). WHO now recommends this as a "highly efficacious . . . and simplest regimen" for prevention of MTCT of HIV-1 for women who are not markedly immuno-suppressed and therefore who do not require chronic ARV treatment for themselves (WHO, 2004b). Ironically, this dual regimen is most efficacious in the most immuno-suppressed women, precisely the group which would qualify for HAART in most African countries (Shapiro et al., 2005). Another Thai study (Chalermchokcharoenkit et al., 2005) in which the same ARV combination was used, but AZT was begun later, at 34 weeks to 36 weeks, showed a transmission rate of 4.6% (Table 17.4). However, the addition of sdNVP to pregnant women already receiving combination ARVs or the long course of PACTG 076, does not improve efficacy (Dorenbaum et al., 2002).

It is worth remembering that these studies were conducted in nonbreastfeeding women. In breastfeeding populations the recent results are also impressive.

4.4. Recent Trials of Combination ARVs in Breastfeeding Populations

In a West African study (Dabis et al., 2005), combination of AZT (from 36 weeks) and sdNVP resulted in a rate of 6.5% (95%CI 3.9%–9.1%) at 4–6weeks, and the transmission was 4.7% (2.4%–7.0%) with the addition of 3TC from 36 weeks to the regimen. These MTCT rates can be compared to historical data from the same site which showed figures of 14.7% at 4–6 weeks with the use of AZT alone (LeRoy et al., 2002).

The shortcomings of the above West African trial are the small proportions of breastfeeding women (46%–54%), and the very early end point at 4–6 weeks for establishing the transmission rates. In many parts of Africa and Asia women breastfeed for longer periods and transmission of HIV occurs as long as they breastfeed.

In the MASHI trial in Botswana (Shapiro et al., 2004), antenatal AZT from 34 weeks provided the backbone for a number of subsequent interventions. Breastfed infants were given AZT for 6 months and there was an 86% adherence to this prophylactic drug. Median duration of breastfeeding was 5.8 months, and 18% of the women breastfed exclusively, whilst 50% gave mixed feeding. The MTCT rates in breast- and formula fed infants are given in Table 17.6. The incremental transmission due to breastfeeding (i.e. between 1–7 months) with AZT was 4.5%; there was no comparison group of breastfed infants who had not been given AZT. In order to provide a rough idea of the benefits of the AZT, the transmission rates obtained in recent breastfeeding studies without ARV interventions, can be used: the transmission was 4.2% between 6 weeks and 6 months in BHITS (BHITS Study Group, 2004) and 4.4% between 6 weeks and 6 months in the ZVITAMBO study (Iliff et al., 2005). It would appear therefore that AZT administration to infants as a prophylactic to reduce breastfeeding transmission was ineffective. The larger benefits of the interventions as measured by the proportion of infants surviving and free of HIV infection at 18 months was similar between the formula- and breast-fed infants. The Infant Mortality Rates were also similar between the formula fed and the breastfed groups. In comparing the two feeding groups, the results imply that the disadvantages of HIV transmission through breastfeeding are compensated by increases in survival. Moreover, the benefits of breastfeeding in infants at-risk of HIV infection are demonstrated by the mortality rates at 7 months: these were 9.3% in the formula fed infants and 4.9% in the breastfed infants. Deaths were mostly due to diarrhea and pneumonia. In general, this interpretation of the data is supported by the finding of more hospitalizations and clinical signs in the formula fed infants. The use of AZT in the breastfed infants produced hematological side effects among a proportion of the infants.

In two other unpublished trials, MITRA (Kilewo et al., 2002) and SIMBA (Vyankandondera et al., 2003), there are characteristics of the study design which preclude conclusions on the efficacy of infant ARVs to reduce breastfeeding transmission of HIV-1.

In the open-label MITRA trial in Tanzania where the effect of infant 3TC, with prior course of a week of AZT, was evaluated, the ARV treated children had transmission rates of 1.8% at 6 weeks (n = 114), and only 3.1% at 3 months (n = 78). There was no control group against which to test this effect of 3TC.

In the SIMBA study, conducted in Rwanda and Uganda, all women had received AZT and ddI from 36 weeks of pregnancy, and AZT intrapartum and after delivery for one week. Infants were randomized to 3TC or to NVP for 6 months. The transmission attributable to breastfeeding by 6 months was 2.4% (9/375 infants). There was no statistical difference between the two arms, and the mean duration of breastfeeding was 3.3–3.5 months. This finding was compared within the HIV Prevention Trials Network to unpublished data from the HIVNET012 trial in which sdNVP only was employed; the rate of breastfeeding transmission in HIVNET012, among those women with a viral load of <4.4 log which was comparable to the maternal viral burden in SIMBA, was 2.8%.

It follows that these two trials do not provide any conclusive evidence of the benefits of infant ARVs to reduce breastfeeding transmission.

5. Development of Resistance To Nevirapine

One of the main problems with the use of ARVs in the prevention of mother-to-child-transmission of HIV-1 is that drugs used in these regimens can induce the development of drug resistant strains. In resource-limited settings, the simplest and safest ARV regimen conferring the most efficacious outcome is the sdNVP intrapartum/postpartum regimen (Guay et al., 1999; Jackson et al., 2003). Resistance to NVP may compromise the subsequent use of NVP-containing pMTCT, and HAART regimens in women or infants previously exposed to the drug in pMTCT programs. This is of greater importance to the HIV-infected mothers than to their offspring, because the numbers of HIV infected infants are likely to decrease as pMTCT programs are successfully scaled-up in Africa and Asia. This subject has been extensively reviewed recently (Fowler et al., 2003). In the HIVNET012 trial 19% of the women and 46% of the HIV-infected infants developed resistance by 6 weeks. However, this resistance in both mother and infant was transient and was undetectable at 18 months (Eshleman et al., 2001). More definitive answers are needed to address the question as to whether resistant mutations could be archived in cellular DNA, and whether this resistance dissipates over time or re-emerges on re-exposure to NVP (Servais et al., 2004).

Factors associated with increasing the risk of development of NVP resistance include long half-life of the drug (Muro et al., 2004); high maternal viral load (Jourdain et al., 2004) and low CD4 counts (Eshleman et al., 2001; Jourdain et al., 2004); HIV sub-type (Clade C > A > D) (Eshleman et al., 2005); more maternal NVP doses (Sullivan, 2000); earlier time of blood sampling (Eshleman et al., 2001) and compartment (rate of NVP resistance mutations in breastmilk was higher than plasma) (Lee et al., 2003).

A study by Jourdain et al., (2004) was the first to assess whether ARVs used in pMTCT programs may compromise the efficacy of ARVs used subsequently for the initiation of chronic treatment of HIV-infected mothers who had disease progression after delivery and required NVP containing HAART. Women who received the sdNVP were more likely to experience virologic failure (defined as the failure to achieve <50 viral copies/ml) compared to the women who had not been exposed to NVP; virological failure is more frequent among those women who had viral resistance mutations. Therefore women with prior exposure to sdNVP may have less than optimal virological response to subsequent HAART regimens containing NVP. Despite development of NNRTI resistance mutations, there were no discernable differences in clinical response or CD4 counts after 6 months of HAART (including NVP) between women having prior NVP and those without. However the long term implications of these findings are not known.

A number of interventions are being studied to decrease the risk of resistance; a study employing combination ARVs to cover the period of persisting NVP levels after discontinuation of the drug, appears especially promising (McIntyre et al., 2004). Another approach is to ascertain whether the maternal or the infant dose of NVP can be omitted when combination ARVs are used.

In the Thai PHPT-2 trial (Lallemant et al., 2004) (with a backbone of AZT antenatally) the transmission was 1.1% when NVP was given to both mother and infant; it was 2.1% when the infant dose was omitted (p > 0.05). Therefore where prior AZT has been given during the antenatal period, the maternal dose of NVP may be sufficient and the infant dose of NVP may be unnecessary.

In the NVAZ trial (Taha et al., 2003) the maternal dose maternal NVP was required to obviate the addition of AZT to infant NVP to ensure reduction of transmission by providing ARVs to mothers who may come very late to the antenatal clinic or arrive already in labor.

In the MASHI trial (Shapiro et al., 2005), where antenatal AZT was the backbone to the study, maternal NVP was not needed to enhance efficacy. Indeed, the use of sdNVP in mothers and infants (transmission 0.8%) was not statistically superior (p = 0.7) to Placebo in mothers and infants (transmission 2.5%); except in formula fed infants. Therefore these studies do not assist in the decision on omission of maternal NVP to reduce viral resistance.

On balance, it would be reasonable to exclude the infant dose of NVP if AZT had been provided during the antenatal period. However, a case could be made for simply retaining the sdNVP doses to both mother and infant, as it acts on the intrapartum transmission and may reduce early breastfeeding transmission.

The Table (see Table 17.7) provides data from recent trials on the proportions of subjects who develop resistance to NVP after exposure to the drug in pMTCT programs; these proportions are higher with more sensitive detection systems which have been recently reported.

6. Conclusions

Reductions in the incidence of HIV-1 among women of child-bearing age, and in unintended pregnancies in those already HIV-infected, are sensible and cost-effective measures to decrease the population burden of vertically transmitted HIV infections. The transmission of HIV-1 from mother to her newborn during the intrauterine and intrapartum periods can be successfully prevented. Vertical

Table 17.7. Proportion of Maternal NVP resistance after sdNVP

Study	HIV-Subtype	Regimen	Proportion with NVP Resistant Strains	Sampling Postpartum	Comment
HIVNET012 Uganda (Eshleman et al., 2005)	C	sdNVP	19%	6–8 weeks	Mutations faded by 12–24 months
South Africa (Soweto, Durban) (Morris et al., 2004)	C	sdNVP	39%	7 weeks	Mutations faded to 11% at 24 weeks
Ditrame Plus 1.0 ANRS Cote d'Ivoire (Chaix et al., 2004)	C	AZT 36 wk + sdNVP	28%	4–6 weeks	
Ditrame Plus 1.1 ANRS Cote d'Ivoire (Chaix et al., 2005)	C	32 wk AZT/3TC + sdNVP + 3d ZDV + 3TC postpartum	1.10%	4–6 weeks	8–9% resistance to 3TC
Thailand (Chalermchokcharoenkit et al., 2004)	E/B	AZT 34–36 wk + sdNVP	18%	4 weeks	
PHPT-2 Thailand (Jourdain et al., 2004)	E/B	AZT 28 wk + sdNVP	20%	2 weeks	
MASHI (Shapiro et al., 2005)	C	AZT 34 wk+/− sdNVP	45%	4 weeks	
Ireland (Lyons et al., 2005)	69% non-B	NVP-HAART +5d AZT/3TC postpartum	13%	6 weeks	
PACTG316 (Cunningham et al., 2002) USA	B	70% combination ARV + sdNVP	15%	6 weeks	

transmission rates, which were between 30% to 45% over 24 months without ARVs in developing countries, can now be reduced to around 5% in early infancy with ARVs. Optimum measures to reduce breastfeeding transmission have yet to be established, although exclusive breastfeeding is associated with very low transmission in the first 6 months of life. For women in socio-economically advantaged areas in Africa who can safely formula feed, combination ARVS can reduce transmission to between 1% to 2%. Single-dose Nevirapine is effective, affordable, and safe, and is still useful in many situations, although sdNVP in combination with other ARVs markedly improves efficacy. A broader public health response to the prevention of vertical transmission of HIV requires the early design, promotion and implementation of national policies which sustain family life, advance community activities, and create social welfare safety-nets for the children who will inevitably be rendered vulnerable and made orphans by HIV/AIDS. At the end of 2003 it was estimated that there were already 43.4 million orphans in Sub-Saharan Africa of which 12.3 million alone were orphaned by AIDS (UNAIDS/UNICEF/USAID, 2004c), a figure roughly equivalent to the total number of children in the United Kingdom. The next wave of the AIDS epidemic in Southern Africa, following on the escalation of adult mortality, is the massive increase in numbers of AIDS orphans and vulnerable children. It would be a tragic irony to save children from HIV through effective pMTCT programs, only to crush their hopes and blight their futures in this next wave.

References

Ayouba, A., Tene, G., Cunin, P., Foupouapouagnigni, Y., Menu, E., Kfutwah, A., Thonnon, J., Scarlatti, G., Monny-Lobe, M., Eteki, N., Kouanfack, C., Tardy, M., Leke, R., Nkam, M., Nlend, A.E., Barre-Sinooussi, F., Martin, P.M., and Nerrient, E. – Yaounde European Network for the Study of In Utero Transmission of HIV-1. (2003). Low rate of mother-to-child transmission of HIV-1 after nevirapine intervention in a pilot public health program in Yaounde, Cameroon. *J. Acqui. Immune. Defic. Syndr.* **34**(3), 274–280.

Chaix, M.L., Dabis, F., Ekouevi, D., Rouet, F., Tonwe-Gold, B., Viho, I., Bequet, L., Peytavin, G., Toure, H., Menan, H., LeRoy, V., and Rouzioux, C. (2005). Addition of 3 days of ZDV + 3TC postpartum to a short course of ZDV + 3TC and single-dose NVP provides low rate of NVP resistance mutations and high efficacy in preventing peri-partum HIV-1 transmission: The ANRS DITRAME Plus study, Abijan, Cote d'Ivoire. *12th Conference on Retrovirus and Opportunistic Infections, Boston, February 2005* Abstract 72LB.

Chaix, M.L., Montcho, C., Ekouevi, D.K., Rouet, F., Bequet, L., Viho, I., Fassinou, P., Welfens-Ekra, C., LeRoy, V., and Dabis, F. (2004). Genotypic resistance analysis in women who received intrapartum nevirapine associated with short course of zidovudine to prevent perinatal HIV-1 transmission: The DITRAME Plus ANRS 1201/1201 study, Abijan, Cote d'Ivoire. *11th Conference on Retrovirus and Opportunistic Infections, San Francisco, February 2004* Abstract 657.

Chalermchokcharoenkit, A., Asavapiriyanont, S., Teeraratkul, A., Vanprap, N., Chotpitayasunondh, T., Chaowanachan, T., Mock, P., Wilasrusme, S., Skunodom, N., Simonds, R.J., Tappero, and J.W., and Culnane, M. (2004). Combination short-course zidovudine plus 2-dose nevirapine for the prevention of mother-to-child transmission: safety, tolerance, transmission, and resistance results. *In: Program and abstracts of the 11th Conference of Retrovirus and Opportunistic Infection, San Francisco, February 8–11, 2004* Abstract 96.

Colvin, M., Jackson, D., Levin, J., Chopra, M., Doherty, T., Willumsen, J., and Goga, A. (2005). Risk factors for early transmission of HIV to infants: results from a cohort study across three PMTCT sites in South Africa. *2nd South African AIDS Conference, Durban, 7–10 June 2005* Abstract 462.

Connor, E.M., Sperling, R.S., Gelber, R., Kiselev, P., Scott, G., O'Sullivan, M.J., VanDyke, R., Bey, M., Shearer, W., Jacobson, R.L., Jimenez, E., O'Neill, E., Bazin, B., Delfraissy, J.F., Culnane, M., Coombs, R., Elkins, M., Moye, J., Stratton, P., and Balsley J. – for The Pediatric AIDS Clinical Trials Group Protocol 076 Study Group. (1994). Reduction of maternal-infant transmission of human immunodeficiency virus type 1 with zidovudine treatment. *N. Engl. J. Med.* **331**, 1173–1180.

Cooper, E.R., Charurat, M., Mofenson, L., Hanson, I.C., Pitt, J., Diaz, C., Hayani, K., Handelsman, E., Smeriglio, V., Hoff, R., and Blattner, W.: Women and Infant Transmission Study Group. (2002). Combination antiretroviral strategies for the treatment of pregnant HIV-1 infected women prevention of perinatal HIV-1 transmission. *J. Acquir. Immune. Defic. Syndr.* **29**, 484–494.

Coutsoudis, A., Goga, A.E., Rollins, N., and Coovadia, H.M. (2002). Free formula milk for infants of HIV-infected women: blessing or curse. *Health Policy Plan* **17**, 154–160.

Coutsoudis, A., Pillay, K., Kuhn, L., Spooner, E., Wei-Yann, T., and Coovadia, H.M. for the South African Vitamin A Study Group. (2001). Method of feeding and transmission of HIV-1 from mothers to children by 15 months of age: prospective cohort study from Durban, South Africa. *AIDS* **15**, 379–387.

Creese, A., Floyd, K., Alban, A., and Guiness, L. (2002). Cost-effectiveness of HIV/AIDS interventions in Africa: a systematic review of the evidence. *Lancet* **359**, 635–643.

Cunningham, C.K., Chaix, M.L., and Rekacewicz, C. (2002). Development of resistance mutations in women receiving standard antiretroviral therapy who received intrapartum nevirapine to prevent perinatal human immunodeficiency virus type-1 transmission: a subtudy of Pediatric Aids Clinical Trials Group 316. *J. Infect. Dis.* **186**, 181–188.

Dabis, F. and Ekpini, E.R. (2002). HIV-1/AIDS and maternal and child health in Africa. *Lancet* **359**, 2097–2104.

Dabis, F., Bequet, L., Ekouevi, D.C., Viho, I., Rouet, F., Horo, A., Sakarovitch, C., Becquet R., Fassinou, P., Dequae-Merchadou, L., Welffens-Ekra, C., Rouzioux, C., and Leroy, V.: ANRS 1201/1202 DITRAME Plus Study Group. (2005). Field efficacy of zidovudine, lamivudine and single-dose nevirapine to prevent peripartum HIV transmission. *AIDS* **19**, 309–318.

Dabis, F., Msellati, P., Meda, N., Welffens-Ekra, C., You, B., Manigart, O., LeRoy, V., Simonon, A., Cartoux, M., Combe, P., Ouangre, A., Ramon, R., Ky-Zerbo, O., Monthcho, C., Salamon, R., Rouzioux, C., Van de Perre, P., and Mandelbrot, L.: for the Ditrame Study Group. (1999). 6-month efficacy, tolerance, and acceptability of a short regimen of oral zidovudine to reduce vertical transmission of HIV-1 in breastfed children in Cote d'Ivoire and Burkina Faso: A double blind, placebo controlled multicentre trial. *Lancet* **353**, 786–792.

De Cock, K.M., Fowler, M.G., Mercier, E., de Vincenzi, I., Saba, J., Hoff, E., Alnwick, D.J., Rogers, M., and Shaffer, N. (2000). Prevention of mother-to-child HIV transmission in resource-poor countries: translating research into policy and practice. *JAMA* **283**, 1175–1182.

Duerr, A., Hurst, S., Kourtis, A.P., Rutenberg, N., and Jamieson, D.J. (2005). Integrating family planning and prevention of mother-to-child HIV transmission in resource-limited settings. *Lancet* **366**, 261–266.

Dorenbaum, A., Cunningham, C.K., Gelber, R.D., Clunane, M., Mofenson, L., Britto, P., Rekacewicz, C., Newell, M.L., Delfraissy, J.F., Cunningham-Schrader, B., Mirochnik, M., and Sullivan, J.L.: International PACTG 316 Team. (2002). Two-dose intrapartum/newborn nevirapine and standard antiretroviral therapy to reduce perinatal HIV-1 transmission: a randomized trial. *JAMA* **288**, 189–198.

Eshleman, E., Mracna, M., Guay, L.A., Deseyve, M., Cunningham, S., Mirochnik, M., Musoke, P., Fleming, T., Fowler, M.G., Mofenson, L., Mmiro, F., and Jackson, J.B. (2001). Selection and fading of resistance mutations in women and infants receiving nevirapine to prevent HIV-1 vertical transmission (HIVNET012). *AIDS* **15**, 1951–1957.

Eshleman, S.H., Hoover, D.R., Chen, S., Hudelson, S.E., Guay, L.A., Mwatha, A., Fiscus, S.A., Mmiro, F., Musoke, P., Brooks, J.B., Kumwenda, N., and Taha, T. (2005). Nevirapine (NVP) resistance in women ith HIV-1 subtype C, compared with subtypes A and D, after administration of single-dose NVP. *J. Infect. Dis.* **192**, 30–36.

Fowler, M.G., Mofenson, L., and McConnell, M. (2003). The interface of perinatal HIV prevention, antiretroviral drug resistance and antiretroviral treatment: What do we really know. *J. Acquir. Immune. Defic. Syndr.* **34**, 308–311.

Guay, L.A., Musoke, P., Fleming, T., Bagenda, D., Allen, M., Nakabiito, C., Sherman, J., Bakaki, P., Ducar, C., Deseyve, M., Emel, L., Mirochnick, M., Fowler, M.G., Mofenson, L., Miotti, P., Dransfield, K., Bray, D., Mmiro, F., and Jackson, J.B. (1999). Intrapartum and neonatal single-dose nevirapine compared with zidovudine for prevention of mother-to-child transmission of HIV-1 in Kampala, Uganda: HIVNET012 randomised trial. *Lancet* **354**, 795–802.

Iliff, P.J., Piwoz, E.G., Tavengwa, N.V., Zunguza, C.D., Marinda, E.T., Nathoo, K.J., Moulton, L.H., and Ward, B.J., the ZVITAMBO study group and Humphrey, J.H. (2005). Early exclusive breast-feeding reduces the risk of postnatal HIV-1 transmission and increases HIV-free survival. *AIDS* **19**, 699–708.

Institute of Medicine of the National Academies. Review of the HIVNET012 perinatal HIV prevention study. April 7 2005. National Academies Press, Washington DC. www.nap.edu.

Jackson, J.B., Musoke, P., Fleming, T., Guay, L.A., Bagenda, D., Allen, M., Nakabiito, C., Sherman, J., Bakaki, P., Maxensia, O., Ducar, C., Deseyve, M., Mwatha, A., Emel, L., Duefield, C., Mirochnik, M., Fowler, M.G., Mofenson, L., Miotti, P., and Gigliotti, M. (2003). Intrapartum and neonatal single-dose nevirapine compared with zidovudine for prevention of mother-to-child transmission of HIV-1 in Kampala, Uganda: 18 month follow-up to the HIVNET012 randomized trial. *Lancet* **362**, 59–68.

Jourdain, G., Ngo-Giang-Huong, N., Le Couer, S., Bowonwatanuwong, G., Kantipong, P., Leechanachai, P., Ariyadej, S., Leenasirimakul, P., Hammer, S., and Lallemant, M. (2004). Intrapartum exposure to nevirapine and subsequent maternal responses to nevirapine-based antiretroviral therapy. *N. Engl. J. Med.* **351**, 229–240.

Kilewo, C. (2002). MITRA Study. 2nd Multifactorial AIDS Conference in Tanzania, 2002 Abstract 061.

Kreitchmann, R., Fuchs, S.C., Suffert, T., and Prussler, G. (2004). Perinatal HIV transmission among low income women participants in the HIV/AIDS control program in Southern Brazil: a cohort study. *B. Jog.* **111(6)**, 579–584.

Lallemant, M., Jourdain, G., Le Coeur, S., Mary, J.Y., Ngo-Giang-Huong, N., Keotsawang, S., Kanshana, S., McIntosh, K., and Thaineua, V. (2004). Single-dose perinatal nevirapine plus standard zidovudine to prevent mother-to-child transmission of HIV-1 in Thailand. *N. Engl. J. Med.* **351**, 217–228.

Lallemant, M., Jourdain, G., Le Coeur, S., Kim, S., Keotsawang, S., Comeau, A.M., Phoolcharoen, W., Essex, M., McIntosh, K., and Vithayasai, V. (2000). A trial of shortened zidovudine regimens to prevent mother-to-child transmission of human immunodeficiency virus type 1. Perinatal HIV Prevention Trial (Thailand) Investigators. *N. Engl. J. Med.* **343**, 982–991.

Lee, E., Kantor R., Johnston, E., Mateta, P., Zijenah, L., and Katzenstein, D. (2003). Breastmilk shedding of drug resistant subtype C HIV-1 and among women receiving single-dose nevirapine. *In: Program and Abstracts of 10th Conference on Retrovirus and Opportunistic Infections, Boston, February 2003* Abstract 96.

Leroy, V., Karon, J.M., Alioum, A., Ekpini, E.R., Meda, N., Greenberg, A.E., Msellati, P., Hudgens, M., Dabis, F., and Wiktor, S.Z. (2002). Twenty-four month efficacy of maternal short-course zidovudine regimen to prevent mother-to-child transmission of HIV-1 in West Africa. *AIDS* **16**, 631–641.

Leroy, V., Sakarovitch, C., Cortina-Borja, M., McIntyre, J., Coovadia, H., Dabis, F., and Newell, M.L., on behalf of the ghent Group on HIV in Women and Children. (2005). Is there a difference in the efficacy of peripartum antiretroviral regimens in reducing mother-to-child transmission of HIV in Africa? *AIDS* **19**, 1865–1875.

Lyons, F.E., Coughlan, S., Byrne, C.M., Hopkins, S.M., Hall, W.W., and Mulcahy, F.M. (2005). Emergence of antiretroviral resistance in HIV-positive women receiving antiretroviral therapy in pregnancy. *AIDS* **19**, 63–67.

McIntyre, J., Martinson, N., Boltz, V., Palmer, S., Coffin, J., Mellors, J., Hopley, M., Kimura, T., Robinson, P., and Mayers, D. (2004). Addition of short course Combivir (CBV) to single dose viramune (sdNVP) for prevention of mother-to-child transmission (MTCT) of HIV-1 can significantly decrease the subsequent development of maternal NNRTI-resistant virus. *XV International AIDS Conference, Bangkok, 11–14 July 2004* Abstract LbOrB09.

Moodley, D., Moodley, J., Coovadia, H., Gray, G., McIntyre, J., Hofmeyer, J., Nikodem, X., Hall, D., Gigliotti, M., Robinson, P., Boshoff, L., and Sullivan, J.L. The South African intrapartum nevirapine trial (SAINT). (2003). A multicentre randomized, controlled trial of nevirapine compared to a combination of zidovudine and lamivudine to reduce intrapartum and early postpartum mother-to-child transmission of human immunodeficiency virus type-1. *J. Infect. Dis.* **18**, 725–735.

Morris, L., Martinson, N., Pillay, C., Moodley, D., Chezzi, C., Lupondwana, P., Ntshala, M., Cohen, S., Puren, A., Sullivan, J., Gray, G., and McIntyre, J. (2004). Persistence of nevirapine resistance mutations 6 months following single-dose nevirapine. *XV Internation AIDS Conference, Bangkok 11–14 July 2004* Abstract ThOrB1353.

Muro, E., Droste, J., ter Hofstede, H., Bosch, M., Dolmans, W., and Burger, D. (2004). Nevirapine plasma concentrations are still detectable after more than 2 weeks in the majority of women receiving single-dose nevirapine: implications for intervention studies. *In Program and Abstracts of 11th Conference on Retrovirus and Opportunistic Infections, San Francisco, February 2004* Abstract 891.

Nduati, R., John, G., Mbori-Ngacha, D., Richardson, B., Overbaugh, J., Mwatha, D., Ndinya-Achola, J., Bwayo, J., Onyango, F.F., Huhes, J., and Kreiss, J. (2000). Effect of breastfeeding and formula feeding on transmission of HIV-1: a randomized clinical trail. *JAMA* **283**, 1167–1174.

Peters, V., Liu, K.L., Dominguez, K., Frederick, T., Melville, S., Hsu, H.W., Ortiz, I., Rakusan, T., Gill, B., and Thomas, P. (2003). Missed opportunities for perinatal HIV prevention among HIV-exposed infants born 1996–2000, pediatric spectrum of HIV disease cohort. *Pediatrics* **111**, 1186–1191.

Petra Study Team. (2002). The efficacy of three short-course regimens of zidovudine and lamivudine in preventing early and late transmission of HIV-1 from mother to child in Tanzania, South Africa, and Uganda (Petra study): a randomised, double-blind placebo-controlled trial. *Lancet* **359**, 1178–1186.

Quaghebeur, A., Mutunga, L., Mwanyumba, F., Mandaliya, K., Verhofstede, C., and Temmerman, M. (2004). Low efficacy of nevirapine (HIVNET012) in preventing perinatal HIV-1 transmission in a real-life situation. *AIDS* **18**, 1854–1856.

Santora, M. New York Times (30 January 2005). US is close to eliminating AIDS in infants, officials say.

Servais, J., Lambert, C., Karita, E., Vanhove, D., Fischer, A., Baurith, T., Schmith, J.T., Schneider, F., Hemmer, R., and Arendt, V. (2004). HIV type-1 pol gene diversity and archived nevirapine resistance mutation in pregnant women in Rwanda. AIDS Res. Human Retroviruses 20, 279–283.

Shaffer, N., Chuuachoowong, R., Mock, P.A., Bhadrakom, C., Siriwasin, W., Young, N.L., Chotpitayasunondh, T., Chearskul, S., Roongpisuthipong, A., Chionayon, P., Karon, J., Mastro, T.D., and Simonds, R.J. (1999). Short-course zidovudine for perinatal HIV-1 transmission in Bangkok, Thailand: a randomized controlled trial. Bangkok Collaborative Perinatal HIV Transmission Study Group. *Lancet* **353**, 773–780.

Shapiro, R., Thior, I., Gilbert, P., Lockman, S., Wester, C., Smeaton, L., Stevens, L., Ndung'u, T., Novitsky, V., van Widenfelt, E., Mazonde, P., Lee, T.H., Marlink, R., Lagakos, S., Essex, M., and the Mashi Study Group. (2005). Maternal single-dose nevirapine may not be needed to reduce mother-to-child HIV transmission in the setting of maternal and infant zidocudine and infant single-dose nevirapine: results of a randomized clinical trail in Botswana. *In: Program and abstracts of the 12th Conference on Retrovirus and Opportunistic Infections, Boston, February 22–25, 2005* Abstract 74LB.

Sullivan, J.L. (2002). South African intrapartum nevirapine trial: selection of resistance mutations. *XIV International AIDS Conference, Barcelona, July 2002* Abstract LbPeB9024.

Sweat, M.D., O'Reilly, K.R., Schmid, G.P., Denison, J., and de Zoysa, I. (2004). Cost-effectiveness of nevirapine to prevent mother-to-child HIV transmission to eight African Countries. *AIDS* **18(12)**, 1661–1671.

Taha, T.E., Kumwenda, N., Gibbons, A., Broadhead, R.L., Fiscus, S., Lema, V., Liomba, G., Nkhoma, C., Miotti, P.G., and Hoover, D.R. (2003). Short post-exposure prophylaxis in newborn infants to reduce mother-to-child transmission of HIV-1: NVAZ randomized clinical trial. *Lancet* **362**, 1171–1177.

Temmerman, M., Quaghebeur, A., Mwanyumba, F., and Mandaliya, K. (2003). Mother-to-child transmission in resource poor settings: how to improve coverage. *AIDS* **17**, 1239–1242.

The Breastfeeding and HIV Transmission Study Group (The BHITS Group). (2004). Late postnatal transmission of HIV-1 in breast-fed children: an individual patient data meta-analysis. *J. Infect. Dis.* **189**, 2154–66.

The Ghent IAS Working Group. (2004). Use of antiretroviral drugs to prevent HIV-1 transmission through breast-feeding: from animal studies to randomized clinical trials. *J. Acquir. Immune. Defic. Syndr.* **35(2)**, 178–186.

UNAIDS and World Health Organization. (2003). AIDS epidemic update, Geneva, UNAIDS. http://www.unaids.org/Unaids/EN/Resources/Publications/corporate+publications/aids+epidemic+update+december+2003.asp)

UNAIDS and World Health Organization. December (2004a). AIDS epidemic update. Geneva, UNAIDS. (http://www.unaids.org/epidemic_update/report_dec01/index.html)

UNAIDS. (2004b). "3 by 5" Progress Report: The way forward: challenges in 2005. WHO Geneva. UNAIDS. 45–47.

UNAIDS, UNICEF and USAID (2004c). Children on the Brink 2004: A joint report of new orphan estimates and a framework for action. 4th edition Page3. www.unaids.org, www.unicef.org and www.usaid.gov

Vyankandondera, J., Luchters, S., Hassink, E., Pakker, N., Mmiro, F., Okong, P., Kituuka, P., Ndugwa, C., Mukanka, N., Beretta, A., Imperiale, M. Jr, Loeliger, E., Giuiliano, M., and Lange, J. (2003). Reducing the risk of HIV-1 transmission from mother to infant through breasftfeeding using antiretroviral prophylaxis in infants (SIMBA Study). *2nd IAS Conference on Pathogenesis and Treatment, Paris, France 13–16 July 2003* Abstract LB7.

World Health Organization. (2004a). HIV transmission through breastfeeding: Review of available evidence. WHO, Geneva. Page iv.

World Health Organization (2004b). Antiretroviral drugs for treating women and preventing HIV infection in infants: Guidelines on care, treatment and support for women living with HIV/AIDS and their children in resource-constrained settings. WHO, Geneva. Pages 3–4.

Wiktor, S., Ekpini, E., Karon, J., Nkengasong, J., Maurice, C., Severin, S.T., Roets, T.H., Kouassi, M.K., Lackritz, E.M., Coulibaly, I., and Greenberg, A.E. (1999). Short-course oral zidovudine for prevention of mother-to-child transmission of HIV-1 in Abijan, Cote d'Ivoire: A randomized trial. *Lancet* **353**, 781–785.

18

Practical Aspects of Antiretroviral Treatment in Children

Dr Sam Walters

1. Introduction

In the mid 1990s effective treatment of HIV with combination therapy became possible in industrialised countries. The success of such combinations in children has been demonstrated by significant declines in deaths, new AIDS diagnoses and hospital admissions even though numbers of infected children have been rising (Gibb et al., 2003). However, benefit from these medications can only be achieved if children are able to take the drugs regularly and consistently. This chapter focuses on the practical aspects of delivering antiretroviral therapy to children successfully.

2. Aims of Therapy

The aims of therapy are to preserve or restore health, to enable normal somatic and neurological development of the child and to prevent the occurrence of encephalopathy. This is achieved through suppression of HIV replication leading to immune preservation or restoration. HIV is a rapidly mutating virus and a number of basic principles need to be borne in mind. It is necessary to maintain drug levels sufficient to prevent the development of viral resistance. Past experience indicates that at least three drugs active against the virus are needed to prevent development of resistance; perhaps future drugs will be effective singly or in two-drug combinations. The current short-term "target" for therapy is to achieve an undetectable plasma viral load using an assay with a limit of detection of 50 RNA copies/ml.

Hot Topics in Infection and Immunity in Children, edited by Andrew J. Pollard and Adam Finn. Springer, New York, 2006

3. Selection of the Combination

3.1. Potency Versus Practicality

Selecting the combination for the child depends upon a balance between its potency against the virus and its practicality. Potency is determined both by the characteristics of the virus and the drugs, whereas practicality is determined only by the characteristics of the drugs and depends upon the preferences and circumstances of the child and parents. A drug's characteristics are delineated through *in vitro* experiments and *in vivo* practice (mainly from adult experience) and may change profoundly when used in combination with other drugs. The virus' characteristics relate to its innate sensitivity to the drugs and to its resistance profile. This resistance profile can be predicted from the child's drug history (and for vertical HIV infection also the mother's drug history) and can be determined through resistance assays.

3.2. Resistance Assays

Two different types of assay are available.

1. Phenotypic resistance; whereby the reverse transcriptase and the protease genes are inserted into an HIV construct and growth *in vitro* is assayed in the presence of the drug. This is an expensive, slow and difficult assay and consequently is not widely available.
2. Genotypic resistance; since several mutations or combinations of mutations are known to be associated with resistance to specific drugs or drug classes, sequencing of the reverse transcriptase gene and the protease gene enables the resistance profile to be predicted. These assays are now widely used.

The value of resistance assays is limited because they only assess the predominant circulating viral species at the time of sampling and mutant species present at low copy number may be missed. If resistance has developed and treatment stopped, fully sensitive "wild type" virus will often emerge as the predominant circulating virus and the resistant mutations are "archived". If drugs are recommenced that favour these "archived" mutations they will rapidly re-emerge as the predominant circulating virus. Therefore, both the detailed history of drug exposure as well as the resistance pattern are required to interpret resistance assay results and plan appropriate choices of antiretrovirals.

3.3. Practicality

The practicality of a combination depends upon many factors. For several drugs pediatric dosing is problematic because there is no liquid preparation and the drug is only available in "adult-sized" un-scored non-dispersible tablets or capsules rendering them unusable in young children. Pediatric pharmacokinetic data profiles for many drugs are limited and dosing is based on adult studies. Such extrapolation is insecure and associated with uncertainty as to whether to use calculation based on surface area or body weight. A drug's side effects, both short-term and known long-term, profoundly influence its practicality. The taste of the preparation when in liquid formulation can sometimes be masked by other flavours but some anti-retrovirals (or the excipients required to hold them in liquid form) have an extremely unpleasant taste that cannot be adequately masked. Known drug-drug interactions with other antiretrovirals must be considered along with interactions with other medications which the child is receiving, in particuar TB treatments. Knowledge of drug interactions continues to increase rapidly and use of one of several websites that frequently update information about them is recommended when considering addition of other drugs (e.g., http://www.hiv-druginteractions.org/index.asp).

Children's parents may have had previous drug therapy, sometimes with worrying side effects and may have a prejudice for or against specific drugs. Similarly, if parents are currently on therapy and doing well, they may have knowledge and a favourable opinion of their medications. This can be important when considering what to recommend for the child. The probability of resistance developing to particular drugs or classes of drug when adherence is sub-optimal is variable. It is thought that some drugs (e.g., lopinavir/ritonavir) are more "forgiving" than others. This may influence prescribing practice if difficulties with adherence are anticipated e.g. in many teenagers. For some children who have previously been treated and have developed highly resistant virus, access to new medications may be required and can sometimes only be achieved through enrolment into therapeutic trials.

3.4. Convenience

The convenience of a drug combination (perhaps more appropriately the level of the inconvenience) also depends on a number of factors. If food restrictions are required at the time of dosing this can be difficult, especially for babies. The storage of the preparations can be important especially if refrigeration is required, and this can be particularly difficult for families living in temporary accommodation who share a fridge with other families. Also for families going on holidays or wishing to visit relatives in other countries, maintaining access to a fridge can be difficult. A short "shelf life" of a preparation can be a problem for those who find it difficult to attend clinic frequently. The "pill burden" (and the volume of liquid for a baby) is important. The fewer the number of pills the more easily this is likely to fit in with lifestyle. For older children especially during "sleepovers" at friends' homes or on school trips where the child would wish to be discreet about taking their medication, the fewer the number of pills the more likely it is that they are able to maintain their adherence.

3.5. Choice of Drugs

With potency and practicality in mind the clinician must select a combination which should best suit the child. There are three main classes of antiretroviral medications, NRTIs (nucleoside reverse transcriptase inhibitors), NNRTIs (non-nucleoside reverse transcriptase inhibitors) and PIs (protease inhibitors). Some of these medications come in a combined form (usually two drugs) enabling reduction of "pill burden" and greater convenience. At the time of writing, widely-used first line combinations would be a combination of two NRTIs plus one PI or two NRTIs plus one NNRTI. For initial therapy of infected babies who typically have an extremely high viral load some pediatricians favour a four drug combination initially of three NRTIs plus one NNRTI (since good pharmacokinetic data for PI usage in babies are lacking). Choice of the specific drugs for these combinations is not within the remit of this chapter but frequently up-dated guidelines are published (Sharland et al., 2004).

3.6. Principles and Practicality

Starting treatment is never an urgent requirement and clinical benefit is seen only after several weeks. It is important to optimise the chance of adequate absorption and tolerance of these drugs. It is usually wrong to initiate treatment in a very ill child with a life threatening illness where gastrointestinal function may be compromised and the child is receiving many other concomitant medications. One should also consider the duration of other treatments (e.g., for tuberculosis) and delay the initiation of HIV therapy until the number of other medications can be reduced whenever possible. It is equally important to optimise the chance of good adherence to the regimen once it is commenced.

When starting therapy the parents and child must be given adequate information and must be made fully aware of the importance of adherence. Realism is essential and the family should know what the combination involves, what to expect particularly during the first few weeks of the regimen and know exactly what to do if there are any problems (e.g., the child vomits, develops a rash, refuses to swallow the medicine). If possible give some choice to the child in the timing of the dosing, the child could taste preparations prior to determining the combination or between pills or liquid. There are several aids to adherence which can be used including a "star chart" where the young child attaches coloured stickers each time they take their medicine, dosset boxes as a reminder to families whether each dose has been taken, pill crushers and pill cutters to enable smaller children to avoid liquid preparations and pill charts with clearly labelled photographs of each of the drugs to help prevent confusion over dosing. Some children (and their parents) prefer a gastrostomy tube or "button" and in this way avoid the taste of the medications. Despite initial reluctance, this approach has been very successful among those families who have elected to try it in our clininc. It is important to stress to families that this is not permanent, it is a completely reversible and relatively safe procedure and it is envisaged that after a number of months or a year or two the child will learn to swallow medication.

3.7. Short Term Side Effects

The family should be warned beforehand about likely or serious short-term side effects. A management plan should have been discussed beforehand so the family know exactly what to do and when and how to contact the clinic. There should also be a plan of who to contact at night or at weekends. The family must know what they should not do such as stopping the medication without discussion with the clinic or waiting for the next scheduled appointment to discuss the problem.

3.8. Specific Side Effects

3.8.1. Rash

A low threshold should be exercised for seeing the child at the first signs of a rash particularly if the child's combination contains drugs with a known serious side effect of rash such as nevirapine (Stevens Johnson) and abacavir (Hypersensitivity). The rash may be trivial (an intercurrent illness is very likely to be the cause) but could be the beginning of a serious reaction. If the child is taking nevirapine and a rash develops, antihistamines can help, steroids may be of benefit but if Stevens-Johnson syndrome does develop substitution of the drug is indicated.

Abacavir hypersensitivity manifests as progressively worsening symptoms over several days, usually with rash, fever, respiratory symptoms and the child becoming increasingly unwell. Its occurrence is rare in black racial groups and appears to be associated with the HLA-B*5701 genotype. The importance of this hypersensitivity is that if present the drug must be stopped and re-challenge never attempted. In the literature there are documented cases of fatal outcome from re-challenge with this drug. The family must be warned in advance that if rash develops they should *not* stop the abacavir but contact the clinic immediately and the child should be seen promptly. Management of this child is usually more safely conducted by admission to the ward for close observation, antihistamines or steroids may be tried and if despite this the rash and symptoms progress, the Abacavir should be stopped. Clinic staff and the family (and older children) must be fully aware that abacavir must never be used again.

3.8.2. Gastrointestinal Symptoms

Diarrhoea is a relatively common side effect when starting a new combination of drugs and the family should be warned to expect its occurrence, especially with protease inhibitors. The family and child should be reassured that it usually settles down and often can be lessened if the drug can be taken with food. If diarrhoea occurs and is interfering with lifestyle loperamide may be helpful and this can usually be stopped after a few weeks. Similarly, nausea is relatively common and also would be expected to settle down but ondansetron may be needed to help some children through the initial few weeks of treatment. Some medications are particularly associated with hepatitis and extra monitoring of blood liver function tests may be indicated over the first few weeks.

3.8.3. Mitochondrial Toxicity

Mitochondrial toxicity (which can also be a long-term side effect) is extremely rare. It has protean clinical manifestations depending upon which organ system is affected. There are no tests that reliably predict its occurrence (blood lactate levels appear to be unhelpful). However, any child presenting unwell with unexplained symptoms should be tested for lactic acidemia and pancreatitis.

4. Maintaining Therapy

4.1. Adherence

Experience has shown that in any chronic illness doctors cannot accurately predict the degree of adherence in their patients. Adherence changes over time for an individual patient and is not related to level of intelligence, educational background, severity of disease, and for adults; income, social class or occupation. Factors, which encourage adherence are easy incorporation of medication into the patient's lifestyle, dosing not affected by food or fluid intake, convenient and simple dosing and good tolerability. Frequency of dosing is critical and once or twice daily medication regimens achieve much better adherence than three or four times a day regimens, especially if there are easy to swallow tablets with a manageable side effect profile.

4.2. Follow-Up Appointments

Follow-up appointments for those on therapy should always include discussion of adherence in a non-confrontational way and display a genuine appreciation of how difficult it is to keep taking these drugs. Where there are admitted difficulties, children should be praised for being honest about these difficulties and ways to help the child and family should be explored. Maintaining treatment requires dose adjustments in a growing child. For some anti-HIV medications whether this adjustment is best based on weight or surface area calculation is not known and therapeutic drug monitoring has an increasingly important role to play in these dose adjustments.

4.3. Long-Term Side Effects

Long-term side effects cannot all be known. Some side effects are known to be associated with specific drugs or classes of drugs and some of these side effects are potentially irreversible.

4.3.1. Lipid Disturbance

Because of the need for long-term suppressive therapy for HIV and because it is envisaged that children will need to take these medications for longer than infected adults, the issue of abnormal lipid metabolism is extremely important. There are two main abnormalities, lipodystrophy (both lipoatrophy and lipohypertrophy) and biochemical lipid disturbance. Mitochondrial toxicity and insulin resistance are thought to play key roles in the development of these problems. Lipohypertrophy may be due mainly to insulin resistance and lipoatrophy may be due to mitochondrial toxicity. Either can cause serious cosmetic disability and the changes can be irreversible. Lipodystrophy has been most frequently linked with stavudine (d4T), didanosine and some protease inhibitors. If present or developing, changing the medication is indicated when possible. Biochemical lipid disturbance manifests as raised triglycerides and raised cholesterol. It is particularly seen with some protease inhibitors and is potentially important because it may lead to an increased risk of cardiovascular disease. In adult practice lipid lowering agents have been tried with some success but very few data exist for pediatric practice. Again, changing the medication should be considered.

4.4. Changing Drugs

Indications for changes in therapy include virological failure, intolerable or unmanageable side effects (both short-term and long-term) or switch to an easier and more convenient regimen as new drugs and alternative formulations become available.

5. Treatment Interruption

It has been suggested that for those on treatment with fully suppressed virus, treatment interruption permitting virus replication so that the immune system can "see" the virus may provide a beneficial boost to the immune system's ability to suppress HIV replication. So far, data supporting this hypothesis have not been compelling. Nevertheless, given the potential for long-term side effects, the inconvenience of daily medications and the high cost of these drugs, interruption of treatment, if it does not result in a rapid and dangerous decline in immune competence, is an attractive option. It may be that the child is able to stay off treatment for many months or even years. Balanced against this is the potential for promoting resistant virus through repeated interruptions, although, in principle, if discontinued in a planned way, drugs should retain their potency. Since the potential benefits balanced against the hazards of this option are still not known, entry into controlled treatment interruption trials is encouraged (PENTA 11, 2005).

6. Conclusion

For those with access to antiretroviral medication, HIV is now a treatable condition. Long-term viral suppression can be achieved and the expectation is maintenance of health. One of the main and most challenging tasks for clinical staff is to help children to continue to take these difficult medications on a daily basis throughout childhood and into adult life. If that can be achieved, an optimistic future should be predicted with a long life and near-normal lifestyle.

References

Gibb, D.M., Duond, T., Tookey, P.A., et al. (2003), Decline in mortality, AIDS, and hospital admissions in perinatally HIV-1 infected children in the United Kingdom and Ireland. Br. Med. J. **327**, 1019

Paediatric European Network for the Treatment of AIDS, 2005. PENTA trials (October 17, 2005) http://www.ctu.mrc.ac.uk/penta/trials.htm

Sharland, M., Blanche, S., Castelli, G., et al. (2004), PENTA guidelines for the use of antiretroviral therapy, 2004. HIV Med. **5**, 61–86

The University of Liverpool, 2005. HIV Drug Interactions (August 25, 2005) http://www.hiv-druginteractions.org/index.asp

19

Antibiotic Treatment for Acute Otitis Media in Children

Matthew J. Thompson and Paul Glasziou

1. Introduction

Acute otitis media (AOM) is common in children in whom it causes considerable morbidity, however in developed countries it is only rarely associated with serious complications or mortality (Stool, 1989). The highest incidence of AOM occurs in children between 6 months and 2 years of age (Teele et al., 1989; Klein, 1989). Indeed, by the age of 1 year, nearly 60% of children will have had at least one episode of AOM, by 2 years this rises to 70%, and by 3 years 80% (Sipila et al., 1987; Alho et al., 1991; Teele et al., 1989; Rosenfeld et al., 1999; Ruuskanen et al., 1994). The reported incidence of AOM appears to be falling in some countries such as the UK (Fleming et al., 2003). This may be a real reduction, due to decreases in environmental risk factors for AOM such as exposure to cigarette smoke and routine childhood vaccination, or may simply reflect changes in health seeking behaviour by parents of children with suspected AOM. However, in other countries such as the USA, the incidence of AOM has continued to rise (Auinger et al., 2003).

AOM is defined as an effusion in the middle ear, with an abnormal appearing tympanic membrane in a child with symptoms and signs of an acute infection. AOM develops when the eustachian tube becomes congested, usually as a result of a viral upper respiratory tract infection. Negative pressure develops within the middle ear cavity and causes secretions to accumulate which then become infected with bacteria and/or viruses, resulting in an effusion. The most common bacteria found in AOM are *Streptococcus pneumoniae*, *Haemophilus influenzae* and *Branhamella catarrhalis*. However infection or co-infection with viruses is common and occurs in 30–40% of infections (Heikkinen et al., 1999; Ruuskanen et al., 1994; Arola et al., 1990). The effusion causes release of inflammatory mediators which contribute to the local and systemic clinical features of an acute infection. In some children, the effusion may persist after the symptoms and signs of the acute infection have resolved, that is, chronic otitis media with effusion.

Hot Topics in Infection and Immunity in Children, edited by Andrew J. Pollard and Adam Finn. Springer, New York, 2006

2. Risk Factors for AOM

AOM occurs most frequently in winter months, and is usually preceded by a viral upper respiratory tract infection (Henderson et al., 1982; Arola et al., 1990). Apart from young age (<2 years), and a history of a previous episode of AOM, other risk factors for AOM include the risk factors for viral upper respiratory infections: attending day care outside the home, exposure to cigarette smoke, having at least one sibling, and possibly use of a dummy (pacifier) (Uhari et al., 1996; Owen et al., 1993). The protective effect of breast feeding is unclear, with studies showing both a beneficial effect, as well as either no effect or increased incidence (Rovers et al., 2004).

3. Diagnosis of AOM

Accurately diagnosing AOM is challenging as young children have limited ability to describe symptoms, and physical examination may be limited by narrow external auditory canals and poor cooperation. Doctors may be uncertain of their diagnosis of AOM in up to 40% of cases (Rosenfeld et al., 2002; Karma et al., 1989). This can lead to over-diagnosis of AOM, resulting in potential iatrogenic harm due to increased use of antibiotics or referral to ENT specialists (in the case of recurrent infections), as well as potentially failing to identify the correct cause of the child's illness (Gonzalez-Vallejo et al., 1998). Diagnosis can be improved by having the parent correctly hold and restrain the child, and removing wax obscuring the tympanic membrane using blunt curettes or ceruminolytic agents (Schwartz et al., 1983; Singer et al., 2000). The mobility of the tympanic membrane can be assessed simply by insufflating a small amount of air using a bulb attachment to an otoscope (Takata et al., 2003). Immobile tympanic membranes imply the presence of middle ear effusions – this may be confirmed using handheld tympanometry or acoustic reflectometry devices which measure the amount of sound energy reflected back from a probe inserted in the external canal.

Although many different diagnostic criteria have been proposed for AOM, a systematic review of the literature regarding the accuracy of clinical features of AOM identified only 4 studies of symptoms, all of which were potentially limited by incorporation bias (Rothman et al., 2003). Ear pain appeared to be the only useful symptom in diagnosis, with a positive likelihood ratio (LR) of 3.0–7.3, but was only present in 50–60% children with AOM. Ear rubbing (or pulling at the ears) had a positive LR of 3.3 (95% CI 2.1–5.1). The presence of fever was variably useful, while most other symptoms were of little value. A further study of clinical signs, which used tympanocentesis to confirm diagnosis, found that the most useful signs were a tympanic membrane that was cloudy (adjusted positive Likelihood Ratio, LR, of 34), distinctly red in colour (LR 8.4), bulging (LR 51) or retracted (LR 3.5) in position, with mobility that was distinctly (LR 31), or slightly (LR 4.0) impaired.

4. Treatment of AOM

Almost all AOM resolves spontaneously, usually within 7 days with only the rare child developing mastoiditis or chronic suppurative otitis (Rosenfeld et al., 1994). In most countries analgesics and antipyretics are used to control pain and fever, while in others decongestants are routinely advised. However, a Cochrane review of randomized trials showed that antihistamines and decongestants had little if any benefit (Flynn et al., 2002), but there have been no randomized trials of the use of various types of analgesics or antipyretics in children with AOM.

Until recently, there has been no global consensus on the appropriate use of antibiotics for children with AOM. Consequently, the rates of use of antibiotics for this condition vary widely from 7–45% in the Netherlands to 98–100% in the USA and UK – depending in some cases on the child's age and severity of presentation (Froom et al., 2001). However more recent studies have shown that rates of antibiotic prescribing to children in general in the UK has fallen since 1993 (Figure 19.1).

Several outcomes need to be addressed in determining the effects of antibiotics on children with AOM; 1) effects on the short term (<7 days) outcomes, such as pain, fever, school attendance, and the rates of AOM in the contralateral ear, 2) effects on medium term (1–2 month) outcomes such as persistence of the middle ear effusion, or recurrence/relapse of infection, 3) whether serious sequelae of AOM such as mastoiditis and its complications are reduced, 4) adverse effects from anti-

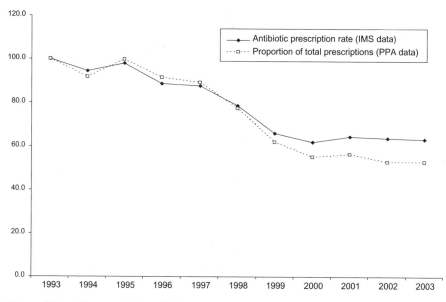

Figure 19.1. Time trends in antibiotic prescribing to children in United Kingdom, 1993–2003, showing declines in antibiotic prescription rate from general practice (IMS data) and proportion of drugs issued by pharmacists (PPA = Prescription Pricing Authority) (Reprinted with permission from Sharland, BMJ 2005, 311:328–9).

biotics such as diarrhoea, rash, allergic reactions, emergence of resistant flora and pathogens, as well as effects on health-seeking behavior of caregivers. Results of recent systematic reviews and other studies allow most of these to be addressed.

4.1. Effects on Short-Term Outcomes

Most evidence on the effects of antibiotics for acute AOM address their effects on short-term outcomes. A 1994 metaanalysis of 33 randomized, controlled trials involving 5400 children (Rosenfeld et al., 1994) found that antibiotics increased complete clinical resolution within 7–14 days after the start of treatment by 13.7% (95% CI 8.2–19.2%). Rosenfeld's metaanalysis included studies which used various outcomes, not all of which would be directly relevant to patients or their caregivers. A recent and more stringent Cochrane review (search updated in March 2003) identified ten randomized trials comparing antibiotics with placebo (Glasziou et al., 2004). Only the eight studies which assessed patient-relevant outcomes (e.g., pain, fever) involving 2,287 children were analysed in the review. Most children with AOM recovered rapidly regardless of whether antibiotics were prescribed – two-thirds were better by 24 hours from the start of treatment, and 80% at 2–7 days. However, antibiotic treatment did result in a further 30% relative reduction (95% CI 19–40%) in pain at two to seven days. Nevertheless, since most children had already recovered, the absolute reduction in pain was only 7%, i.e. 7% fewer children had pain at 2–7 days and 15 (95% CI 11–24) children would need to be treated with antibiotics to prevent one child having pain at 2–7 days. Antibiotics did reduce the rates of AOM in the contralateral ear, however this was non-significant in the random effects model due to heterogeneity of the results.

Since the publication of this Cochrane review, one further randomized, double-blind, placebo-controlled trial of 10 days of amoxicillin for AOM in children 6 months to 5 years of age has been published (Le Saux et al., 2005). In this study of 512 children, those who received placebo experienced significantly more pain and fever in the first two days, and required more doses of analgesics in the first 3 days than those who received antibiotic. However, by 14 days clinical resolution of symptoms had occurred in 84.2% of the children who received placebo and 92.8% who received antibiotics (absolute difference −8.6%, 95% CI −14.4 to −3.0%). Of the 512 children included in the data analysis, 415 (81.1%) had moderate disease, based on clinical symptoms and signs at recruitment. If the Le Saux trial is added to the metaanalysis of the eight studies included in the Cochrane review (Figure 19.2), pain at 24 hours occurred in 39.8% of children receiving placebo compared to 35.7% of those receiving antibiotics, and at 2–7 days pain occurred in 22% and 15.8% of children respectively. Thus, despite the addition of this further large trial that included mostly children with moderate AOM, the overall results of the Cochrane review are unchanged.

4.2. Effects on Medium-Term Outcomes

Reducing the proportion of children who develop a chronic effusion (otitis media with effusion) is important clinically, as persistence of the effusion may pre-

Review: Antibiotics for acute otitis media in children (Version 02)
Comparison: 01 Antibiotic versus Placebo
Outcome: 01 Pain

Study or sub-category	Treatment n/N	Control n/N	RR (fixed) 95% CI
01 Pain at 24 hours			
vanBuchem 1981b	17/48	10/36	
vanBuchem 1981a	13/47	11/40	
Burke 1991	53/112	56/117	
Thalin 1985	58/159	58/158	
Le Saux 2005	82/258	106/254	
Subtotal (95% CI)	624	605	
Total events: 223 (Treatment), 241 (Control)			
Test for heterogeneity: Chi² = 4.12, df = 4 (P = 0.39), I² = 2.9%			
Test for overall effect: Z = 1.37 (P = 0.17)			
02 Pain at 2-7 days			
Halsted 1968	17/62	7/27	
vanBuchem 1981a	6/46	10/38	
Appelman 1991	11/67	10/54	
vanBuchem 1981b	10/48	11/35	
Thalin 1985	15/158	25/158	
Mygind 1981	15/72	29/77	
Burke 1991	20/111	29/114	
Kaleida 1991	19/488	38/492	
Le Saux 2005	43/253	53/246	
Damoiseaux 2000	69/117	89/123	
Subtotal (95% CI)	1422	1364	
Total events: 225 (Treatment), 301 (Control)			
Test for heterogeneity: Chi² = 7.07, df = 9 (P = 0.63), I² = 0%			
Test for overall effect: Z = 4.63 (P < 0.00001)			

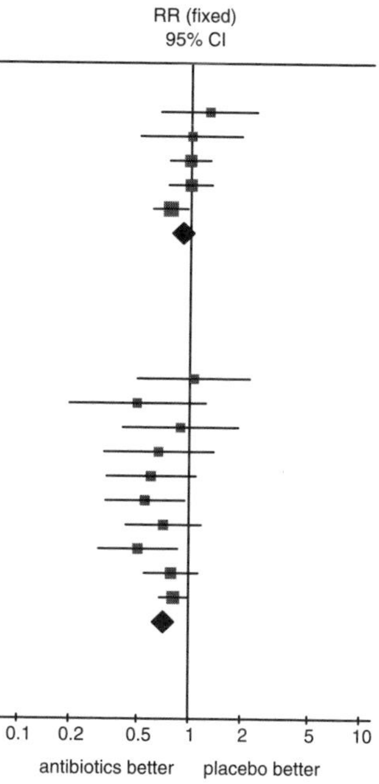

0.1 0.2 0.5 1 2 5 10
antibiotics better   placebo better

Figure 19.2. Results of meta-analysis of the effects of antibiotic compared to placebo for relief of pain in children with acute otitis media.

dispose to recurrent AOM, cause hearing impairment and may result in surgical interventions such as tympanostomy tube placement. However, in the Cochrane review, antibiotics had no statistically significant effect in reducing persistent effusion as measured by tympanometry 1–3 months after the acute infection (Figure 19.3). However audiometry was only performed in two studies, neither of which found significant differences between groups. The Cochrane review found no effect on perforation in the two studies which had assessed this outcome.

4.3. Effects on Serious Sequelae of AOM

In the studies in the Cochrane review, serious outcomes were rare, with only one child (who had received antibiotic not placebo) developing mastoiditis. In the UK where the rates antibiotic prescribing in children has fallen by 37% from 1993–2003 (Figure 19.1) (Sharland et al., 2005), rates of mastoiditis have been relative stable (Figure 19.4). However, the primary studies were all trials in industrialized countries, therefore the results may only apply to settings where the more serious complications of AOM, such as mastoiditis and its sequelae such as subperiosteal abscess, facial palsy, meningitis or brain abscesses are rare, and where

Review: Antibiotics for acute otitis media in children (Version 02)
Comparison: 01 Antibiotic versus Placebo
Outcome: 02 Abnormal Tympanometry

Study or sub-category	Treatment n/N	Control n/N	Peto OR 95% CI	Weight %	Peto OR 95% CI
01 1 Month					
Appelman 1991	21/51	25/45		11.57	0.57 [0.25, 1.26]
Mygind 1981	23/72	25/77		15.70	0.98 [0.49, 1.94]
Burke 1991	41/111	41/116		25.24	1.07 [0.62, 1.84]
Le Saux 2005	68/233	77/222		47.49	0.78 [0.52, 1.15]
Subtotal (95% CI)	467	460		100.00	0.84 [0.64, 1.10]

Total events: 153 (Treatment), 168 (Control)
Test for heterogeneity: Chi² = 2.06, df = 3 (P = 0.56), I² = 0%
Test for overall effect: Z = 1.24 (P = 0.21)

02 3 Months					
Mygind 1981	18/72	18/77		18.71	1.09 [0.52, 2.31]
Burke 1991	20/110	31/111		26.86	0.58 [0.31, 1.08]
Le Saux 2005	58/228	47/210		54.43	1.18 [0.76, 1.83]
Subtotal (95% CI)	410	398		100.00	0.96 [0.70, 1.33]

Total events: 96 (Treatment), 96 (Control)
Test for heterogeneity: Chi² = 3.50, df = 2 (P = 0.17), I² = 42.8%
Test for overall effect: Z = 0.24 (P = 0.81)

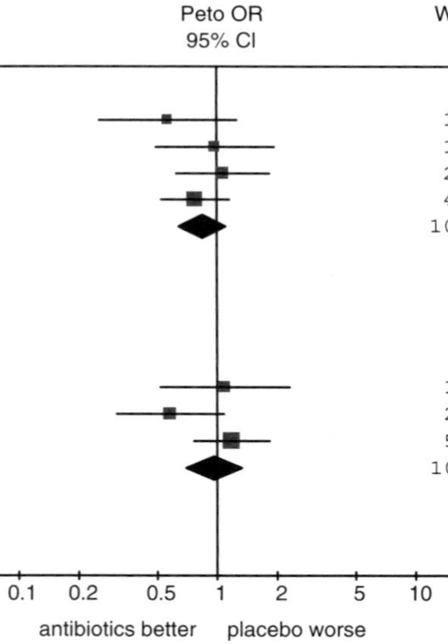

0.1 0.2 0.5 1 2 5 10

antibiotics better   placebo worse

Figure 19.3. Results of meta-analysis of the effects of antibiotic compared to placebo on abnormal tympanometry in children with acute otitis media.

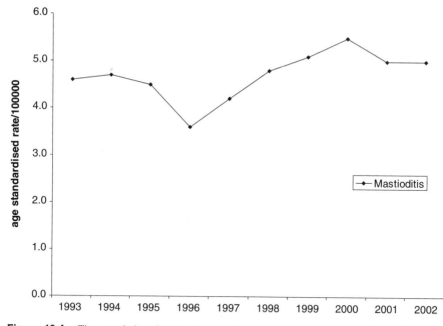

Figure 19.4. Time trends in episodes of mastoiditis or simple mastoidectoy in United Kingdom, 1993–2003. Age standardized rates of hospital admission in children ages 0–14 years of age (Reprinted with permission from Sharland, BMJ 2005, 311:328–9).

children are able to re-attend when clinical improvement fails to occur. An earlier semi-randomised study from Sweden in the 1950s reported a rate of mastoiditis of 17% in the untreated group vs none in the children who received penicillin (Rudberg, 1954). Indeed, in developing countries mastoiditis may be more common: for example in Uganda mastoiditis occurred in 18% of children and adolescents presenting to a hospital ENT clinic (Berman, 1995).

4.4. Adverse Effects of Antibiotics

Children who received antibiotics in the studies reviewed in the Cochrane review were significantly more likely to experience vomiting, diarrhoea or rash (16.6% in antibiotic group vs. 10.5% in placebo group). However in the more recent LeSaux study, there were no statistically significant differences in the occurrence of diarrhoea, rash or vomiting between children in the placebo and control groups. This may be because the majority of children in this latter study had moderate AOM, and hence more likely to have systemic symptoms already. However, these findings are important since the risk of adverse effects from antibiotics may weigh parents' and clinicians' decision against using antibiotics. Emergence of antibiotic resistance in normal as well as pathogenic bacteria is a further potential side effect of using antibiotics, particularly with longer courses or lower doses of antibiotics (Guillemot et al., 1998). Routine prescribing of antibiotics also has a significant role in determining parents' health seeking behaviour for children with possible AOM. Advising parents to use symptomatic measures and providing accurate information on AOM in children with mild disease can reduce the prescribing of antibiotics, without adversely affecting the doctor-patient relationship (Cates, 1999; Little et al., 2001).

5. Duration and Type of Antibiotics Used in AOM

The effect of the type and length of course of antibiotics on outcomes of AOM as well as rates of adverse effects have been explored with both a systematic review and a later Cochrane review (Kozyrskyj et al., 1998; Kozyrskyj et al., 2000). Most studies have compared antibiotic courses of 5 days (short course), to those of 8–10 days (long course). In the 12 trials that reported outcomes at 1 month or less, the summary odds ratio (OR) for treatment failure in 1549 children treated for 5 days compared to 1569 treated for 8–10 days was 1.38 (95% CI 1.15–1.66) when evaluated at any time up to 30 days. The OR for treatment failure increased to 1.52 (95% CI 1.17–1.98) when children were evaluated at 8–19 days, with weighted mean failure rates of 19% in the short course and 13.7% in the long course: a summary risk difference of 7.8% (95% CI 4–11.6%). At an early evaluation point, 8–19 days after starting treatment, short course slightly increased the risk of treatment failure (ie child experiencing clinical features of AOM, or relapse or reinfection) (OR 1.52, 95% CI 1.17–1.98). However, by 30 days there is no difference in outcome between the long and short course (OR 1.22, 95% CI 0.98–1.54). However, there is evidence that longer courses of antibiotics (>5 days), and those at a lower daily dose may cause selective pressure to increase the risk of pharyngeal carriage of resistant bacteria, such as *Streptococcus pneumoniae* (Guillemot et al., 1998). The 1994 meta-analysis of 33 randomized, controlled trials involving 5400 children (Rosenfeld et al., 1994) found no significant differences in the comparative efficacy of various antibiotics used.

6. Limitations of Current Evidence

The Cochrane review by Glasziou et al. used reduction in pain at any point between 2 and 7 days after the onset of treatment as the primary outcome. Since most children with AOM will have recovered spontaneously by 7 days regardless of treatment used, it is possible that such a wide range may mask effects on treatment of children at the lower end of this timescale. There is some evidence that the findings of the Cochrane and other reviews may not apply equally to children of different ages, or with varying severity of disease. Indeed there is evidence that younger children and those with more severe disease may benefit more from antibiotics. Two trials (Burke et al., 1991; Appleman et al., 1991) found higher rates of

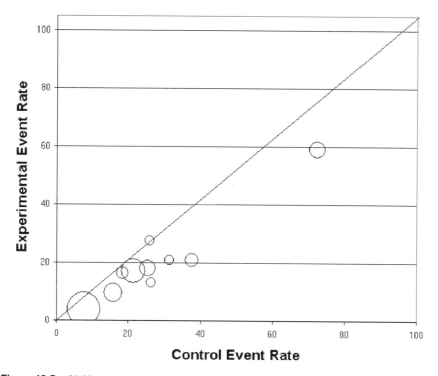

Figure 19.5. L'Abbe plot of studies of antibiotics versus placebo for pain relief at 2–7 days in children with acute otitis media.

failure of placebo in children with bilateral AOM, and those younger than 2 years of age. This may be because younger children have a longer duration of AOM than older children (Damoiseaux et al., 2000). The L'Abbe plot from the Cochrane review, as well as from a subgroup analysis of one additional trial (Little et al., 2001) suggests that children with severe disease, or those with high fevers or vomiting may benefit the most from antibiotics (Figure 19.5).

Given the difficulty in diagnosing AOM, it is possible that several of the studies included in the above reviews included a proportion of children who did not have true AOM. Indeed, in the studies where tympanometry was performed at the time of initial diagnosis, an effusion was confirmed in only approximately two-thirds of cases. Overdiagnosis is often reported in AOM, but in the context of intervention studies, is likely to reduce any beneficial effects of antibiotics on clinical resolution (Figure 19.6). However, in LeSaux's study, if only those children with effusion indicated by abnormal tympanometry were included, symptom resolution at 14 days was 83% (placebo) vs 93.2% in the amoxicillin group (OR 2.8, 95% CI 1.31–5.98), which was similar to the overall results.

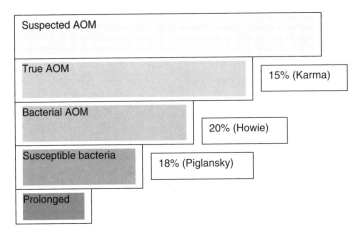

Figure 19.6. A cascade of explanations for the limited effects of antibiotics in children with acute otitis media (AOM). Of children with suspected AOM, only approximately 85% will actually have the condition (Karma et al., 1989). Only approximately 80% of children with true AOM have a bacterial infection (Howie et al., 1972), in which only approximately 82% of the bacteria are susceptible to the antibiotics used (Piglansky et al., 2003).

7. Prevention of AOM

Preventive approaches include reducing environmental risk factors, vaccination, antibiotic prophylaxis, tympanostomy tubes and adenoidectomy (Giebink, 2001). Prevention of frequent recurrences of AOM includes the possible use of daily antibiotic prophylaxis. The effects of this may be limited, and are associated with emergence of resistant bacteria, adverse drug reactions and suppression of symptoms without affecting the actual disease process (Williams et al., 1993). The addition of the heptavalent pneumococcal vaccine to the routine childhood series in the USA has led to a fall in infections caused by the pneumococci strains included in the vaccine (Veenhoven et al., 2003; Kilpi et al., 2003). There is some evidence that the normal flora in the upper respiratory tract may have a role in preventing infection with pathogenic bacteria by inhibiting growth of common pathogenic bacteria. One randomized placebo controlled trial of a spray containing alpha streptococci to prevent recurrence of AOM in 108 children with recurrent AOM, did show a significant reduction in recurrence of AOM compared to placebo (Roos et al., 2001). Xylitol is a naturally occurring polyol sugar that inhibits the growth of pneumococci. In two trials of xylitol chewing gum or ingested orally reduced the incidence of AOM in healthy children by approximately 30–40% (Uhari et al., 1996; Uhari et al., 1998).

8. Conclusions

For the majority of children, AOM is a mild, self-limiting infection for which antibiotics offer little benefit. Most clinicians would agree that treating children with common analgesic antipyretic medications should be the primary management, despite the lack of evidence in AOM, and the theoretical effects of suppressing the febrile response to infection. However, for a subset of children with more severe illness, such as those experiencing high fever or other systemic symptoms, it is possible that antibiotics will give modest benefit in terms of reducing these systemic symptoms in the initial few days after treatment onset. Regardless of initial severity and the modest benefit in the first few days, there appear to be no differences in the rates of persistence of effusions, or rare complications (e.g., mastoiditis). Identifying AOM, especially in younger infants can be very challenging, and even in research situations it is apparently overdiagnosed in about one third of children.

One technique to reduce use of antibiotics is a "delayed" prescription: a prescription is given to the patient (or caregiver) but they are instructed only to use if the patient remains unwell after a set period of time. A trial of antibiotics for children with AOM in primary care randomized 315 children to either an immediate prescription, or a delayed prescription that could be collected at the caregiver's discretion after 72 hours if the child remained unwell (Little et al., 2001). Only a quarter (36/150) of the children in the delayed group actually used their antibiotic prescription – confirming that in the majority of children AOM resolves spontaneously. Children who received immediate antibiotics had approximately one day shorter illness, less night disturbance and needed less paracetamol as compared to those who received the delayed prescription. However, there was no difference in school absence, pain or distress between the groups. The benefit of the immediate prescription occurred after the first 24 hours, when symptoms of distress or pain may have been settling.

Clinicians need to be able to make correct diagnosis, assess the severity of the illness, and in conjunction with the caregiver, make a decision on the use of symptomatic as well as antibiotic treatment. Reductions in the use of antibiotics for AOM have been achieved in many settings. However, providing parents with a "safety net" of a further clinical evaluation if a child fails to improve or deteriorates clinically is important so that the decision to prescribe antibiotics can be reconsidered if necessary.

References

Appelman, C.L., Claessen, J.Q., Touw Otten, F.W., Hordijk, G.J., de Melker, R.A. (1991). Co-amoxiclav in recurrent acute otitis media: placebo controlled study. *BMJ*. 303:1450–1452

Arola, M., Ruuskanen, O., Ziegler, T., Mertsola, J., Nanto-Salonen, K., Putto-Laurila, A., Viljanen, M.K., Halonen, P. (1990). Clinical role of respiratory virus infection in acute otitis media. *Pediatrics*. 86:848–855

Auinger, P., Lanphear, B.P., Kalkwarf, H.J., Mansour, M.E. (2003). Trends in otitis media among children in the United States. *Pediatrics*. 112:514–520

Berman, S. (1995). Otitis media in developing countries. *Pediatrics*. 96:126–131

Burke, P., Bain, J., Robinson, D., Dunleavey, J. (1991). Acute red ear in children: controlled trial of non-antibiotic treatment in general practice. *BMJ*. 305:558–562

Cates, C. (1999). An evidence based approach to reducing antibiotic use in children with acute otitis media: controlled before and after study. *BMJ*. 318:715–716

Damoiseaux, R.A.M.J., van Balen, F.A.M., Hoes, A.W., Verheij, T.J.M., de Melker, R.A. (2000). Primary care based randomized, double blind trial of amoxicillin versus placebo for acute otitis media in children aged under 2 years. *BMJ*. 320:350–354

Fleming, D.M., Ross, A.M., Cross, K.W., Kendall, H. (2003). The reducing incidence of respiratory tract infection and its relation to antibiotic prescribing. *Br. J. Gen. Pract.* 53:778–783

Flynn, C.A., Griffin, G., Tudiver, F. (2002). Decongestants and antihistamines for acute otitis media in children. *Cochrane Database Syst. Rev.* 1:CD001727

Froom, J., Culpepper, L., Green, L., de Melker, R.A., Grob, P., Heeren, T., van Balen, F. (2001). A cross-national study of acute otitis media: risk factors, severity, and treatment at initial visit. Report from the International Primary care Network (ICPN) and the Ambulatory Sentinel Practice Network (ASPN). *J. Am. Board Fam. Pract.* 14:406–417

Giebink, G.S. (2001). Otitis media prevention: non-vaccine prophylaxis. *Vaccine.* 19:S129–133

Glasziou, P.P., Del Mar, C.B., Sanders, S.L., Hayem, M. (2004). Antibiotics for acute otitis media in children. *Cochrane Database Syst Rev.* 1:CD000219

Gonzalez-Vallejo, C., Sorum, P.C., Stewart, T.R., Chessare, J.B., Mumpower, J.L. (1998). Physicians' diagnostic judgements and treatment decisions for acute otitis media in children. *Med. Decis. Making.* 18:149–162

Guillemot, D., Carbon, C., Balkau, B., Geslin, P., Lecoeur, H., Vanzelle-Kervroedan, F., Bouvenot, G., Eschwege, E. (1998). Low dosage and long treatment duration of beta-lactam: risk factors for carriage of penicllin-resistant Streptococcus pneumoniae. *JAMA* 279:394–395

Heikkinen, T., Thint, M., Chonmaitree, T. (1999). Prevalence of various respiratory viruses in the middle ear during acute otitis media. *N. Engl. J. Med.* 340:260–264

Henderson, F.W., Collier, A.M., Sanyal, M.A. (1982). A longitudinal study of respiratory virus and bacteria in the etiology of acute otitis media with effusion. *N. Engl. J. Med.* 306:1377–1383

Howie, V.M., Ploussard, J.H. (1972). Efficacy of fixed combination antibiotic versus separate components in otitis media: effectiveness of erythromycin estolate, triple sulfonamide, ampicillin, erythromycin estolate-triple sulfonamide, and placebo in 280 patients with acute otitis media under two and one-half years of age. *Clin. Pediatrics.* 11:205–214

Karma, P.H., Penttila, M.A., Sipila, M.M., Kataja, M.J. (1989). Otoscopic diagnosis of middle ear effusion in acute and non-acute otitis media. I. The value of different otoscopic findings. *Int. J. Pediatr. Otorhinolaryngol.* 17:37–49

Kilpi, T., Ahman, H., Jokinen, J., Lankinen, K.S., Palmu, A., Savolainen, H., Gronholm, M., Leinonen, M., Hovi, T., Eskola, J., Kayhty, H., Bohidar, N., Sadoff, J.C., Makela, P.H., Finnish Otitis Media Study Group. (2003). Protective efficacy of a second pneumococcal conjugate vaccine against pneumococcal acute otitis media in infants and children: randomized, controlled trial of a 7-valent pneumococcal polysaccharide-meningococcal outer membrane protein complex conjugate vaccine in 1666 children. *Clin. Infect. Dis.* 37:1155–1164. Epub 2003 Oct 7

Klein, J.O. (1989). Epidemiology of otitis media. *Ped. Infect. Dis. J.* 8(Suppl 1):8

Kozyrskyj, A.L., Hildes-Ripstein, E., Longstaffe, S.E., Wincott, J.L., Sitar, D.S., Klassen, T.P., Moffat, M.E. (1998). Treatment of acute otitis media with a shortened course of antibiotics. A meta-analysis. *JAMA.* 279:1736–1742

Kozyrskyj, A.L., Hildes-Ripstein, E., Longstaffe, S.E., Wincott, J.L., Sitar, D.S., Klassen, T.P., Moffat, M.E. (2000). Short course antibiotics for acute otitis media. *Cochrane Database Syst. Rev.* 2: CD001095

Le Saux, N., Gaboury, I., Baird, M., Klassen, T.P., MacCormick, J., Blanchard, C., Pitters, C., Sampson, M., Moher, D. (2005). A randomized, double-blind, placebo-controlled noninferiority trial of amoxicillin for clinically diagnosed acute otitis media in children 6 months to 5 years of age. *CMAJ.* 172(3):335–341

Little, P., Gould, C., Williamson, I., Moore, M., Warner, G., Dunleavey, J. (2001). Pragmatic randomized controlled trial of two prescribing strategies for childhood acute otitis media. *BMJ.* 322:336–342

Owen, M.J., Baldwin, C.D., Swank, P.R., Pannu, A.K., Johnson, D.L., Howie, V.M. (1993). Relation of infant feeding practices, cigarette smoke exposure, and group child care to the onset and duration of acute otitis media with effusion in the first two years of life. *J. Pediatr.* 123:702–711

Piglansky, L., Leibovitz, E., Raiz, S., Greenberg, D., Press, J., Leiberman, A., Dagan, R. (2003). Bacteriologic and clinical efficacy of high dose amoxicillin for therapy of acute otitis media in children. *Pediatr. Inf. Dis. J.* 22:936–937

Roos, K., Håkansson, E.G., Holm, S. (2001). Effect of recolonisation with "interfering" α streptococci on recurrences of acute and secretory otitis media in children: randomized placebo controlled trial. *BMJ.* 322:1–4

Rosenfeld, R.M., Vertrees, J.E., Carr, J., Cipolle, R.J., Uden, D.L., Giebink, G.S., Canafax, D.M. (1994). Clinical efficacy of antimicrobial drugs for acute otitis media: metaanalysis of 5400 children from thirty-three randomized trials. *J. Pediatr.* 124:355–367

Rothman, R., Owens, T., Simel, D.L. (2003). Does this child have acute otitis media? *JAMA.* 290:1633–1640

Rovers, M.M., Schilder, A.G.M., Zielhuis, G.A., Rosenfeld, R.M. (2004). Otitis media. *Lancet.* 363:465–473

Rudberg, R.D. (1954). Acute otitis media: comparative therapeutic results of sulphonamide and penicillin administered in various forms. *Acta Otolaryngol.* 113(Suppl):1–79

Ruuskanen, O., Heikkinen, T. (1994). Viral-bacterial interaction in acute otitis media. *Pediatr. Infect. Dis. J.* 13:1047–1049

Schwartz, R.H., Rodriguez, W.J., McAveney, W., Grundfast, K.M. (1983). Cerumen removal. How necessary is it to diagnose acute otitis media? *Am. J. Dis. Child.* 137:1064–1065

Sharland, M., Kendall, H., Yeates, D., Randall, A., Hughes, G., Glasziou, P., Mant, M. (2005). Antibiotic prescribing in general practice and hospital admissions for peritonsillar abscess, mastoiditis, and rheumatic fever in children: time trend analysis. *BMJ.* 331:328–329

Singer, A.J., Sauris, E., Viccellio, A.W. (2000). Ceruminolytic effects of docusate sodium: a randomized, controlled trial. *Ann. Emerg. Med.* 36:228–232

Sipila, M., Pukander, J., Karma, P. (1987). Incidence of acute otitis media up to the age of 1 years in urban infants. *Acta Otolaryngol. (Stockh).* 104:138–145

Stool, S.E., Field, M.J. (1989). The impact of otitis media. *Ped. Inf. Dis. J.* 1(Suppl):11–14

Takata, G.S., Chan, L.S., Morphew, T., Mangione-Smith, R., Morton, S.C., Shekelle, P. (2003). Evidence assessment of the accuracy of methods of diagnosing middle ear effusion in children with otitis media with effusion. *Pediatrics.* 112:1379–1387

Teele, D.W., Klein, J.O., Rosner, B. (1989). Epidemiology of otitis media during the first seven years of life in children in Greater Boston: a prospective, cohort study. *J. Infect. Dis.* 160:83–94

Uhari, M., Mantysaari, K., Niemela, M. (1996). A meta-analytic review of the risk factors for acute otitis media. *Clin. Infect. Dis.* 22:1079–1083

Uhari, M., Kontiokari, T., Koskela, M., Niemelä, M. (1996). Xylitol chewing gum in prevention of acute otitis media: double blind randomized trial. *BMJ.* 313:1180–1183

Uhari, M., Kontiokari, T., Niemelä, M. (1998). A novel use of xylitol sugar in preventing acute otitis media. *Pediatrics.* 102:879–884

Veenhoven, R., Bogaert, D., Uiterwaal, C., Brouwer, C., Kiezebrink, H., Bruin, J., IJzerman, E., Hermans, P., de Groot, R., Zegers, B., Kuis, W., Rijkers, G., Schilder, A., Sanders, E. (2003). Effect of conjugate pneumococcal vaccine followed by polysaccharide pneumococcal vaccine on recurrent acute otitis media: a randomised study. *Lancet.* 361:2189–2195

Williams, R.L., Chalmers, T.C., Stange, K.C., Chalmers, F.T., Bowlin, S.J. (1993). Use of antibiotics in preventing recurrent acute otitis media and in treating otitis media with effusion. *JAMA.* 270:1344–1351

20

Antibiotic Prophylaxis for the Prevention of Recurrent Urinary Tract Infections in Children

Elliot Long, Samantha Colquhoun, and Jonathan R. Carapetis

1. Introduction

Urinary tract infections (UTI) in children are common, often recurrent, and sometimes severe. The following discussion explores the reasons for and evidence underlying the use of antibiotic prophylaxis for prevention of recurrent urinary tract infections.

2. Why Prevent Urinary Tract Infections in Children?

There are two main reasons for trying to actively prevent UTIs:

a. *To prevent morbidity/ mortality associated with acute cystitis & pyelonephritis.*
Young children are especially at risk of severe infections, with 6% of UTIs in the first year of life associated with bacteraemia, 2% requiring fluid resuscitation at diagnosis, and 1% requiring ventilatory support (Craig et al., 1998). The incidence of death from urinary sepsis was as high as 11% during the 1960s, but is now rare. Because UTIs often recur −12% to 30% of children with their first UTI have another infection during the ensuing 12 months (Winberg et al., 1974) – these occasional severe outcomes may justify offering preventive interventions to young children with UTI.
b. *To prevent long-term sequelae of hypertension and end-stage renal disease.*
The combination of vesicoureteric reflux (VUR) and renal damage from recurrent UTIs has traditionally been thought to lead to end-stage renal

Hot Topics in Infection and Immunity in Children, edited by Andrew J. Pollard and Adam Finn. Springer, New York, 2006

disease (ESRD) (Bailey, 2000). In recent years, however, evidence has emerged that hypertension (HT), renal impairment, and ESRD may not be causally linked with VUR and recurrent UTIs (Figure 20.1) (Wennerstrom et al., 2000b; Yeung et al., 1997; Pope et al., 1999; Marr et al., 2004). Long-term sequelae of HT and renal dysfunction were addressed in a Swedish cohort of children who were examined 16–26 years after their first symptomatic UTI (Wennerstrom et al., 2000a). Sixty-eight participants with renal parenchymal defects on dimerccaptosuccinic acid (DMSA) scan at the time of childhood UTI diagnosis were matched for age and gender with 51 controls (who had UTI with normal DMSA scans). Hypertension was found in 9% of the group with defects and 6% of the group without defects (difference not statistically significant). Glomerular filtration rate (GFR) was well preserved in both study groups, and neither had substantial rates of proteinuria (Wennerstrom et al., 2000a). Although the sample size was small, these data suggest that childhood UTI is not associated with high rates of renal impairment or HT, even if an abnormal DMSA scan is present initially.

These results stand in contrast to previous case-series, in which 18–25% of patients with evidence of renal "scarring" developed hypertension on follow-up 15–30 years after their first UTI (Goonasekera et al., 1996; Smellie et al., 1998). However, those studies selected patients with severe renal parenchymal defects and used conventional and not ambulatory BP measurements, which may account for the discrepancy in results. Jacobson et al. found three of 30 patients followed up 27 years after the detection of non-obstructive pyelonephritic renal scarring had ESRD, and the remainder had significantly lower GFR than matched controls (Jacobson et al., 1989). Martinell et al. followed 54 female patients with renal scarring continuously for 15 years (Martinell et al., 1996). GFR was reduced compared to controls only in those with severe scarring. In summary, there is no clear evidence that renal scarring is a risk factor for the future development of HT, but severe renal scarring may be a risk factor for future reduction in GFR.

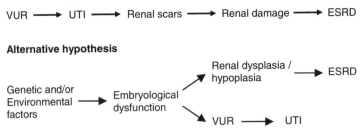

VUR, vesico-ureteric reflux; UTI, urinary tract infection; ESRD, end-stage renal disease

Figure 20.1. Traditional and alternative hypotheses of the link between vesico-ureteric reflux, urinary tract infections, and long-term renal damage.

The natural history of renal parenchymal defects as detected by DMSA scans is resolution over time. Forty percent of DMSA scans performed at the time of acute UTI will demonstrate defects. This reduces to 5% three years later (Stokeland et al., 1998). Many of the defects found at the time of a UTI may have been present prior to the infection, possibly due to congenital renal dysplasia (Figure 20.1) (Polito et al., 2000; Wennerstrom et al., 2000b). This may be diagnosed antenatally and may be progressive despite medical and surgical intervention.

3. Which Children are at Risk of Recurrent UTI?

Data from a prospective hospital based cohort study in children less than 5 years of age who were identified following their first symptomatic UTI implicates young age and presence of high grade VUR as risk factors for recurrence of UTI (Panaretto et al., 1999). Vesico-ureteric reflux of grade 3 to 5 is an independent risk factor for recurrence (odds ratio: 3.5), as is age less than 6 months (OR: 2.9). Recurrence of UTI is also a predictor of future recurrence. After 2 UTIs, the risk of recurrence within one year is ~60%, and after 3 UTIs the risk of recurrence within one year is ~75% (Hellerstein, 1982).

4. Can Recurrent Urinary Tract Infections be Prevented?

A Cochrane meta-analysis of studies comparing antibiotic prophylaxis with antibiotic prophylaxis combined with surgical re-implantation of ureters demonstrated that there may be a small reduction in recurrent UTIs in the surgery plus antibiotic group (Wheeler et al., 2004). However, the effect size was such that nine re-implantations would be needed to prevent one febrile UTI after 5 years. No difference in GFR or abnormalities on DMSA scan was demonstrated between groups. Importantly, none of the studies included a control group that received no treatment.

Williams et al. performed a meta-analysis of 5 existing randomised controlled trials of antibiotic prophylaxis in the prevention of recurrent UTIs (Williams et al., 2001). Three trials (n = 392) examined long-term antibiotic treatment (2–6 months) and subsequent off-treatment development of UTI. These trials included patients with asymptomatic bacteriuria. Two trials (n = 72) examined antibiotic prophylaxis for prevention of recurrent UTI during treatment. Overall, long-term antibiotic treatment reduced the risk of UTI while on treatment (relative risk 0.31, 95% confidence interval 0.10–1.00), but there was no sustained benefit once antibiotic treatment had ceased. The quality of included trials was poor due to lack of blinding and unclear definitions for UTI. The methods of randomization were not described, and allocation concealment was unclear. In addition, the trials included very few boys, infants, and children with abnormal renal tracts. The authors concluded that there remains considerable uncertainty regarding the effectiveness of antibiotic prophylaxis in preventing recurrent UTIs in children.

A systematic review of trials examining antimicrobial prophylaxis for prevention of recurrent UTIs in children with normal urinary tracts was undertaken by Le Saux et al. (Le Saux et al., 2000). They identified two cross-over studies and one randomised controlled trial comparing antibiotic prophylaxis with placebo or no treatment. The quality of the included trials was also poor, with unclear allocation concealment and small sample sizes (mean 28 children). The one randomised trial included was not blinded. There was considerable heterogeneity in rates of recurrence of UTI in each trial, ranging from 0 to 4.0 per 10 patient years in the treatment group and from 4.0 to 16.7 per 10 patient years in the control groups. The authors concluded that the treatment effect observed may have been significantly skewed in favor of antimicrobial prophylaxis by systematic bias.

A Cochrane review of three trials (n = 151) comparing antibiotics with placebo demonstrated a reduction in risk of recurrent UTI with antibiotic prophylaxis (RR 0.36, CI 0.25–0.92) (Williams et al., 2001). The rate of recurrent UTI in the placebo group was 63% overall. Improper allocation concealment, lack of blinding, and lack of intention-to-treat analysis were thought to bias these trials in favor of overestimating any treatment effect. Additionally, the criteria for diagnosis of UTI differed between studies and two studies did not state their method of randomization. The authors concluded that the evidence for effectiveness of antibiotic prophylaxis in preventing recurrent UTIs is unreliable, and the small number of studies performed to date are of poor quality. One trial (n = 120) found nitrofurantoin to be more effective than trimethoprim at preventing recurrent UTI (RR 0.48; CI 0.25–0.92), although nitrofurantoin was poorly tolerated due to gastrointestinal side-effects. Patients who received nitrofurantoin were three times more likely to stop their antibiotic than those receiving trimethoprim because of nausea, vomiting, and abdominal pain.

There is considerable evidence that chemoprophylaxis has a significant failure rate. One quarter of children with moderate to severe VUR in the Birmingham Reflux Study and one third of children in the International Reflux Study developed breakthrough UTI regardless of treatment with antibiotics alone or antibiotics and surgery (Birmingham Reflux Study Group, 1987; Hjalmas et al., 1982). Neither study was placebo controlled.

5. Is There Any Harm Associated with the Long-Term Use of Antibiotic Prophylaxis?

Antibiotic prophylaxis is not always a benign intervention. The presence of *E. coli* resistant to trimethoprim has been demonstrated in up to 66% of children on antibiotic prophylaxis following breakthrough infection, suggesting that prophylaxis may play a role in inducing the emergence of antibiotic-resistant organisms (Braendstrup et al., 1990). Moreover, there is an 8–10% risk of adverse reactions with antimicrobial prophylaxis (Uhari et al., 1996). Trimethoprim-sulfamethoxazole has been associated with nausea and vomiting (5%), skin reactions (2%), diarrhea (1%), and blood dyscrasias (<0.5%). Nitrofurantoin has been associated with abdominal pain limiting adherence.

6. Do Non-antibiotic Interventions Have a Role in Preventing Recurrent Urinary Tract Infections?

Cranberry juice has been examined as a UTI prophylactic in a Cochrane meta-analysis. No difference in incidence of bacteriuria between treatment and control groups was found, although small numbers may have underestimated any treatment effect (Jepson et al., 2004). The authors concluded that this safe intervention requires larger trials to evaluate its efficacy.

Addressing voiding dysfunction with bladder training programs with or without the additional use of anticholinergic agents may be an alternative to prophylactic chemotherapy. The anticholinergic oxyphenomium bromide was found to be effective at decreasing detrusor contractility and the degree of reflux in children with these conditions (Scholtmeijer and Van Mastrigt, 1991), although it has not been proven to decrease the incidence of UTI.

Circumcision has been demonstrated to reduce the incidence of UTI in boys 10-fold in a retrospective cohort study (Schoen et al., 2000). This intervention may not be appealing as a first-line preventive strategy, but may benefit boys with multiple recurrent UTIs.

7. Recommendations for the Use of Antibiotic Prophylaxis

Current clinical guidelines regarding the use of antimicrobial prophylaxis for recurrent UTIs in children highlight the uncertainty that surrounds their efficacy. For example, the American Academy of Pediatrics currently recommends the use of prophylactic antibiotics for all children with UTIs in whom documented VUR exists (American Academy of Pediatrics, 1999), whereas Swedish guidelines limit prophylaxis to high grades of reflux (4–5) only (Jodal and Lindburg, 1999).

We know that UTIs may be severe, particularly in infants, and that frequent recurrent UTIs are associated with, but not necessarily the cause of, permanent renal damage in a small percentage of cases. There is a limited amount of poor-quality evidence that recurrent UTIs may be prevented by long-term prophylactic antibiotics, although it is not known if the benefits of this approach (if any) outweigh the risks. It is very difficult to determine if antibiotic prophylaxis can prevent the development of long-term renal damage. In Australia, where reflux nephropathy accounts for approximately 4% of all ESRD cases, there has been no apparent reduction in reflux-nephropathy associated ESRD over the past 20 years, during which time antibiotic prophylaxis for children with UTI and VUR has been a common practice (Craig et al., 2000). Therefore, until we have better data as to whether UTI is linked with long-term renal damage, and whether antibiotic prophylaxis can prevent recurrent UTI, it will be difficult to provide clinicians with clear evidence-based recommendations. It is our practice to consider a voiding cystourethrogram in infants, particularly those aged less than six months, presenting with their first UTI and to offer antibiotic prophylaxis to those with grade 3 or more reflux. We also consider antibiotic prophylaxis in other children at increased risk of severe or recurrent UTI (Figure 20.2). We prescribe a daily dose of trimethoprim-sulphamethoxazole (2 mg

Infants, particularly age <6 months

Grade 3 or higher vesico-ureteric reflux

A history of a previous UTI

A family history of recurrent UTI or vesico-ureteric reflux

Figure 20.2. Risk factors for recurrent urinary tract infection or vesico-ureteric reflux in children with urinary tract infection.

trimethoprim – 10 mg sulphamethoxazole per kg) for six to twelve months or until the child is no longer wearing nappies, before a trial off prophylaxis. Given the uncertainties in this area, other clinicians will have different indications for the use of prophylaxis, and many will not use it at all. Hopefully the next few years will see better evidence emerge so that recommendations can be revised with confidence.

Acknowledgements

The authors would like to thank Dr. Jonathan Craig for his assistance with the preparation of this manuscript.

References

American Academy of Pediatrics (1999) *Practice Parameter: the diagnosis, treatment, and evaluation of the initial urinary tract infection in febrile infants and young children.*, Pediatr. Inf. Dis. J., **103**, 843–853.

Bailey, R. (2000) *The relationship of vesicoureteral reflux to urinary tract infection and chronic pyelonephritis nephropathy.*, Clin. Nephrol., **1**, 132–141.

Birmingham Reflux Study Group (1987) *Prospective trial of operative versus non-operative treatment of vesicoureteric reflux in children: five year's observation.*, Br. Med. J., **295**, 237–241.

Braendstrup, L., Hjelt, K., Petersen, K., Petersen, S., Andersen, E.A., Daugbjerg, P., Stagegaard, B.R., Nielsen, O.H., Vejlsgaard, R. and Schon, G. (1990) *Nitrofurantoin versus trimethoprim prophylaxis in recurrent urinarytract infection in children.*, Acta Paediatr. Scand., **79**, 1225–1234.

Craig, J., Irwig, L., Knight, J., Sureshkumar, P. and Roy, L.P. (1998) *Symptomatic urinary tract infection in Australian preschool children: clinical features, microbiology, renal tract abnormalities, and short-term outcomes.*, J. Paediatr. Child Health., **34**, 154–159.

Craig, J.C., Irwig, L.M., Knight, J.F. and Roy, L.P. (2000) *Does treatment of vesicoureteric reflux in childhood prevent end-stage renal disease attributable to reflux nephropathy?*, Pediatr. Inf. Dis. J., **105**, 1236–1241.

Goonasekera, C., Shah, V., Wade, A.M., Barratt, T.M. and Dillon, M.J. (1996) *15-year follow-up of renin and blood pressure in reflux nephropathy.*, Lancet., **347**, 640–643.

Hellerstein, S. (1982) *Recurrent urinary tract infections in children.*, Paediatr. Inf. Dis., **1**, 271–281.

Hjalmas, K., Lohr, G., Tamminen-Mobius, T., Seppanen, J., Olbing, H. and Wikstrom, S. (1982) *Surgical Results in the International Reflux Study in Children (Europe).*, J. Urol., **148**, 1657–1661.

Jacobson, S.H., Eklof, O., Erikson, C.G., Linsl, E., Tidgren, B. and Winberg, J. (1989) *Development of hypertension and uremia after pyelonephritis in childhood: 27 years of follow-up.*, Br. Med. J., **299**, 703–706.

Jepson, R., Mihalejvic, L. and Craig, J. (2004) *Cranberries for preventing urinary tract infections. Cochrane Database Syst. Rev.,* **1:CD001321**.

Jodal, U. and Lindburg, U. (1999) *Guidelines for management of children with urinary tract infection and vesico-ureteral reflux. Recommendations from a Swedish state-of-the-art conference., Acta Paediatr. Suppl.,* **431,** 87–89.

Le Saux, N., Pham, B. and Moher, D. (2000) *Evaluating the benefits of antimicrobial prophylaxis to prevent urinary tract infections in children: a systematic review., Can. Med. Assoc. J.,* **163,** 523–529.

Marr, G., Oppezzo, C., Ardissino, G., Dacco, V., Testa, S., Avolia, L., Taioli, E. and Sereni, F. (2004) *Severe vesicoureteral reflux and chronic renal failure: a condition peculiar to male gender? Data from the ITALKID project., J. Pediatr.,* **144,** 677–681.

Martinell, J., Linden-Jansen, O., Jagenburg, G., Sivertsson, R., Claesson, I. and Jodal, U. (1996) *Girls prone to urinary tract infections followed into adulthood: indices of renal disease., Pediatr. Nephrol.,* **10,** 139–142.

Panaretto, K., Craig, J., Knight, J., Giles, R., Sureshkumar, P. and Roy, L. (1999) *Risk factors for recurrent urinary tract infection in preschool children., J. Paediatr. Child Health.,* **35,** 454–459.

Polito, C., La Manna, A., Rambaldi, P.F., Nappi, B., Mansi, L. and Di Toro, R. (2000) *High incidence of a generally small kidney and primary vesicoureteral reflux., J. Urol.,* **164,** 479–482.

Pope, J.C., Brock, J.W., Adams, M.C., Stephens, F.D. and Ichikawa, I. (1999) *How they begin and how they end: classic and new theories for the development and deterioration of Congenital Anomalies of the Kidney and Urinary Tract, CAKUT., J. Am. Soc. Nephrol.,* **10,** 2018–2028.

Schoen, E., Colby, C. and Ray, G. (2000) *Newborn circumcision decreases incidence and costs of urinary tract infection in the first year of life., Pediatr. Inf. Dis. J.,* **105,** 7789–7793.

Scholtmeijer, R. and Van Mastrigt, R. (1991) *The effect of Oxyphenonium Bromide and Oxybutynin Hydrochloride on Detrusor Contractility and Reflux in Children with Vesicoureteral reflux and Voiding Dysfunction., J. Urol.,* **146,** 660–662.

Smellie, J., Prescod, N.P., Shaw, P.J., Risdon, R.A. and Bryant, T.N. (1998) *Childhood reflux and urinary tract infection: a follow-up of 10–41 years in 226 adults., Pediatr. Nephrol.,* **12,** 727–736.

Stokeland, E., Hellstrom, M., Jacobsson, S., Jodal, U. and Sixt, R. (1998) *Evaluation of DMSA scintigraphy and urography in assessing both acute and permanent renal damage in children., Acta Radiol.,* **39,** 447–452.

Uhari, M., Nuutinen, M. and Turtinen, J. (1996) *Adverse reactions in children during long-term antimicrobial therapy., Pediatr. Inf. Dis. J.,* **15,** 404–408.

Wennerstrom, M., Hansson, M., Jodal, U., Sixt, R. and Stokland, E. (2000a) *Renal function 16 to 26 years after the first urinary tract infection in childhood., Arch. Pediatr. Adol. Med.,* **154,** 339–345.

Wennerstrom, M., Hansson, S., Jodal, U. and Stokland, U. (2000b) *Primary and acquired renal scarring in boys and girls with urinary tract infection., J. Pediatr.,* **136,** 30–34.

Wheeler, D.M., Vimalachandra, D., Hodson, E.M., Smith, G.H. and Craig, J.C. (2004) *Interventions for primary vesicoureteric reflux., Cochrane Database Syst. Rev.,* **3:CD001532**.

Williams, G., Lee, A. and Craig, J. (2001) *Long-term antibiotics for preventing recurrent urinary tract infection in children., Cochrane Database Syst. Rev.,* **4:CD001534**.

Winberg, J., Andersen, H.J., Bergstrom, T., Jacobsson, B., Larson, H. and Lincoln, K. (1974) *Epidemiology of symptomatic urinary tract infections in childhood., Acta Paediatr. Scand.,* **Suppl. 252,** 1–20.

Yeung, C.K., Godley, M.L., Dhillon, H.K., Gordon, I., Duffy, P.G. and Ransley, P.G. (1997) *The characteristics of primary vesico-ureteric reflux in male and female infants with pre-natal hydronephrosis., Br. J. Urol.,* **80,** 319–327.

21

Human Metapneumovirus: An Important Cause of Acute Respiratory Illness

Adilia Warris and Ronald de Groot

1. Introduction

Acute respiratory infections are an important cause of morbidity and mortality in children worldwide. The World Health Organization ranks respiratory diseases as the second-leading cause of death worldwide in children <5 years. Bronchiolitis and pneumonia are the most common clinical manifestations of viral respiratory pathogens. Respiratory viruses are also associated with exacerbations of asthma and recurrent wheezing.

In 2001, a new infectious agent, human metapneumovirus (hMPV), was isolated from nasopharyngeal aspirates of young children with respiratory tract illness from The Netherlands (van den Hoogen et al., 2001). Since then, numerous reports have described the detection of hMPV in clinical specimens from children, adults and the elderly, both immunocompetent and immunocompromised patients, diagnosed with an acute respiratory illness all over the world. Serological surveys indicate that the virus has been circulating in humans for at least 50 years. It has a high prevalence in the Dutch population: all children tested were seropositive before the age of 6 years (van den Hoogen et al., 2001).

2. Characterization of the Virus

The *Paramyxovirinae* and *Pneumovirinae* subfamilies of the *Paramyxoviridae* family, a group of negative stranded RNA viruses, include several major pathogens of humans and animals. The *Pneumovirinae* are taxonomically divided in the *Pneumovirus* and the *Metapneumovirus* genera. The human pathogen respiratory syncytial virus (hRSV) belongs to the genus *Pneumovirus*. On the basis of the

Hot Topics in Infection and Immunity in Children, edited by Andrew J. Pollard and Adam Finn. Springer, New York, 2006

organisation of the viral genome and sequence identity to the *Metapneumovirus* avian pneumovirus, also known as turkey rhinotracheitis virus, hMPV seemed to be a tentative member of the *Metapneumovirus* genus. Genetic analysis of the N (nucleocapsid RNA binding protein), M (matrix protein), P (phosphoprotein) and F (fusion glycoprotein) genes revealed that hMPV showed a higher sequence homology to the *Metapneumovirus* genus (average of 66%) as compared to the genus *Pneumovirus* (average of 30%) (van den Hoogen et al., 2001; Bastien et al., 2003a).

hMPV is an enveloped virus with a genome that is a single strand of RNA of approximately 13 kb (van den Hoogen et al., 2001). A nearly complete genome sequence was determined for the prototype Netherlands 00-1 strain of hMPV, and complete genome sequences were determined for two Canadian strains, CAN97-83 and CAN98-75, which represent the two proposed hMPV genetic subgroups (van den Hoogen et al., 2002; Biacchesi et al., 2003). These studies also confirmed that the 3′ to 5′ hMPV gene order is N-P-M-F-M2,1 (transcription elongation factor)-M2,2 (RNA synthesis regulatory factor)-SH (small hydrophobic surface protein)-G (major attachment protein)-L (major polymerase subunit). None of these proteins have identified or characterized by direct biochemical means, and their functions remain to be confirmed [Crowe, 2004]. In contrast to the genomic organization of the pneumoviruses, metapneumoviruses lack the NS1 and NS2 genes and have a different positioning of the other common genes, i.e., the N, P, M, F, L, G, M2, and SH. The absence of open reading frames (ORFs) between the M and F genes in this virus and the lack of NS1 and NS2 genes is in agreement with it being the first identified non-avian member of the *Metapneumovirus* genus (van den Hoogen et al., 2002; Boivin et al., 2002).

Although genetically not closely related, hMPV shares many biologic properties with hRSV. The hMPV isolates replicate slowly in tertiary monkey kidney (tMK) and rhesus monkey kidney (LLC-MK2) cells, very poorly in Vero cells and A549 cells, and could not be propagated in Madin Darby canine kidney (MDCK) cells or chicken embryo fibroblasts (CEF) (van den Hoogen et al., 2001). The cytopathic effects are indistinguishable from those caused by hRSV, although occurred slightly later, 10–17 days post inoculation. Electron microscopy reveals paramyxovirus-like pleiomorphic particles in the range of 150–600 nm, with short envelope projections in the range of 13–17 nm, indistinguishable from hRSV (van den Hoogen et al., 2001). hMPV is chloroform-sensitive and replicates optimally in a trypsin-dependent manner, in contrast to hRSV, in tMK cells. No hemagglutinating activity with turkey, chicken or guinea pig erythrocytes was displayed. These combined virological data indicate that the hMPV is indeed a member of the *Paramyxoviridae* family.

RT-PCR analyses using primer sets for specific paramyxoviruses (parainfluenza virus, mumps virus, measles virus, hRSV, simian virus type 5, Senday virus and Newcastle disease virus) did not react with the newly identified hMPV, indicating no close genetic relatedness to these viruses. hMPV-specific antisera did not react in immune fluorescence assays (IFA) with cells infected with a panel of paramyxoviruses and orthomyxoviruses (parainfluenza viruses, influenza virus A and B, hRSV) (van den Hoogen et al., 2001).

3. Molecular Epidemiology

Genetic studies on hMPV have demonstrated the presence of 2 distinct hMPV groups each divided in two subgroups (Boivin et al., 2002; Peiris et al., 2003; Madhi et al., 2003; Bastien et al., 2003a; van den Hoogen et al., 2004b).

Bastien and colleagues (2003a) determined the complete nucleotide sequences of the N, P, M, and F genes of Canadian hMPV isolates. Comparison of the deduced amino acid sequences for the N, M, and F genes of the different isolates revealed that all three genes were well conserved with 94.1%–97.6% identity between the two distinct clusters. The P gene showed more diversity with 81.6%–85.7% amino acid identity for isolates between the two clusters, and 94.6%–100% for isolates within the same cluster. The Canadian cluster 1 isolates showed over 96% amino acid identity with the NDL00-1 Dutch isolates for all the viral proteins analyzed (van den Hoogen et al., 2002).

Analysis of the F and G protein genes of the 4 subgroups showed that the F protein was highly conserved and demonstrated low variability within the four groups (van den Hoogen et al., 2004b). With an amino acid identity of 93–96% between the subgroups, the F protein becomes attractive as a principal target of protective antibodies. In contrast, the G gene showed high sequence diversity. Furthermore, these phylogenetic analyses showed that the hMPV strains obtained from different years and from different countries were randomly distributed over all four sublineages. To address the antigenic relationship between the different lineages, virus neutralization assays were performed showing a difference in antigenicity between lineage A and B. On the basis of these results it was proposed to define the two main lineages of hMPV as serotypes A and B. Although each serotype can be divided into two genetic subgroups, these subgroups did not reflect major antigenic differences.

To characterize the extent of genetic diversity among hMPV strains in Australia and worldwide, comparative nucleotide- and predicted amino acid-sequence studies were performed with the N gene and with the P gene [Mackay, 2004]. Comparison of aligned sequences revealed an 11.9%–17.6% nucleotide variation, which divided the viral strains into 2 main lineages. In addition, 2 distinct subtypes were apparent within each lineage, which were defined as hMPV types A1, A2, B1, and B2. The variability of the P gene permitted the reliable classification of hMPV into its 4 subtypes, providing the P gene to be a valuable target for phylogenetic studies. To confirm that this classification, based on both the N and the P gene, agreed with that proposed in other studies, similar sequencing and analyses of all available M, F, G, and L gene sequences were performed. The same lineages were found and thus the P gene seems a useful single target for genotyping and for the creation of a global classification scheme for hMPV.

A large community-based phylogenetic study of hMPV for both surface glycoproteins F and G provides the evidence for the presence of multiple genotypes within each subgroup of hMPV [Ludewick, 2005]. This evidence comes from the topology of the phylogenetic trees and bootstrap values in which sequences were arbitrarily considered a genotype if they clustered together with bootstrap values of 70% to 100%. This resulted in 9 genotypes and 6 possible genotypes in the 4 subgroups together.

Strains from both hMPV groups may cocirculate in a particular year as shown in South Africa, but at the same time not all 4 subgroup viruses are detected in a single year [Ludewick, 2005]. Limited data indicate that both hMPV groups can circulate in a single season with the possibility of the predominant group switching in successive seasons [Boivin, 2002; Mackay, 2004]. The circulation of multiple lineages and the changes of the dominant group of virus may suggest an attempt at evasion of preexisting immunity, as has been seen also for hRSV. Studies performed in very different geographic areas showed that specific strains coexist across geographic areas [Mackay, 2004; Ludewick, 2005].

4. Clinical Epidemiology

hMPV has circulated in humans for at least 50 years; a 100% seroprevalence was found in 72 serum samples obtained from individuals 8 to 99 years old, collected in 1958 in the Netherlands [van den Hoogen, 2001]. The few seroprevalence studies available indicate that virtually all children are infected in early childhood. Since the first description of hMPV infection in children by van den Hoogen et al. [2001], hMPV has been found in most parts of the world (Table 21.1): North America, Europe, Asia, and Australia. The virus has also been identified in HIV-infected and non-immunocompromised children from South-Africa [Madhi, 2003]. hMPV infections account for at least 4 to 8% of the RTI in hospitalized children, although some studies report much higher prevalences (see Table 21.1). In the general community hMPV infections account for at least 3% of children who visit a general practitioner for RTI. The relative role of hMPV in respiratory syndromes of adults has not been well studied.

In a large study of patients with RTI, the diagnostic outcome for 685 specimens sent specifically for respiratory pathogen testing were compared. hRSV was detected most frequently, in 126 (18%) of 685 samples obtained with patients with RTI. hMPV was the second-most-detected viral pathogen in 7% of the samples and was isolated more frequently than parainfluenza viruses, adenovirus, rhinovirus, and influenza viruses types A and B [Peiris, 2003]. The mode of transmission is not formally studied, but is likely by large particle respiratory secretions and fomites, based on its relatedness to other pneumoviruses. Nosocomial transmission does occur and warrants contact isolation and scrupulous hand washing by health care providers.

5. Pathogenesis

An experimental infection model in cynomolgus macaques (Macaca Fasicularis) has confirmed that hMPV is a primary pathogen of the respiratory tract in primates [Kuiken, 2004]. The hMPV-infected macaques showed mild clinical signs of rhinorrhea corresponding with a suppurative rhinitis at pathological examination. In addition, mild erosive and inflammatory changes in the mucosa and submucosa of conducting airways, and an increased number of alveolar macrophages in

Table 21.1.   Incidence of hMPV infections in several studies

	country	study period	population	method	hMPV-positive/total number of patients	prevalence	peak age
HOSPITAL							
König, 2004	Germany	Nov. 99–Oct. 01	<3 yr; RTI <6mo; apneu admitted to ICU	PCR	15/87	18%	
McAdam, 2004	USA	Oct. 00–Sept. 02	≤18 yr	RT-PCR(1)	54/868	6.2%	3–24 months
Jartti, 2004	Finland	Sept. 00–June 02	3 mo.–16yr; acute expiratory wheezing	RT-PCR(2)	12/291	4%	3–11 months
Døllner, 2004	Norway	Nov. 02–Apr. 03	children; RTI	PCR(2)	50/236	21%	≤12 months
Mullins, 2004	USA	Aug. 00–Sept. 01	<5 yr; RTI	RT-PCR(2)	26/641	4%	6–24 months
Esper, 2004	USA	Nov. 01–Nov. 02	<5 yr	RT-PCR(3)	54/668	8.1%	<12 months
Madhi, 2003	South-Africa	Mar. 00–Oct. 00	infants	RT-PCR(3)	14/196	7.1%	
vd Hoogen, 2003	Netherlands	Oct. 00–Feb. 02	all ages; RTI	RT-PCR(1)	48/685*	6.5%	4–6 months
Viazov, 2003	Germany	Jan. 02–May 02	<2yr; RTI	RT-PCR(2)	11/65	17.5%	
Maggi, 2003	Italy	Jan. 00–May 02	<2yr; RTI	RT-PCR(4)	23/90	25%	≤3 months
Peiris, 2003	Hong-Kong	Aug. 01–Mar. 02	≤18yr; RTI	RT-PCR(2)	32/587	5.5%	
Thanasugarn '03	Thailand	Mar. 01–Sept. 02	<14yr; RTI	RT-PCR(2)	5/120	4.2%	
Rawlinson, 2003	Australia	2 summers & 2 winters 00–02	<12yr; URTI	PCR(2)	9/150	6%	
			<17yr; asthma	PCR(2)	3/179	2%	
Falsey, 2003	USA	Nov. 99–Apr. 00	high risk adults; RTI	RT-PCR(2)	7/238	2.9%	
		Nov. 00–Apr. 01	nursing homes; RTI	serology	2/37	5.4%	
			hospitalized; RTI		20/309	6.5%	
Freymuth, 2003	France	Nov. 00–Mar. 01 Nov. 01–Feb. 02	Children	RT-PCR(3)	19/337*	6.6%	<1 yr

Table 20.1. *Continued*

	country	study period	population	method	hMPV-positive/total number of patients	prevalence	peak age
GENERAL COMMUNITY							
König, 2004	Germany	Oct. 00–Apr. 01	<3 yr; RTI <6 mo; apneu	PCR(3)	2/620	<1%	
Laham, 2004	Argentina	June 02–Sept. 02	<1 yr; RTI	RT-PCR(4)	22/373	6%	
Principi, 2004	Italy	Nov. 02–Apr. 03	<15 yr; RTI	PCR	41/1331	3.1%	
Williams, 2004	USA	1976–2001	<5 yr; RTI or AOM	RT-PCR(3)	49/248	20%	6–12 months
Bastien, 2003	Canada	Oct. 01–Apr. 02	RTI	RT-PCR(1)	66/445	14.8%	<5 yr, >50 yr
Falsey, 2003	USA	Nov. 99–Apr. 00	fit elderly >65 yr; RTI	RT-PCR(2)	4/233	1.7%	
		Nov. 00–Apr. 01	young adults; RTI	serology	11/167	6.6%	

(U)RTI: (upper) respiratory tract infection; AOM: acute otitis media.
1 all respiratory specimens obtained; 2 nasopharyngeal aspirates; 3 on common respiratory viruses negative nasopharyngeal aspirates; 4 nasal swabs; * number of samples.

bronchioles and pulmonary alveoli were observed. A close association between these laesions and the specific expression of hMPV antigen was shown by immunohistochemistry. Based on the antigen expression, viral replication mainly took place at the apical surface of ciliated epithelial cells throughout the respiratory tract. Pharyngeal excretion of hMPV showed a peak at day 4 after infection decreasing to zero by day 10, concomittant with a reduction in the number of infected epithelial cells. The mild upper respiratory tract disease as observed in these macaques corresponds to that in immunocompetent adults. Due to the fact the hMPV can replicate in the lower respiratory tract of cynomolgus macaques, more severe disease can be expected in immunocompromised patients. Similar histopathological findings of the respiratory tract were reported in an experimental cotton rat model of hMPV-infection [Wyde, 2005]. These experimental models of hMPV-infection show similarities with the pathogenesis, as far as studied, of hRSV infection in humans. With the results of the above described experimental hMPV-infection, the Rivers' modified Koch's postulates for viral disease are fulfilled. These modified postulates propose that (1) the virus has to be isolated from a diseased host, (2) the virus has to be cultivated in host cells, (3) the virus withstands the proof of filterability, (4) the isolated virus produces the disease in original or related host species, (5) the virus has to be re-isolated after experimental infection, and (6) a specific immune response upon infection can be detected.

hMPV, in contrast with hRSV, seems to be a poor inducer of inflammatory cytokines at the respiratory epithelial surfaces but elicits identical symptoms of similar severity [Laham, 2004]. Levels of the inflammatory cytokines Interleukin (IL)-1β, TNF-α, IL-6, IL-8, IL-10, and IL-12 in respiratory secretions of infants <1 year with an acute respiratory tract infection, were 2–6-fold lower in those infected with hMPV compared to hRSV. Levels of IL-8 and RANTES (regulated upon activation, normal T cell expressed and secreted) in nasal secretions of children <16 year admitted with acute expiratory wheezing were different from that reported in infections with hRSV [Jartii, 2002]. Patients with hRSV-infection had high concentrations of RANTES and varying levels of IL-8, whereas children with hMPV-infection had lower concentrations of RANTES and higher levels of IL-8. It seems that other mechanisms than known for hRSV elicit symptomatic disease after infection with hMPV. Other mechanisms may included, although not limited to, (1) direct viral damage to the airways; (2) Th1 vs. Th2 polarization of the pulmonary immune response, leading to different clinical symptoms; and (3) chemokine-mediated inflammation. Further research is needed to elicudate the exact mechanisms of illnesses caused by hMPV.

6. Diagnosis

For a virus that is not easily detected by virus isolation in the laboratory, it will be of great importance that rapid, sensitive and reproducible diagnostic tests be developed. The identification of the two hMPV serotypes, A and B, with each serotype divided into genetic sublineages, 1 and 2, has implications for the development of RT-PCR assays and serologic diagnostic tests. Because of the unavailability of rapid antigen detection tests and because its fastidious growth in cell cultures, RT-PCR has become the method of choice. RT-PCR procedures proved to be more sensitive than virus isolation, and can detect genetically distinct hMPV strains [van den Hoogen, 2003].

The cytopathic effect is variable with hRSV-like syncytia formation or focal rounding and cell destruction. The search by van den Hoogen et al. [2004a] for a cell line with similar susceptibility for the 4 hMPV lineages and with enhanced detection of the virus by cytopathic effects, resulted in the generation of a subclone of Vero cells (Vero cell clone 118). This cell line is now used routinely for virus isolation in the Netherlands. Commercially available antibodies are not yet available. Monoclonal antibodies recognising conserved epitopes will be useful for rapid viral diagnostics by use of IF or direct IF techniques as currently used for diagnosing hRSV. Confirmation of hMPV causing the cytopathic effect is achieved by RT-PCR testing of the viral culture.

Most RT-PCR protocols reported to date have relied on amplification of the L, N, or F gene with primer sequences mainly derived from the prototype strain 001 from the Netherlands. A comparative evaluation of RT-PCR assays performed in a LightCycler instrument for detection of hMPV in infected cell cultures showed positivity rates of 100%, 90%, 75%, 60%, and 55% using primers for the N, L, M, P, and F genes [Cote S, 2003]. A second evaluation in the same study on nasopharyngeal aspirates positive for the hMPV N gene, the PCR positivity rate for the L, M, P, and F genes were 90%, 60%, 30% and 80%, respectively. From this study it can be concluded that RT-PCR assays aimed at amplifying the N and L genes, which are coding for two internal viral proteins, appear particularly suitable to detect hMPV from both lineages [Coté, 2003; Mackay, 2003; van den Hoogen, 2003]. The N and L genes seem to be the more conserved regions of the hMPV genome. Rapid and sensitive RT-PCR assays for the N gene (detection limit of 100 copies) has been developed allowing rapid amplification and detection of hMPV sequences directly from clinical samples in <2 h [Coté, 2003; Mackay, 2003]. It might be clear that if inadequate primers are selected for PCR amplification, the hMPV detection might be underestimated.

Serological testing only permits a retrospective diagnosis. Because infection is almost universal in childhood, a seroconversion or a ≥4-fold increase in antibody titers must be demonstrated to confirm recent infection. The serologic survey performed in the Netherlands was based on an indirect immunofluoresence assay using hMPV-infected cells [van den Hoogen, 2001]. A home-made ELISA method has also been developed using cell lysates of hMPV [Falsey, 2003]. To conduct large serological surveys, simpler ELISA tests using viral proteins possibly derived from the 2 serotypes will be needed.

7. Clinical Characteristics

The first description of hMPV in children with lower RTI has been reported by a Dutch group that identified the virus in respiratory secretions [van de Hoogen, 2001]. Clinical symptoms were similar to those caused by hRSV, ranging from upper RTI, severe bronchiolitis and pneumonia during the winter season. All 28 children were <5 years old, whereas 46% were <1 year old. Asymptomatic carriage seems to be rare in children; no hMPV was detected in 400 infants without respiratory symptoms [van den Hoogen, 2001].

The prevalence and clinical symptoms of hMPV-infected patients, identified by RT-PCR in respiratory samples obtained from patients in a university hospital, indicated that the prevalence and clinical severity due to hMPV infections are slightly lower than those of hRSV infections during the winter season [van den Hoogen, 2003]. Most of the hMPV-positive patients were children <2 years old who did not have any underlying illnesses. hMPV was found significantly less frequently in children <2 months old than was hRSV. Of the 31 hMPV-positive children <2 years old, only 4 (31%) were <2 months old, whereas 43 (35%) of the 122 hRSV-positive children <2 years old were also <2 months old. Others have found that the mean age of patients infected with hMPV was slightly lower than that compared to hRSV [Williams, 2004]. Of the hMPV-positive patients who were >5 years old, most had other diseases (e.g. cystic fibrosis, leukemia, and non-Hodgkin lymphoma) or had recently received bone-marrow or kidney transplantation, indicating an association with immunosuppression. Two severely immuncompromised patients died due to progressive respiratory failure with hMPV as the sole pathogen detected [Hamelin, 2004]. In studies involving young and elderly adults, hMPV caused more severe disease in fragile elderly than in healthy elderly or young adults [Boivin, 2002; Falsey, 2003]. Clinical symptoms in children due to hMPV infection (Table 21.2) include cough (82%), rhinitis (67%), fever (72%), respiratory distress (71%), wheezing (59%), and retractions (54%) [van den Hoogen, 2003; Hamelin, 2004; Døllner, 2004; Mullins, 2004; Esper, 2004; Williams, 2004; Laham, 2004]. Specific clinical syndromes caused by hMPV seem to differ from that caused by other respiratory viruses (Table 21.3). These data are derived from Williams et al. [2004] who tested respiratory specimens over a 25-year period in the US from previously healthy children. Infection due to hMPV was more likely to be associated with bronchiolitis and less likely to be associated with croup than infection due to (para)influenzavirus. hMPV infection was less likely to be associated with pneumonia than was infection with hRSV or influenzavirus.

Table 21.2.   Clinical features of hMPV-infected children <10 years of age*

Features	Percentages	Patients/total patients
male sex	55%	98/117
cough	82%	194/238
rhinorrhea/rhinitis	67%	160/238
fever	72%	151/209
respiratory distress	71%	119/167
retractions	54%	43/79
wheezing	59%	125/212

*data derived from van den Hoogen 2003, Hamélin 2004, Døllner 2004, Mullins 2004, Esper 2004, Williams 2004, Laham 2004.

Table 21.3. Mean age at onset and clinical diagnosis of lower respiratory tract infections caused by hMPV as compared with other respiratory viruses*

| | n | mean age (mo) | P-value | Diagnosis (%) | | | | |
				bronchiolitis	croup	pneumonia	exacerbation of asthma	P-value
hMPV	49	11.6	–	59	18	8	14	–
hRSV	103	13.0	0.24	65	11	21	3	0.009
PIV	58	15.1	0.008	28	64	7	2	<0.001
InV	32	20.0	0.005	22	41	28	0	<0.001
AdV	28	10.9	0.13	61	21	14	4	0.48

PIV = parainfluenzavirus, InV = influenzavirus, AdV = adenovirus.
*data derived from Williams 2004 with permission.

Studies examining the role of hMPV with respect to exacerbations of asthma have yielded conflicting results [Jarrti, 2002; Williams, 2004; Rawlinson, 2003]. Although there is no doubt that some patients with asthmatic exacerbations have hMPV infection, whether or not the virus is associated more frequently than other respiratory viruses with these exacerbations is not yet clear. Remarkably, a history of asthma or a family member with asthma was more often associated with hMPV (16% and 67%, resp.) than with hRSV (0% and 30%, resp.) [Peiris, 2003]. It is currently not known whether hMPV infection leads to an increased susceptibility to secondary bacterial infections.

The similar seasonality and susceptible population shared by several respiratory viral infections will result in prevalent co-infection of hMPV with other respiratory viruses. This might lead to an underestimation of the percentage of hMPV-positive samples identified in studies in which only samples negative for other respiratory viruses were tested (see also Table 21.1). Co-infection rates of 5–10% with 1 or more respiratory viruses has been demonstrated in several studies searching for the causative pathogen of RTI. Because the epidemic seasonality for hRSV coincides with that for hMPV, the potential exists for hRSV/hMPV co-infections. Several studies have identified cases of LRTI in which evidence for the presence of both hRSV and hMPV has been detected [Williams, 2004; Maggi, 2003; Semple, 2005]. One study observed that dual infection with hRSV and hMPV was more frequent in infants with severe disease (i.e. those who needed supplementary oxygen) and even more frequent in infants with severe disease admitted to the intensive care unit for mechanical ventilation [Semple, 2005]. In contrast, others found the same rate of bronchopneumonia in infants infected with hMPV alone as in dual infections [Maggi, 2003]. On the other hand, the seasonal distribution of hHMP and hRSV may differ in specific geographic areas as demonstrated in studies from Argentina and Hong-Kong where co-infections were not or infrequent observed [Peiris, 2003; Laham, 2004]. The peak of hMPV in these countries becomes prevalent in late winter and early spring.

It is likely that by the development of more sensitive detection methods, dual or mixed infections will be increasingly recognized, and do not necessarily result in more severe infection. A positive RT-PCR test result does not differentiate between active infection or prolonged shedding after a recent acute infection that has been terminated.

8. Therapy

No vaccines, chemotherapeutic agents, or antibodies (monoclonal or polyclonal) are currently licensed for use to prevent or treat hMPV infections. However, ribavirin and polyclonal antibody preparations (IVIG), used in the therapy and prevention of hRSV-infections in children, are known to have broad-spectrum activity and can inhibit different viruses. In tissue culture-based assays the antiviral potential of both compounds was assessed and found to have equivalent activity against hMPV and hRSV [Wyde, 2003]. These findings would suggest that ribavirin and/or IVIG could be useful for treatment of severe hMPV-infections and therefor, these compounds should undergo clinical evaluation.

9. Prevention

The circulation of two serotypes of hMPV might have implications for the development of vaccines. Studies in cynomolgous macaques showed that reinfection is suppressed by high titers of virus neutralization antibodies against the homologous virus and far less by heterologous virus neutralization antibodies [van den Hoogen, 2004b]. Others report cross-protection and reciprocal cross-neutralization studies in experimental models of hMPV infection showing that cross-protection is induced at a high level, consistent with a single serotype [reviewed by Crowe, 2004]. The most relevant test of the importance of genetic diversity is whether or not viruses of one genotype induce greater protection against the homologous virus than against the heterologous one. Although difficult to assess, the extent of cross-protection is important to estimate to provide information for the development of a monovalent or bivalent vaccine formulation. One of the difficulties in assessing cross-protection is the occurrence of reinfection. Virus neutralization antibody titers in children >5 years of age are higher than in those 1–2 years of age, which suggests that reinfection may occur frequently.

Before the discovery of hMPV in 2001, several groups were working with molecular systems that allow the generation of recombinant paramyxoviruses from plasmid DNA copies of virus genes and virus genome. This technique, referred to as reverse genetics, was rapidly used to study the replication of hMPV and to generate recombinant hMPV strains. Foreign genes such as the reporter gene for green fluorescence protein were inserted into the hMPV genome and expressed, which effectively defined the transcription start and gene end signals [Biacchesi, 2004]. Reverse genetics has been used to rescue both strains from Canada and the Netherlands entirely from complementary DNA (cDNA). Because the viruses are made from DNA copies, chimeric viruses can be made with the use of the antigenic protein of one virus inserted into the genome of another virus. Neutralizing antibody responses can be induced by such a chimeric virus protecting the host against challenge with hMPV strains.

10. Conclusions

The epidemiology and clinical manifestations associated with hMPV have been found to be reminiscent of those of the hRSV, with most severe RTI occuring in young infants, elderly subjects, and immunocompromised hosts. The seasonal distribution resembles that of hRSV and influenza virus infections, with recurrent epidemics during the winter months. hMPV is the second most important cause, after hRSV, of viral lower RTI in children. hMPV infections account for at least 4 to 8% of the RTI in hospitalized children. In the general community hMPV infections account for at least 3% of patients who visit a general practitioner for RTI. Interestingly, the rates of detection of hMPV have been generally higher in retrospective than prospective studies, an observation consistent with some selection bias. Larger prospective studies, not limited to the typical respiratory virus season, not limited to testing respiratory samples negative for the other respiratory viruses, and using appropriate controls need to be conducted. Diagnosis is made by RT-PCR assays aimed at amplifying the N, L, or P genes. Additional research to define the pathogenesis of this viral infection and the host' specific immune response will enhance our knowledge and guide the search for preventive and therapeutic strategies.

References

Bastien, N., Normand, S., Taylor, T., Ward, D., Peret, T.C.T., Boivin, G., Anderson, L.J., and Li, Y. (2003a). Sequence analysis of the N, P, M and F genes of Canadian human metapneumovirus strains. *Virus Res* 93:51–62.

Bastien, N., Ward, D., Van Caeseele, P., Brandt, K., Lee, S.H.S., McNabb, G., Klisko, B., Chan, E., and Li, Y. (2003b). Human metapneumovirus infection in the Canadian population. *J Clin Microbiol* 41:4642–4646.

Biacchesi, S., Skiadopoulos, M.H., Boivin, G., Hanson, C.T., Murphy, B.R., Collins, P.L., and Buchholz, U.J. (2003). Genetic diversity between human metapneumovirus subgroups. *Virology* 315:1–9.

Biacchesi, S., Skiadopoulos, M.H., Tran, K.C., Murphy, B.R., Collins, P.L., and Buchholz, U.J. (2004). Recovery of human metapneumovirus from cDNA: optimization of growth in vitro and expression of additional genes. *Virology* 321:247–259.

Boivin, G., Abed, Y., Pelletier, G., Ruel, L., Moisan, D., Côté, S., Peret, T.C.T., Erdman, D.D., and Anderson, L.J. (2002). Virological features and clinical manifestations associated with human metapneumovirus: a new paramyxovirus responsible for acute respiratory-tract infections in all age groups. *J Infect Dis* 186:1330–1334.

Côté, S., Abed, Y., and Boivin, G. (2003). Comparative evaluation of real-time PCR assays for detection of the human metapneumovirus. *J Clin Microbiol* 41:3631–3635.

Crowe, J.E. (2004). Human metapneumovirus as major cause of human respiratory tract disease. *Pediatr Infect Dis J* 23:S215–S221.

Døllner, H., Risnes, K., Radtke, A., and Nordbø, S.A. (2004). Outbreak of human metapneumovirus infection in Norwegian children. *Pediatr Infect Dis J* 23:436–440.

Esper, F., Martinello, R.A., Boucher, D., Weibel, C., Ferguson, D., Landry, M.L., and Kahn, J.S. (2004). A 1-year experience with human metapneumovirus in children aged <5 years. *J Infect Dis* 189:1388–1396.

Falsey, A.R., Erdman, D., Anderson, L.J., and Walsh, E.E. (2003). Human metapneumovirus infections in young and elderly adults. *J Infect Dis* 187:785–790.

Freymuth, F., Vabret, A., Legrand, L., Eterradossi, N., Lafay-Delaire, F., Brouard, J., and Guillois, B. (2003). Presence of the new human metapneumovirus in french children with bronchiolitis. *Pediatr Infect Dis J* 22:92–94.

Hamelin, M.-E., Abed, Y., and Boivin, G. (2004). Human metapneumovirus: a new player among respiratory viruses. *Clin Infect Dis* 38:983–990.

van den Hoogen, B.G., de Jong, J.C., Groen, J., Kuiken, T., de Groot, R., Fouchier, R.A.M., and Osterhaus, A.D.E.M. (2001). A newly discovered human pneumovirus isolated from young children with respiratory tract disease. *Nat Med* 7:719–724.

van den Hoogen, B.G., Bestebroer, T.M., Osterhaus, A.D.E.M., and Fouchier, R.A.M. (2002). Analysis of the genomic sequence of a human metapneumovirus. *Virology* 295:119–132.

van den Hoogen, B.G., van Doornum, G.J.J., Fockens, J.C., Cornelissen, J.J., Beyer, W.E.P., de Groot, R., Osterhaus, A.D.E.M., and Fouchier, R.A.M. (2003). Prevalence and clinical symptoms of human metapneumovirus infection in hospitalised patients. *J Infect Dis* 188:1571–1577.

van den Hoogen, B.G., Osterhaus, A.D.E.M., and Fouchier, R.A.M. (2004a). Clinical impact and diagnosis of human metapneumovirus infection. *Pediatr Infect Dis J* 23:S25–S32.

van den Hoogen, B.G., Herfst, S., Sprong, L., Cane, P.A., Forleo-Neto, E., de Swart, R.L., Osterhaus, A.D.E.M., and Fouchier, R.A.M. (2004b). Antigenic and genetic variability of human metapneumoviruses. Emerg Infect Dis 10:658–666.

Jartti, T., van de Hoogen, B., Garofalo, R.P., Osterhaus, A.D.E.M., and Ruuskanen, O. (2002). Metapneumovirus and acute wheezing in children. *Lancet* 360:1393–1394.

Jartti, T., Lehtinen, P., Vuorinen, T., Österback, R., van de Hoogen, B., Osterhaus, A.D.E.M., and Ruuskanen, O. (2004). Respiratory picornaviruses and respiratory syncytial virus as causative agents of acute expiratory wheezing in children. *Emerg Infect Dis* 10:1095–1101.

Kuiken, T., van den Hoogen, B.G., van Riel, D.A.J., Laman, J.D., van Amerongen, G., Sprong, L., Fouchier, R.A.M., and Osterhaus, A.D.E.M. (2004). Experimental human metapneumovirus infection of cynomolgus macaques (*Macaca fascicularis*) results in virus replication in ciliated epithelial cells and pneumocytes with associated lesions throughout the respiratory tract. *Am J Pathol* 164:1893–1900.

König, B., König, W., Arnold, R., Werchau, H., Ihorst, G., and Forster, J. (2004). Prospective study of human metapneumovirus infection in children less than 3 years of age. *J Clin Microbiol* 42:4632–4635.

Laham, F.R., Israele, V., Casellas, J.M., Garcia, A.M., Lac Prugent, C.M., Hoffman, S.J., Hauer, D., Thumar, B., Name, M.I., Pascual, A., Taratutto, N., Ishida, M.T., Balduzzi, M., Maccarone, M., Jackli, S., Passarino, R., Gaivironsky, R.A., Karron, R.A., Polack, N.R., and Polack, F.P. (2004). Differential production of inflammatory cytokines in primary infection with human metapneumovirus and with other common respiratory viruses of infancy. *J Infect Dis* 189:47–56.

Ludewick, H.P., Abed, Y., van Niekerk, N., Boivin, G., Klugman, K.P., and Madhi, S.A. (2005). Human metapneumovirus genetic variability, South Africa. *Emerg Infect Dis* 11:1074–1078.

Mackay, I.M., Jacob, K.C., Woolhous, D., Waller, K., Syrmis, M.W., Whiley, D.M., Siebert, D.J., Nissen, M., and Sloots, T.P. (2003). Molecular assays for detection of human metapneumovirus. *J Clin Microbiol* 41:100–105.

Mackay, I.M., Bialasiewicz, S., Waliuzzaman, Z., Chidlow, G.R., Fegredo, D.C., Laingam, S., Adamson, P., Harnett, G.B., Rawlinson, W., Nissen, M., and Sloots, T.P. (2004). Use of the P genes to genotype human metapneumovirus identifies 4 viral subtypes. *J Infect Dis* 190:1913–1918.

Madhi, S.A., Ludewick, H., Abed, Y., Klugman, K.P., and Boivin, G. (2003). Human metapneumovirus-associated lower respiratory tract infections among hospitalized human immunodeficiency virus type 1 (HIV-1)-uninfected African infants. *Clin Infect Dis* 37:1705–1710.

Maggi, F., Pifferi, M., Vatteroni, M., Fornai, C., Tempestini, E., Anzilotti, S., Lanini, L., Andreoli, E., Ragazzo, V., Pistello, M., Specter, S., and Bendinelli, M. (2003). Human metapneumovirus associated with respiratory tract infections in a 3-year study of nasal swabs from infants in Italy. *J Clin Microbiol* 41:2987–2991.

McAdam, A.J., Hasenbein, M.E., Feldman, H.A., Cole, S.E., Offermann, J.T., Riley, A.M., and Lieu, T.A. (2004). Human metapneumovirus in children tested at a tertiary-care hospital. *J Infect Dis* 190:20–26.

Mullins, J.A., Erdman, D.D., Weinberg, G.A., Edwards, K., Hall, C.B., Walker, F.J., Iwane, M., and Anderson, J.J. (2004). Human metapneumovirus infection among children hospitalized with acute respiratory illness. *Emerg Infect Dis* 10:700–705.

Peiris, J.S.M., Tang, W.-H., Chan, K.-H., Khong, P.-L., Guan, Y., Lau, Y.-L., and Chiu, S.S. (2003). Children with respiratory disease associated with metapneumovirus in Hong Kong. *Emerg Infect Dis* 9:628–633.

Principi, N., Esposito, S., and Bosis, S. (2004). Human metapneumovirus and lower respiratory tract disease in children. Correspondence. *New Engl J Med* 350:1788.

Rawlinson, W.D., Waliuzzaman, Z., Carter, I.W., Belessis, Y.C., Gilbert, K.M., and Morton, J.R. (2003). Asthma exacerbations in children associated with rhinovirus but not human metapneumovirus infection. *J Infect Dis* 187:1314–1318.

Semple, M.G., Cowell, A., Dove, W., Greensill, J., McNamara, P.S., Halfhide, C., Shears, P., Smyth, R.L., and Hart, C.A. (2005). Dual infections of infants by human metapneumovirus and human respiratory syncytial virus is strongly associated with severe bronchiolitis. *J Infect Dis* 191:382–386.

Thanasugarn, W., Samransamruajkit, R., Vanapongtipagorn, P., Prapphal, N., van den Hoogen, B., Osterhaus, A.D.E.M., and Poovorawan, Y. (2003). Human metapneumovirus infection in Thai children. *Scand J Infect Dis* 35:754–756.

Viazov, S., Ratjen, F., Scheidhauer, R., Fiedler, M., and Roggendorf, M. (2003). High prevalence of human metapneumovirus infection in young children and genetic heterogeneity of the viral isolates. *J Clin Microbiol* 41:3043–3045.

Williams, J.V., Harris, P.A., Tollefson, S.J., Halburnt-Rush, L.L., Pingsterhaus, J.M., Edwards, K.M., Wright, P.F., and Crowe, J.E. (2004) Human metapneumovirus and lower respiratory tract disease in otherwise healthy infants and children. *N Eng J Med* 350:443–450.

Wyde, P.R., Chetty, S.N., Jewell, A.M., Boivin, G., and Piedra, P.A. (2003). Comparison of the inhibition of human metapneumovirus and respiratory syncytial virus by ribavirin and immune serum globulin in vitro. *Antiviral Res* 60:51–59.

Wyde, P.R., Chetty, S.N., Jewell, A.M., Schoonover, S.L., and Piedra, P.A. (2005). Development of a cotton rat-human metapneumovirus (hMPV) model for identifying and evaluating potential hMPV antivirals and vaccines. *Antiviral Res* 66:57–66.

Index